To my children
Tomás, Rolando and Ana Cristina   *To Allison and Mallory*

Ricardo Estrada   Ram P. Kanwal

Ricardo Estrada
Ram P. Kanwal

ASYMPTOTIC ANALYSIS
A Distributional Approach

1994

Birkhäuser

Boston • Basel • Berlin

Ricardo Estrada
Escuela de Matemática
Universidad de Costa Rica
San José, Costa Rica

Ram P. Kanwal
Department of Mathematics
The Pennsylvania State University
University Park, PA 16802

Library of Congress Cataloging In-Publication Data

Estrada, Ricardo, 1956-
    Asymptotic analysis : a distributional approach / Ricardo Estrada
and Ram P. Kanwal.
        p.   cm.
    Includes bibliographical references and index.
    ISBN 0-8176-3716-8 (acid free)
    1. Asymptotic expansions.   2. Theory of distributions (Functional
analysis)   I. Kanwal, Ram P., 1924-   .   II. Title.
    QA295.E83   1994   93-36473
    515'.234--dc20   CIP

Printed on acid-free paper
© Birkhäiser Boston 1994

ISBN 0-8176-3716-8
ISBN 3-7643-3716-8

Typeset in TEX by the authors.
Printed and bound by Quinn-Woodbine, Woodbine, NJ.
Printed in the U.S.A.

9 8 7 6 5 4 3 2 1

CONTENTS

CHAPTER 4   The Asymptotic Expansion of Multidimensional Generalized Functions

CHAPTER 5   The Asymptotic Expansion of Certain Series Considered by Ramanujan

CHAPTER 6   Series of Dirac Delta Functions

PREFACE

Asymptotic analysis is an old subject that has found applications in various fields of pure and applied mathematics, physics and engineering. For instance, asymptotic techniques are used to approximate very complicated integral expressions that result from transform analysis. Similarly, the solutions of differential equations can often be computed with great accuracy by taking the sum of a few terms of the divergent series obtained by the asymptotic calculus. In view of the importance of these methods, many excellent books on this subject are available [19], [21], [27], [67], [90], [91], [102], [113].

An important feature of the theory of asymptotic expansions is that experience and intuition play an important part in it because particular problems are rather individual in nature. Our aim is to present a systematic and simplified approach to this theory by the use of distributions (generalized functions). The theory of distributions is another important area of applied mathematics, that has also found many applications in mathematics, physics and engineering. It is only recently, however, that the close ties between asymptotic analysis and the theory of distributions have been studied in detail [15], [43], [44], [84], [92], [112]. As it turns out, generalized functions provide a very appropriate framework for asymptotic analysis, where many analytical operations can be performed, and also provide a systematic procedure to assign values to the divergent integrals that often appear in the literature.

The book is suitable for a one semester graduate course in mathematical and physical sciences. We have offered courses inspired by this material at our respective universities. Although the book is based on our own research work, we have attempted to make the material self-contained. With that goal in mind we have incorporated one chapter on the introduction of classical asymptotic analysis and one chapter on the basic principles of generalized functions as needed in the sequel.

The material is divided into six chapters. In Chapter 1, we explain the classical principles of asymptotic theory. The subject matter of this chapter is designed to motivate and prepare the reader for the next chapters. Since generalized functions, or distributions, permeate our entire presentation, we explain all the basic principles of this fascinating field in Chapter 2. We have taken great care in explaining all the function spaces that we need in the presentation of the theory.

Chapter 3 is devoted to the theme and methodology of our theory. We start with the *moment asymptotic expansion* of generalized functions. As we show, the moment expansion holds in a wide variety of situations, par-

ticularly for distributions of fast decay and for distributions of rapid oscillation. The moment asymptotic expansion immediately yields the asymptotic
development of several integrals and series. We illustrate these ideas with
many examples, deriving in particular, results such as Watson's lemma for
the expansion of Laplace transforms and Stirling's approximation of $n!$.
Use of the notion of change of variables in distributions allows us to obtain
the expansion of more complicated distributional kernels, which in turn
provide the most important methods for the asymptotic development of
integrals, namely, the Laplace formula, the method of stationary phase and
the method of steepest descent. We also consider the situation when the
moment expansion does not hold and show that the expansions should be
given in terms of homogeneous and associated homogeneous functions. In
order to make the presentation easily accessible, we limit ourselves to the
one-dimensional theory in this chapter.

In Chapter 4 we extend the analysis of the previous chapter to the corresponding multidimensional problems. As in Chapter 3, we start with the
moment asymptotic expansion and, using the notion of change of variables
in distributions, obtain the Laplace asymptotic formula in several variables.
We also use these ideas to derive the asymptotic development of oscillatory
integrals, the so-called Fourier type integrals. Besides, we employ the theory of topological tensor products to obtain the asymptotic expansion of
vector valued generalized functions. This helps us in developing the theory
of partial asymptotic expansions. Partial expansions are a powerful tool,
providing a way to obtain rather sharp multidimensional developments.
Moreover, we illustrate the use of partial expansions with an application to
quantum mechanical twisted products. Another interesting feature of this
chapter is the derivation of the far-field behavior of potential and scattering
fields. We find that there arise many non-classical terms, in addition to the
previously known classical terms.

In Chapter 5 we present the asymptotic development for certain functions defined by series containing a small parameter that are of importance
in number theory. Series of this kind were first studied by Ramanujan [95]
who gave their asymptotic expansion. However, as in much of Ramanujan's
work, these results are given without any proofs. Proofs of some of these
results have been provided by appealing to Mellin transforms [7]. By using
our methods, we are able to present proofs of Ramanujan's results and of
many interesting generalizations for a wide class of functions. We study
not only the one-dimensional series but the multidimensional series as well.
Among many other concepts we introduce the notion of a distributionally
small sequence and relate it to the classical principles of summability.

Chapter 6 surveys the use of divergent series of delta functions in various contexts. Series of Dirac delta functions appear rather frequently in
the previous chapters since they form the basic blocks in the moment as-

ymptotic expansion of generalized functions. Furthermore, they have been used as formal tools in other branches of applied mathematics, such as the solution of differential and functional equations, the construction of weight functions for orthogonal polynomials and the solution of moment problems. Our aim is to give a rigorous interpretation of these formal methods in the framework of the asymptotic analysis. We also show how problems in singular perturbations can be studied by the use of such divergent series of generalized functions.

We would like to express our thanks to Berny Alvarado for his careful typing of the manuscript and to Professor B.K. Sachdeva for checking the manuscript and to the staff at Birkhäuser for their patience and cooperation.

CHAPTER 1

Basic Results in Asymptotics

1.1 Introduction

In many problems of engineering and the physical sciences we attempt to write the solutions as infinite series of functions. The simplest series representation is the power series. Given a function $f(x)$ of a real variable x containing a number x_0 in its domain of definition, we try to find a power series of the form

$$\sum_{j=0}^{\infty} a_j (x - x_0)^j, \tag{1.1.1}$$

which provides a valid representation of $f(x)$ in the interval I of convergence of the power series. It emerges that if $f(x)$ has uniformly bounded derivatives of all orders at each point in I, the above series is uniquely determined and

$$a_j = \frac{f^{(j)}(x_0)}{j!},$$

where $f^{(j)}(x_0)$ is the j-th derivative of $f(x)$ evaluated at x_0. Then the series (1.1.1) is called the Taylor series.

The so-called remainder term in the Taylor expansion plays a crucial role. When we write the series (1.1.1) as

$$f(x) = \sum_{j=0}^{n} \frac{f^{(j)}(x_0)}{j!} (x - x_0)^j + R_n(x), \tag{1.1.2}$$

the remainder $R_n(x)$ is given by

$$R_n(x) = \frac{f^{(n+1)}(\hat{x})}{(n+1)!} (x - x_0)^{n+1}, \tag{1.1.3}$$

where $\hat{x}$ is a point between x_0 and x. If M denotes a uniform bound of $f^{(n+1)}(x)$ in I, that is, $\mid f^{(n+1)}(x) \mid \leq M$, $x \in I$, then the error introduced by using the Taylor polynomial

$$f_n(x) = \sum_{j=0}^{n} \frac{f^{(j)}(x_0)}{j!} (x - x_0)^j \tag{1.1.4}$$

for $f(x)$ is the same order of magnitude as the first term which is neglected in the Taylor series. Also observe that in this case

$$\lim_{n\to\infty} \mid f(x) - f_n(x) \mid = 0. \tag{1.1.5}$$

The important feature of the Taylor polynomial $f_n(x)$ as given by (1.1.4) is that it is a function $g(n, x)$ of two independent variables. The convergent series approach is to consider x fixed and determine the behavior of $g(n, x)$ as n increases. Accordingly, the approximation is considered adequate if the error in using the Taylor polynomial can be made sufficiently small by choosing n appropriately large. Even for convergent series this becomes a formidable task.

The concept of an asymptotic series reverses the role of n and x in $g(n, x)$. That is, the approximation is considered adequate if the error can be made sufficiently small, for any fixed number of terms, by using values of x sufficiently close to some value. Of course, some series may be both convergent and asymptotic. However, very often, the series is divergent but still asymptotic. It is for this reason that the asymptotic series is frequently referred to as divergent. Although this is erroneous, the fact remains that taking more terms in the asymptotic series may not improve the approximation. Indeed, for a large number of problems, one or two terms will be adequate.

We shall meet functions for which both an asymptotic expansion and a convergent expansion are available and for which the accuracy achieved by three or four terms of the asymptotic series can only be obtained by taking several hundreds of terms of the convergent series.

We devote this chapter to considering the basic notions of asymptotic analysis. We also present some simple methods for the approximation of integrals and sums.

1.2 Order Symbols

In this section, we define and study the basic properties of the order symbols. They play a very important role in the study of asymptotic expansions. This useful notation was introduced by E. Landau.

Let X be a topological space and let $x_0 \in X$. We assume that x_0 is not an isolated point. A *pointed neighborhood* of x_0 is a set of the form $V \backslash \{x_0\}$, where V is a neighborhood of x_0 . Pointed neighborhoods are also called deleted neighborhoods.

The basic examples we want to consider are:

(a) $X = [a, b]$ and $x_0 \in X$. Usually x_0 is one of the endpoints, namely, $x_0 = a$ or $x_0 = b$.

(b) X is a sector in the complex plane of the form

$$S(\alpha, \beta) = \{re^{i\theta} : r \geq 0, \alpha < \theta < \beta\}, \tag{1.2.1}$$

and $x_0 = 0$.

(c) $X = [a, \infty] = [a, \infty) \cup \{\infty\}$, with the natural topology at ∞, and with $x_0 = \infty$.

(d) $X = S(\alpha, \beta) \cup \{\infty\}$ with $x_0 = \infty$.

(e) $X = \mathbb{N} \cup \{\infty\}$ with $x_0 = \infty$.

Our analysis applies equally well to any of these situations.

Definition. *Let $f(x)$ and $g(x)$ be functions defined in a pointed neighborhood of x_0. We say that $f(x)$ is " big O " of $g(x)$ as $x \to x_0$ and write*

$$f(x) = O(g(x)), \quad as \ x \to x_0, \tag{1.2.2}$$

if there exists a pointed neighborhood V of x_0 and a constant $M > 0$ such that

$$| f(x) | \leq M \, | \, g(x) \, |, \quad x \in V. \tag{1.2.3}$$

Observe that if $g(x)$ does not vanish near x_0, then the relation $f(x) = O(g(x))$, as $x \to x_0$ is equivalent to the condition

$$\overline{\lim_{x \to x_0}} \left| \frac{f(x)}{g(x)} \right| < \infty. \tag{1.2.4}$$

Here $\overline{\lim}_{x \to x_0}$ denotes the limit superior as $x \to x_0$.

Another useful concept is the little o symbol, defined as follows.

Definition. *Let $f(x)$ and $g(x)$ be functions defined in a pointed neighborhood of x_0. We say that $f(x)$ is " little o " of $g(x)$ as $x \to x_0$, and write*

$$f(x) = o(g(x)), \quad as \ x \to x_0, \tag{1.2.5}$$

if for each $\varepsilon > 0$ there is a pointed neighborhood $V = V(\varepsilon)$ of x_0 such that

$$| f(x) | \leq \varepsilon \, | \, g(x) \, |, \quad x \in V. \tag{1.2.6}$$

If $g(x)$ does not vanish near x_0, the condition $f(x) = o(g(x))$, as $x \to x_0$, is equivalent to the vanishing of the limit

$$\lim_{x \to x_0} \frac{f(x)}{g(x)} = 0. \tag{1.2.7}$$

Clearly if $f = o(g)$, as $x \to x_0$, then $f = O(g)$, as $x \to x_0$, but not conversely.

Some useful properties of the order symbols o and O are given in Theorem 1. The notation is the one usually used in asymptotic analysis; when we write $O(g)$ we mean *some* function f that is $O(g)$. Thus, for instance, the result $O(g) + O(g) = O(g)$ means that if $f_1 = O(g)$ and $f_2 = O(g)$ then also $f_1 + f_2 = O(g)$.

Theorem 1.

    (a) If c_1, c_2 are constants then $c_1 O(g) + c_2 O(g) = O(g)$.

    (b) If c_1, c_2 are constants then $c_1 o(g) + c_2 o(g) = o(g)$.

    (c) $O(O(g)) = O(g)$

    (d) $O(o(g)) = o(O(g)) = o(o(g)) = o(g)$.

    (e) $O(f)O(g) = O(f\,g)$.

    (f) $O(f)o(g) = o(f\,g)$.

Incidentally, the function g in the relations $f = O(g)$ or $f = o(g)$ is called a *gauge function*: it is the function against which the behavior of $f(x)$ is gauged.

Both types of order symbols suppress information since they indicate only the orders of magnitude. Indeed, the expression $f(x) = O(g(x))$, as $x \to x_0$, indicates only that $\dfrac{f(x)}{g(x)}$ has a finite bound as $x \to x_0$. Writing $f(x) = o(g(x))$, as $x \to x_0$, also suppresses information since it only indicates that $\dfrac{f(x)}{g(x)}$ tends to zero as $x \to x_0$, but it does not say how fast it tends to zero. For example $\sin x = O(x)$, as $x \to 0$ and $\sin x = O(8\,x)$ are both correct. Similarly, $\dfrac{1}{1+x^2} = o(1)$, as $x \to \infty$, but also $\dfrac{1}{1+x^2} = o\left(\dfrac{1}{x}\right)$, as $x \to \infty$. Thus the information given by the order symbols is quite vague. Fortunately, it is precisely this vagueness that allows us to employ these symbols in a variety of situations.

We shall also make use of the notation of asymptotic equivalence, defined as follows.

Definition. *The functions $f(x)$ and $g(x)$ are called asymptotically equivalent as $x \to x_0$ if*

$$f(x) - g(x) = o(g(x)), \quad as \ x \to x_0. \tag{1.2.8}$$

In this case we write

$$f(x) \sim g(x), \quad as \ x \to x_0. \tag{1.2.9}$$

The relation $\sim$ is symmetric since actually $f \sim g$ as $x \to x_0$, if and only if in a neighborhood of x_0 the zeros of f and g coincide and

$$\lim_{x \to x_0} \frac{f(x)}{g(x)} = 1. \tag{1.2.10}$$

Sometimes we shall also employ the notation " $<<$ " , which reads as "much smaller than." The notation $\varepsilon << 1$ means $\varepsilon \to 0$, while $\lambda \gg 1$ means $\lambda \to \infty$. Similary, the notation $f(x) << g(x)$ is another way of expressing that $f(x) = o(g(x))$.

Example 1. If $f(x)$ is continuous at $x = x_0$ then

$$f(x) = f(x_0) + o(1), \quad as \ x \to x_0.$$

Example 2. If $f(x)$ is differentiable at $x = x_0$ then

$$f(x) = f(x_0) + (x - x_0)f'(x_0) + o(|\ x - x_0\ |), \quad as \ x \to x_0.$$

In the case where $f'(x_0) \neq 0$ this can be rewritten as

$$f(x) - f(x_0) \sim (x - x_0)f'(x_0), \quad as \ x \to x_0.$$

Example 3. We have

$$\sin x = O(1), \quad as \ x \to \infty,$$

but

$$\sin x \neq o(1), \quad as \ x \to \infty,$$

and

$$1 \neq O(\sin x), \quad as \ x \to \infty.$$

Example 4. The famous Stirling approximation of $n!$ can be written as

$$n! = \sqrt{2\pi}\, n^{\frac{1}{2}+n}\, e^{-n}(1 + o(1)), \quad \text{as } n \to \infty.$$

We shall discuss this and related formulas in Section 1.8 and in Chapter 3.

Example 5. The series $\sum_{k=1}^{\infty} \frac{1}{k}$ is divergent. We have, however, the following approximation for its partial sums

$$\sum_{k=1}^{n} \frac{1}{k} = \ln n + \gamma + o(1), \quad \text{as } n \to \infty.$$

The asymptotic approximation of series is discussed in Section 1.7 and in Chapter 5.

We would also like to indicate that the order symbols can be integrated. Let $g(x)$ be continuous and positive for $x > 0$. Then if f is locally integrable in $[a, \infty)$ and

$$f(x) = O(g(x)), \quad \text{as } x \to \infty, \tag{1.2.11}$$

then

$$\int_{a}^{x} f(t)dt = O\left(\int_{a}^{x} g(t)dt\right), \quad \text{as } x \to \infty. \tag{1.2.12}$$

When $\int_{a}^{\infty} g(x)dx < \infty$, we also have

$$\int_{x}^{\infty} f(t)dt = O\left(\int_{x}^{\infty} g(t)dt\right), \quad \text{as } x \to \infty. \tag{1.2.13}$$

The proof of (1.2.12) and (1.2.13) is straightforward.

Similarly, when

$$f(x) = o(g(x)), \quad \text{as } x \to \infty, \tag{1.2.14}$$

then if $\int_{a}^{\infty} g(t)dt = \infty$, we have

$$\int_{a}^{x} f(t)dt = o\left(\int_{a}^{x} g(t)dt\right), \tag{1.2.15}$$

while if $\int_{a}^{\infty} g(t)dt < \infty$, then

$$\int_{x}^{\infty} f(t)dt = o\left(\int_{x}^{\infty} g(t)dt\right). \tag{1.2.16}$$

To prove (1.2.15) observe that if $\varepsilon > 0$ there exists $b > 0$ such that $\mid f(t) \mid \le \varepsilon g(t)$ for $t \ge b$. Thus

$$\varlimsup_{x \to \infty} \frac{\mid \int_a^x f(t)dt \mid}{\int_a^x g(t)dt} \le \varlimsup_{x \to \infty} \frac{\mid \int_a^b f(t)dt \mid}{\int_a^x g(t)dt} + \varlimsup_{x \to \infty} \frac{\mid \int_b^x f(t)dt \mid}{\int_a^x g(t)dt} \le \varepsilon,$$

since $\lim_{x \to \infty} \int_a^x g(t)dt = \infty$. But ε is arbitrary and thus

$$\lim_{x \to \infty} \frac{\int_a^x f(t)dt}{\int_a^x g(t)dt} = 0.$$

In the case where $\int_a^\infty g(t)dt < \infty$, (1.2.16) follows directly from L'Hôpital's rule:

$$\lim_{x \to \infty} \frac{\int_x^\infty f(t)dt}{\int_x^\infty g(t)dt} = \lim_{x \to \infty} \frac{f(x)}{g(x)} = 0.$$

Finally, in the case of

$$f(x) \sim g(x), \quad \text{as } x \to \infty, \tag{1.2.17}$$

it follows that when $\displaystyle\int_a^\infty g(t)dt = \infty$ then

$$\int_a^x f(t)dt \sim \int_a^x g(t)dt, \quad \text{as } x \to \infty, \tag{1.2.18}$$

whereas

$$\int_x^\infty f(t)dt \sim \int_x^\infty g(t)dt, \text{as } x \to \infty, \tag{1.2.19}$$

if $\int_a^\infty g(t)dt < \infty$. Formulas (1.2.18) and (1.2.19) are obtained from our previous analysis since (1.2.17) means that $f - g = o(g)$.

Unfortunately, the behavior of the order symbols under differentiation is not very good. In fact, in general no order relation between $f'(x)$ and $g'(x)$ can be deduced from relations of the forms $f = O(g)$, $f = o(g)$, not even $f \sim g$. For instance if $f(x) = x + \sin x^2$, $g(x) = x$ then $f(x) \sim g(x)$ as $x \to \infty$, but neither $f'(x) = O(g'(x))$ nor $g'(x) = O(f'(x))$ as $x \to \infty$.

1.3 Asymptotic Series

If we are interested in the approximation of a function $f(x)$ in the neighborhood of $x = x_0$, a relation of the type

$$f(x) = c_1 \phi_1(x) + o(\phi_1(x)), \quad \text{as } x \to x_0, \tag{1.3.1}$$

in terms of a known gauge function $\phi_1(x)$ is quite useful. We could attempt a better approximation by trying to write the error $f(x) - c_1\phi_1(x)$ in terms of another gauge function $\phi_2(x)$, as follows:

$$f(x) = c_1\phi_1(x) + c_2\phi_2(x) + o(\phi_2(x)) \text{ as } x \to x_0. \qquad (1.3.2)$$

Improved approximations are obtained if we repeat this process several times. After n steps a development of the type

$$f(x) = c_1\phi_1(x) + \cdots + c_n\phi_n(x) + o(\phi_n(x)), \text{ as } x \to x_0, \qquad (1.3.3)$$

should be obtained.

The approximation (1.3.3) will be helpful as long as the sequence of gauge functions $\phi_1(x), \ldots, \phi_n(x)$ satisfies some natural properties, i.e, the functions should form an *asymptotic sequence.*

Definition. *A finite or infinite sequence of functions $\{\phi_n(x)\}$, defined in a pointed neighborhood of x_0 is called an asymptotic sequence as $x \to x_0$ if the following two conditions are satisfied:*

(a) $\phi_n(x) \neq 0$, for $x \neq x_0$ and $n = 1, 2, 3, \ldots$

(b) $\phi_{n+1}(x) = o(\phi_n(x))$, as $x \to x_0$.

Examples of asymptotic sequences include the following. If $x_0 \in \mathbb{R}$, the most common asymptotic sequence as $x \to x_0$ is $\{(x - x_0)^n\}$. More generally if $\{\lambda_n\}$ is a sequence of complex numbers with $\Re e\, \lambda_n \nearrow \infty$ then $\{(x - x_0)^{\lambda_n}\}$ is also an asymptotic sequence as $x \to x_0$.

Logarithmic scales can also be included, as in the sequence $\{\ln | x - x_0 |, 1, (x - x_0) \ln | x - x_0 |, (x - x_0), \ldots \}$. Among asymptotic sequences as $x \to \infty$ we could mention the sequences $\{x^{-n}\}$ and $\{e^{-nx}\}$.

Approximations of the type (1.3.3) are called *asymptotic expansions* or *asymptotic developments* . The formal definition is as follows.

Definition. *Let $\{\phi_n(x)\}$ be an asymptotic sequence as $x \to x_0$. A function $f(x)$ has an asymptotic development to N terms with respect to the sequence $\{\phi_n(x)\}$ if there exist constants $c_1, \cdots, c_N$ such that*

$$f(x) = c_1\phi_1(x) + \cdots + c_N\phi_N(x) + o(\phi_N(x)), \text{ as } x \to x_0. \qquad (1.3.4)$$

In the case where $f(x)$ has an expansion to N terms for every N then we say that $f(x)$ has an asymptotic expansion in terms of the sequence $\{\phi_n(x)\}$ and write

$$f(x) \sim \sum_{n=1}^{\infty} c_n\phi_n(x), \text{ as } x \to x_0. \qquad (1.3.5)$$

The relation (1.3.4) expresses the fact that when using the approximation $c_1\phi_1(x)+\cdots+c_N\phi_N(x)$, *the error is of smaller order than the last retained term*. In the case of an expansion to all orders, equation (1.3.4) is equivalent to the stronger condition

$$f(x) = c_1\phi_1(x) + \cdots + c_N\phi_N(x) + O(\phi_{N+1}(x)), \text{ as } x \to x_0. \qquad (1.3.6)$$

According to this condition, *the error is of the order of the first neglected term*.

A given function $f(x)$ may or may not have an expansion with respect to a gauge function sequence $\{\phi_n(x)\}$. However, if an expansion exists, it is unique.

Theorem 2. *Let $f(x)$ be a function with an expansion to N terms with respect to the sequence $\{\phi_n(x)\}$, as in (1.3.4). Then the coefficients $c_1, \ldots, c_N$ can be computed recursively as*

$$c_k = \lim_{x \to x_0} \frac{f(x) - \sum_{j=1}^{k-1} c_j\,\phi_j(x)}{\phi_k(x)}. \qquad (1.3.7)$$

Example 6. If a function is N times continuously differentiable near x_0 then we have the Taylor approximation

$$f(x) = \sum_{n=0}^{N} \frac{f^{(n)}(x_0)}{n!}\,(x - x_0)^n + o(|\,x - x_0\,|^N), \text{ as } x \to x_0. \qquad (1.3.8)$$

For a C^∞ function we thus have

$$f(x) \sim \sum_{n=0}^{\infty} \frac{f^{(n)}(x_0)}{n!}\,(x - x_0)^n, \text{ as } x \to x_0. \qquad (1.3.9)$$

We remark that, in general, the asymptotic relation in (1.3.9) cannot be replaced by an equality. Actually, as will follow from Borel's theorem (Theorem 8 of Section 1.5), given an arbitrary sequence a_n of real or complex numbers, there is a C^∞ function $f(x)$ with $f^{(n)}(x_0) = a_n$. Therefore, the series on the right of (1.3.9) will "usually" be divergent. But even if convergent, the Taylor series $\sum_{n=0}^{\infty} \frac{f^{(n)}(x_0)}{n!}\,(x - x_0)^n$ might not converge to $f(x)$. Take for instance the function

$$f(x) = \begin{cases} e^x + e^{-\frac{1}{|x|}}, & x \neq 0, \\ 1, & x = 0. \end{cases}$$

Then

$$f^{(n)}(0) = 1, \quad n = 0, 1, 2, \ldots,$$

but

$$\sum_{n=0}^{\infty} \frac{f^{(n)}(0)}{n!} x^n = \sum_{n=0}^{\infty} \frac{x^n}{n!} = e^x \neq f(x).$$

This example also serves to illustrate that an asymptotic expansion does not determine a function uniquely. Indeed,

$$e^x \sim 1 + x + \frac{x^2}{2!} + \frac{x^3}{3!} + \ldots, \quad \text{as } x \to 0,$$

but also

$$e^x + e^{-\frac{1}{|x|}} \sim 1 + x + \frac{x^2}{2!} + \frac{x^3}{3!} + \ldots, \quad \text{as } x \to 0.$$

Example 7. Sometimes the expansion of a function in terms of an asymptotic expansion does not give much information. In those cases the use of another asymptotic sequence is probably needed. For instance, for the exponential integral function

$$E(x) = \int_x^{\infty} \frac{e^{-y}}{y} \, dy, \quad x > 0, \tag{1.3.10}$$

we have

$$E(x) = o(x^{-n}), \quad \text{as } x \to \infty, \tag{1.3.11}$$

for each $n = 0, 1, 2, 3, \ldots$, as follows from L'Hôpital's rule. Thus the expansion of $E(x)$ in terms of the sequence $\{x^{-n}\}$ becomes

$$E(x) \sim 0(1) + 0(x^{-1}) + 0(x^{-2}) + \ldots, \quad \text{as } x \to \infty. \tag{1.3.12}$$

A much better approximation is obtained in terms of the sequence $e^{-x} x^{-n}$. Indeed, by using integration by parts we obtain

$$E(x) = e^{-x} \left(\frac{1}{x} - \frac{1}{x^2} + \frac{1}{x^3} + \cdots + \frac{(-1)^{n-1}(n-1)!}{x^n} \right)$$
$$+ (-1)^n \int_x^{\infty} \frac{e^{-y}}{y^{n+1}} \, dy.$$

$$\tag{1.3.13}$$

The use of L'Hôpital's rule gives

$$(-1)^n n! \int_x^{\infty} \frac{e^{-y}}{y^{n+1}} \, dy = o\left(\frac{e^{-x}}{x^n} \right), \quad \text{as } x \to \infty,$$

and thus we obtain the infinite expansion

$$E(x) \sim e^{-x} \sum_{n=1}^{\infty} \frac{(-1)^{n-1}(n-1)!}{x^n}, \text{ as } x \to \infty. \tag{1.3.14}$$

Observe that the series $\sum_{n=1}^{\infty} \frac{(-1)^{n-1}(n-1)!}{x^n}$ diverges for all x.

The fact that an asymptotic series might be divergent brings out a rather important distinction between asymptotic and convergent series. Indeed, if

$$f(x) = \sum_{n=1}^{\infty} \psi_n(x), \ x \in I, \tag{1.3.15}$$

is a representation of the function $f(x)$ as a *convergent* series, then for each $x \in I$, the partial sums

$$S_N(x) = \sum_{n=1}^{N} \psi_n(x) \tag{1.3.16}$$

provide better and better approximations as N increases.

On the other hand, if

$$f(x) \sim \sum_{n=1}^{\infty} c_n \phi_n(x), \text{ as } x \to x_0, \tag{1.3.17}$$

is an *asymptotic* representation of $f(x)$ near $x = x_0$, then, in general, the partial sums

$$T_N(x) = \sum_{n=1}^{N} c_n \phi_n(x) \tag{1.3.18}$$

do not converge to $f(x)$. Therefore for each value of x there is an optimal value of N for which the error $| f(x) - T_N(x) |$ is smallest: the inclusion of terms beyond the N-th order does not improve the approximation, rather, it makes it worse.

Thus, for each x there is a fixed level of accuracy that can be achieved by using an asymptotic expansion. Naturally, the closer x is to x_0, the higher the level of accuracy.

Consider for instance the series $\sum_{n=1}^{\infty} \frac{(-1)^{n-1}(n-1)!}{x^n}$. According to the last example, we have

$$e^x E(x) \sim \sum_{n=1}^{\infty} \frac{(-1)^{n-1}(n-1)!}{x^n}, \text{ as } x \to \infty. \tag{1.3.19}$$

Since the series is alternating it is not hard to see that

$$T_{2N}(x) < e^x E(x) < T_{2N-1}(x). \tag{1.3.20}$$

Observe, however, that since the series diverges,

$$\lim_{N\to\infty} T_{2N}(x) = -\infty, \quad \lim_{N\to\infty} T_{2N-1}(x) = +\infty.$$

Intuitively, we could expect the approximation to improve as long as the ratio between successive terms of the series remain less than 1 in absolute value. Thus, the optimal value of N should be about $[\![x]\!]$, the integral part of x.

When $x = 1$, the best approximation is given by $T_1(1) = 1$. However, the actual value $eE(1) = .59634$ is not very close. For $x = 10$ we have the value $e^{10} E(10) = .09156$, while $T_{10}(10) = .09154$, an excellent approximation. Actually, as the table shows,

$T_4(10)$	$T_5(10)$	$T_6(10)$	$T_7(10)$	$T_8(10)$	$T_9(10)$	$T_{10}(10)$
.09140	.09164	.09152	.09159	.09154	.09158	.09154

the approximations are already very good for small N. We remark that according to (1.3.20) it is better to consider the average approximations $\frac{1}{2}(T_{2N}(x) + T_{2N-1}(x))$ or $\frac{1}{2}(T_{2N}(x) + T_{2N+1}(x))$. In this case we have $\frac{1}{2}(T_8(10) + T_9(10)) = .09156$, correct to five decimal places.

For $x = 100$, already the sum $T_4(100) = .0099019$ is correct to seven decimal places!.

The function $e^x E(x)$ also admits a convergent series expansion. Using the notion of principal value for integrals discussed in Chapter 2, the following expansion about $x = 0$ is easily obtained

$$E(x) = \int_x^\infty \frac{e^{-y}}{y} dy = -\ln x - \gamma + \sum_{n=1}^\infty \frac{(-1)^{n+1}x^n}{nn!}, \tag{1.3.21}$$

where γ is the Euler constant, and $\gamma = .5772157\cdots$. Thus,

$$e^x E(x) = e^x \left[-\ln x - \gamma + \sum_{n=1}^\infty \frac{(-1)^{n+1}x^n}{nn!} \right], \tag{1.3.22}$$

a series that converges for all x.

However, although (1.3.22) converges for all x, the convergence is very slow for $x \gg 1$. Indeed, over 40 terms are needed to achieve the accuracy given by $T_6(10)$. For $x = 100$ the results are even more dramatic: while

$T_4(100)$ is already correct to seven decimal places, the sum of 1000 terms in (1.3.22) is only correct to three decimal places.

Summarizing, an asymptotic series may be divergent, and it may be that the level of accuracy is fixed and cannot be improved by taking more terms of the series. Nevertheless, near x_0 even the sum of just a few terms could give excellent approximations.

Next we shall consider a useful generalization, the notion of an *extended asymptotic expansion*. We introduce it by an example.

Example 8. Let

$$S(x) = \int_0^x \frac{\sin t}{t} dt \qquad (1.3.23)$$

be the sine integral function. We are going to consider the behavior of $S(x)$ for large x. Observe first that

$$\lim_{x \to \infty} S(x) = \int_0^\infty \frac{\sin t}{t} dt = \frac{\pi}{2}. \qquad (1.3.24)$$

Next we write $S(x) = \frac{\pi}{2} - \int_x^\infty \frac{\sin t}{t} dt$, and apply integration by parts to obtain

$$S(x) = \frac{\pi}{2} - \int_x^\infty \frac{\sin t}{t} dt$$

$$= \frac{\pi}{2} - \frac{\cos x}{x} + \int_x^\infty \frac{\cos t}{t^2} dt$$

$$= \frac{\pi}{2} - \frac{\cos x}{x} - \frac{\sin x}{x^2} + 2\int_x^\infty \frac{\sin t}{t^3} dt,$$

and more generally,

$$S(x) = \frac{\pi}{2} + \cos x \left(-\frac{1}{x} + \frac{2!}{x^3} - \frac{4!}{x^5} + \cdots + \frac{(2n)!(-1)^{n-1}}{x^{2n+1}} \right)$$

$$+ \sin x \left(-\frac{1}{x^2} + \frac{3!}{x^4} - \frac{5!}{x^6} + \cdots + \frac{(2n-1)!(-1)^n}{x^{2n}} \right)$$

$$+ (2n+1)!(-1)^n \int_x^\infty \frac{\cos t}{t^{2n+2}} dt, \qquad (1.3.25a)$$

and

$$S(x) = \frac{\pi}{2} + \cos x \left(-\frac{1}{x} + \frac{2!}{x^3} - \frac{4!}{x^5} + \cdots + \frac{(2n)!(-1)^{n-1}}{x^{2n+1}} \right)$$

$$+ \sin x \left(-\frac{1}{x^2} + \frac{3!}{x^4} - \frac{5!}{x^6} + \cdots + \frac{(2n+1)!(-1)^{n+1}}{x^{2n+2}} \right)$$

$$+(2n + 2)!(-1)^n \int_x^\infty \frac{\sin t}{t^{2n+3}} dt. \tag{1.3.25b}$$

As far as the remainders are concerned, we remark that

$$\int_x^\infty \frac{\cos t}{t^n} dt = O\left(\frac{1}{x^n}\right), \quad \int_x^\infty \frac{\sin t}{t^n} dt = O\left(\frac{1}{x^n}\right), \quad \text{as } x \to \infty. \tag{1.3.26}$$

We are then tempted to write the infinite expansion

$$S(x) \sim \frac{\pi}{2} - \frac{\cos x}{x} - \frac{\sin x}{x^2} + \frac{2\cos x}{x^3} + \frac{6\sin x}{x^4} - \cdots, \quad \text{as } x \to \infty. \tag{1.3.27}$$

Note however that the sequence $\{\frac{\pi}{2}, \frac{\cos x}{x}, \frac{\sin x}{x^2}, \frac{\cos x}{x^3}, \frac{\sin x}{x^4}, \cdots\}$ is not an asymptotic sequence as $x \to \infty$. Moreover, the remainder in (1.3.25a, b) is big O of x^{-m}, for appropriate m, not big O of $x^{-m} \sin x$ nor $x^{-m} \cos x$. The expansion (1.3.27) is an example of an extended asymptotic expansion, with respect to $\{x^{-m}\}$ in this case.

Definition. *Let $\{\phi_n(x)\}$ be an asymptotic sequence as $x \to x_0$. A function $f(x)$, defined in a pointed neighborhood of x_0, has an extended asymptotic expansion with respect to $\{\phi_n(x)\}$ if we can find functions $f_1(x), f_2(x), \cdots$ with $f_n(x) = O(\phi_n(x))$, such that*

$$f(x) = f_1(x) + \cdots + f_N(x) + o(\phi_N(x)), \quad \text{as } x \to x_0, \tag{1.3.28}$$

for each N. In that case we write

$$f(x) \sim \sum_{n=1}^\infty f_n(x), \ \{\phi_n\}, \ \text{as } x \to x_0. \tag{1.3.29}$$

It is clear that (1.3.27) is then an extended asymptotic expansion of the sine integral function $S(x)$ in terms of the sequence $\{x^{-n+1}\}$.

1.4 Algebraic and Analytic Operations

We shall consider some of the basic operations that can be applied to asymptotic expansions. We do not attempt to give the results for general asymptotic sequences: usually there are so many cases to consider that such general theorems become hard to read and hard to use. Rather we try to illustrate the *methods* involved by considering some particular asymptotic sequences.

The linear combination of asymptotic sequences is simple.

Theorem 3. *If the functions $f(x)$ and $g(x)$ have the asymptotic developments to N terms*

$$f(x) \sim \sum a_i \phi_i(x), \quad g(x) \sim \sum b_i \phi_i(x), \ as \ x \to x_0, \qquad (1.4.1)$$

then for any constants λ, μ the function $\lambda f(x) + \mu g(x)$ has the asymptotic development to N terms.

$$\lambda f(x) + \mu g(x) \sim \sum (\lambda a_i + \mu b_i) \phi_i(x), \ as \ x \to x_0. \qquad (1.4.2)$$

The multiplication of asymptotic developments is somewhat more complicated since, in general, if $\{\phi_n(x)\}$ and $\{\psi_m(x)\}$ are asymptotic sequences as $x \to x_0$, there is no canonical way to obtain an asymptotic sequence out of the double sequence $\{\phi_n(x)\psi_m(x)\}$. When such an asymptotic sequence can be constructed then we can indeed multiply the expansions. The following theorem gives the result for asymptotic power series. Other situations are considered in the examples that follow.

Theorem 4. *Let $f(x)$ and $g(x)$ have the asymptotic developments*

$$f(x) \sim \sum_{n=0}^{\infty} a_n (x - x_0)^n, \quad g(x) \sim \sum_{n=0}^{\infty} b_n (x - x_0)^n, \ as \ x \to x_0. \quad (1.4.3)$$

Then

$$f(x)g(x) \sim \sum_{n=0}^{\infty} c_n (x - x_0)^n, \ as \ x \to x_0, \qquad (1.4.4)$$

where

$$c_n = \sum_{j+i=n} a_j b_i. \qquad (1.4.5)$$

Proof. For every N we have

$$f(x) = \sum_{n=0}^{N} a_n (x - x_0)^n + o((x - x_0)^N),$$

$$g(x) = \sum_{n=0}^{N} b_n (x - x_0)^n + o((x - x_0)^N),$$

thus

$$f(x)g(x) = \left(\sum_{n=0}^{N} a_n(x-x_0)^n\right)\left(\sum_{n=0}^{N} b_n(x-x_0)^n\right) + (o(x-x_0)^N)$$

$$= \sum_{n=0}^{N} c_n(x-x_0)^n + o((x-x_0)^N).$$

∎

Example 9. If

$$f(x) \sim a_1 e^{-x} + a_2 e^{-2x} + a_3 e^{-3x} + \cdots, \quad \text{as } x \to \infty$$

and

$$g(x) \sim b_1 e^{-x} + b_2 e^{-2x} + b_3 e^{-3x} + \cdots, \quad \text{as } x \to \infty,$$

then

$$f(x)g(x) \sim a_1 b_1 e^{-2x} + (a_1 b_2 + a_2 b_1)e^{-3x} + \cdots, \quad \text{as } x \to \infty.$$

Example 10. If

$$f(x) \sim \frac{a_1}{x} + \frac{a_2}{x^2} + \frac{a_3}{x^3+} \cdots, \quad \text{as } x \to \infty,$$

and

$$g(x) \sim b_1 e^{-x} + b_2 e^{-2x} + b_3 e^{-3x} + \cdots, \quad \text{as } x \to \infty,$$

then the terms $b_2 e^{-2x}, b_3 e^{-3x}$, etc., are lost when we multiply the expansions:

$$f(x)g(x) \sim \frac{a_1 b_1 e^{-x}}{x} + \frac{a_2 b_1 e^{-x}}{x^2} + \frac{a_3 b_1 e^{-x}}{x^3} + \cdots, \quad \text{as } x \to \infty.$$

For the division of asymptotic power series we have

Theorem 5. *Let $f(x)$ have the expansion*

$$f(x) \sim a_0 + a_1(x-x_0) + a_2(x-x_0)^2 + \cdots, \quad \text{as } x \to x_0, \qquad (1.4.6)$$

with $a_0 \neq 0$. Then $f(x) \neq 0$ for x near x_0 and

$$\frac{1}{f(x)} \sim b_0 + b_1(x-x_0) + b_2(x-x_0)^2 + \cdots, \quad \text{as } x \to x_0, \qquad (1.4.7)$$

where the b_i are computed by formally solving the equation

$$\left(\sum_{n=0}^{\infty} a_n \, (x - x_0)^n\right) \left(\sum_{n=0}^{\infty} b_n \, (x - x_0)^n\right) = 1, \quad \text{so that}$$

$$b_0 = \frac{1}{a_0}, \quad b_1 = \frac{-a_1}{a_0^2}, \quad b_2 = \frac{a_1^2 - a_0 \, a_2}{a_0^3}, \text{ etc.}$$

Example 11. Suppose

$$f(x) \sim \frac{a_1}{x} + \frac{a_2}{x^2} + \frac{a_3}{x^3} + \cdots, \quad \text{as } x \to \infty.$$

Then

$$\frac{1}{f(x)} \sim \frac{1}{\frac{a_1}{x} + \frac{a_2}{x^2} + \frac{a_3}{x^3} + \cdots} \sim \frac{x}{a_1} - \frac{a_2}{a_1^2}$$
$$+ \left(\frac{a_2^2 - a_1 \, a_3}{a_1^3}\right) \frac{1}{x} + \cdots, \quad \text{as } x \to \infty.$$

The integration of asymptotic developments can be studied from our results on the integration of the order symbols given in Section 1.2. Suppose, for instance, that

$$f(x) \sim a_1 \phi_1(x) + a_2 \phi_2(x) + a_3 \phi_3(x) + \cdots, \quad \text{as } x \to \infty, \qquad (1.4.8)$$

where $\{\phi_n(x)\}$ is an asymptotic sequence of positive functions with $\int_c^{\infty} \phi_n(x)dx < \infty$. Then it follows from (1.2.16) that

$$\int_x^{\infty} f(t)dt \sim a_1 \int_x^{\infty} \phi_1(t)dt + a_2 \int_x^{\infty} \phi_2(t)dt + a_3 \int_x^{\infty} \phi_3(t)dt + \cdots,$$
$$\text{as } x \to \infty, \qquad (1.4.9)$$

where we observe that $\{\int_x^{\infty} \phi_n(t)dt\}$ is also an asymptotic sequence as $x \to \infty$.

If, on the other hand, $\int_x^{\infty} \phi_n(t)dt = \infty$, for each n, then by (1.2.15)

$$\int_c^{x} f(t)dt \sim a_1 \int_c^{x} \phi_1(t)dt + a_2 \int_c^{x} \phi_2(t)dt + a_3 \int_c^{x} \phi_3(t)dt + \cdots, \quad \text{as } x \to \infty,$$

where now $\{\int_c^x \phi_n(t)dt\}$ is an asymptotic sequence as $x \to \infty$.

In the mixed case we have

Theorem 6. *Suppose $\int_c^\infty \phi_n(t)dt = \infty$ for $n < N$, but $\int_c^\infty \phi_N(t)dt < \infty$. Then we have*

$$\int_x^\infty [f(t) - a_1\phi_1(t) - \cdots - a_{N-1}\phi_{N-1}(t)]\, dt$$

$$\sim a_N \int_x^\infty \phi_N(t)dt + a_{N+1} \int_x^\infty \phi_{N+1}(t)dt + \cdots, \qquad \text{as } x \to \infty. \quad (1.4.10)$$

Let us consider an illustration.

Example 12. If

$$f(x) \sim \frac{a_1}{x} + \frac{a_2}{x^2} + \frac{a_3}{x^3} + \cdots, \quad \text{as } x \to \infty,$$

then Theorem 6 yields

$$\int_x^\infty (f(t) - \frac{a_1}{t})dt \sim a_2 \int_x^\infty \frac{dt}{t^2} + a_3 \int_x^\infty \frac{dt}{t^3} + \cdots, \quad \text{as } x \to \infty$$

$$\sim \frac{a_2}{x} + \frac{a_3}{2x^2} + \frac{a_4}{3x^3} + \cdots, \quad \text{as } x \to \infty.$$

Therefore,

$$\int_c^x f(t)dt = a_1(\ln x - \ln c) + \int_c^x (f(t) - \frac{a_1}{t})dt.$$

$$= a_1(\ln x - \ln c) + \int_c^\infty (f(t) - \frac{a_1}{t})dt - \int_x^\infty (f(t) - \frac{a_1}{t})dt$$

Hence

$$\int_a^x f(t)dt \sim a_1 \ln x + A - \frac{a_2}{x} - \frac{a_3}{2x^2} - \frac{a_4}{3x^3} - \cdots, \quad \text{as } x \to \infty, \quad (1.4.11)$$

where

$$A = \int_c^\infty (f(t) - \frac{a_1}{t})dt - a_1 \ln c. \quad (1.4.12)$$

As we shall explain in the next chapter, the constant A is precisely the finite part, in the sense of Hadamard, of the divergent integral $\int_c^\infty f(t)dt$.

The results concerning the differentiation of asymptotic developments are rather weak. Indeed the sequence of derivatives $\{\phi_n'(x)\}$ of an asymptotic sequence $\{\phi_n(x)\}$ might not be an asymptotic sequence. But even if it is, the development of $f'(x)$ need not be that obtained by the differentiation of the development of $f(x)$. Take for instance $f(x) = e^{-x} \sin e^x$; then $f(x) = o(x^{-n})$, as $x \to \infty$, for each n, but $f'(x) = -e^{-x} \sin e^x + \cos e^x$ does not have a development with respect to $\{x^{-n}\}$.

Asymptotic developments that can be differentiated k times are called *strong developments* of order k. In case this holds for every k the asymptotic development is termed *strong*.

Thus asymptotic developments usually are not strong, not even of order 1. A remarkable exception is provided by the asymptotic power series expansions of analytic functions: as we shall see in the Section 1.6 these expansions are always strong.

1.5 Existence of Functions with a Given Asymptotic Expansion

We have already seen that the asymptotic expansion of a function in terms of a given sequence, if it exists, is unique. Also, we have already encountered examples of different functions having the same asymptotic development. In this section we shall study the *existence problem,* namely, given an asymptotic sequence $\{\phi_n(x)\}$, as $x \to x_0$, and constants $c_1, c_2, c_3, \cdots$, is there a function $f(x)$ with asymptotic expansion

$$f(x) \sim \sum_{n=1}^{\infty} c_n \phi_n(x), \quad \text{as } x \to x_0?$$

Let us start with an example.

Example 13. The series

$$\sum_{n=0}^{\infty} (-1)^n n! x^n, \tag{1.5.1}$$

is called the Stieltjes series. It diverges for all $x \neq 0$.

Our aim is to examine whether it is the asymptotic expansion as $x \to 0^+$ of some function. For this purpose, we appeal to the definition of the gamma function

$$\Gamma(n+1) = n! = \int_0^{\infty} e^{-t} t^n dt,$$

and proceed in a formal way so that (1.5.1) can be rewritten as

$$\sum_{n=0}^{\infty} (-x)^n \int_0^{\infty} e^{-t} t^n dt = \int_0^{\infty} e^{-t} \sum_{n=0}^{\infty} (-x\,t)^n dt,$$

where we have interchanged the order of summation and integration (a step which, of course, cannot be justified). But since $\sum_{n=0}^{\infty} (-x\,t)^n = \dfrac{1}{1+x\,t}$ for $|\,x\,t\,| < 1$ and we are interested in the situation when $x \ll 1$, we are led to the integral

$$f(x) = \int_0^\infty \frac{e^{-t}}{1+x\,t}\,dt. \qquad (1.5.2)$$

Observe that although our procedure is purely formal, we have arrived at a function $f(x)$ defined in (1.5.2) by an integral that converges for all $x \geq 0$. Furthermore, as we now show, as $x \to 0^+$ the function $f(x)$ has the asymptotic expansion (1.5.1). Indeed, repeated integration by parts yields

$$f(x) = \int_0^\infty (1+x\,t)^{-1} e^{-t}\,dt$$

$$= 1 - x \int_0^\infty (1+x\,t)^{-2} e^{-t}\,dt$$

$$= 1 - x + 2x^2 \int_0^\infty (1+x\,t)^{-3} e^{-t}\,dt$$

$$\vdots$$

$$= 1 - x + 2!\,x^2 - 3!\,x^3 + \cdots + (-1)^n n!\,x^n + R_n(x),$$

where

$$R_n(x) = (-1)^{n+1}(n+1)!\,x^{n+1} \int_0^\infty (1+x\,t)^{-n-2} e^{-t}\,dt = O(x^{n+1}),$$

as $x \to 0^+$. Therefore we have succeeded in finding a function $f(x)$ with the given asymptotic development (1.5.1).

This example hints that no matter how fast a series $\sum_{n=1}^{\infty} c_n \phi_n(x)$ diverges, it is still possible to find a function $f(x)$ whose asymptotic development as $x \to x_0$ is precisely $\sum_{n=1}^{\infty} c_n \phi_n(x)$. That this is indeed the case even for extended asymptotic expansions is shown in Theorem 7.

For simplicity we deal with a metric space, but the reader can easily show that the argument carries over to normal topological spaces. The basic result we are going to use is that, given two disjoint closed sets H and K of the metric space X, there exists a continuous function $\rho(x)$ that satisfies the following three conditions:

(a) $\rho(x) = 0, \quad x \in H$.

(b) $\rho(x) = 1$, $x \in K$.
(c) $0 \le \rho(x) \le 1$, $x \in \mathbf{X}$.

One such function is given by $\rho(x) = \dfrac{d(x,H)}{d(x,H) + d(x,K)}$, where $d(x,F) =$ $\inf\{d(x,y) : y \in F\}$.

Theorem 7. *Let $\mathbf{X}$ be a metric space and let $\{\phi_n(x)\}$ be an asymptotic sequence of continuous functions as $x \to x_0$. Then if $\{f_n(x)\}$ is a sequence of continuous functions on $\mathbf{X} \setminus \{x_0\}$ that satisfy $f_n(x) = O(\phi_n(x))$ as $x \to x_0$, there exists a continuous function $f(x)$ such that*

$$f(x) \sim \sum_{n=1}^{\infty} f_n(x), \quad \{\phi(x)\}, \quad \text{as } x \to x_0. \tag{1.5.3}$$

Proof. We can construct a sequence of open neighborhoods $\{V_n\}$ of x_0 such that: $(a)\overline{V}_{n+1} \subset V_n$; $(b) \bigcap_{n=1}^{\infty} V_n = \{x_0\}$; (c) if $x \in V_n \setminus \{x_0\}$ then $|f_{n+1}(x)| \le \frac{1}{2}|\phi_n(x)|$ and $|\phi_{n+1}(x)| \le \frac{1}{2}|\phi_n(x)|$. Let ρ_n be continuous functions with $0 \le \rho_n \le 1$, $\rho_n(x) = 1$ if $x \in V_{n+1}$, $\rho_n(x) = 0$ if $x \notin V_n$. Set

$$f(x) = \sum_{n=1}^{\infty} \rho_n(x) f_n(x). \tag{1.5.4}$$

Then $f(x)$ is well defined for any $x \ne x_0$ since the sum (1.5.4) consists of only a finite number of non-zero terms. The continuity also follows since any $x \ne x_0$ has a neighborhood where $f(x)$ is a finite sum of the continuous functions $\rho_n(x) f_n(x)$. To show that $f \sim \sum f_n$, $\{\phi_n\}$ as $x \to x_0$, let N be fixed. Then if $x \in V_{N+1}$,

$$\left| f(x) - \sum_{n=1}^{N} f_n(x) \right| \le \sum_{n=N+1}^{\infty} |\rho_n(x) f_n(x)| \le \{|f_{N+1}(x)|$$

$$+ \left(\sum_{n=N+2}^{\infty} 2^{N-2-n} \right) |\phi_{N+1}(x)|\}$$

$$= o(\phi_N(x)), \quad \text{as } x \to x_0.$$

∎

When $\{\phi_n(x)\}$ and $\{f_n(x)\}$ are of class C^k in a subset X of $\mathbb{R}^n$ then the above construction yields a function $f(x)$ of class C^k since we can take the $\rho_n(x)$ infinitely differentiable. It follows, in particular, that if $x_0 \in \mathbb{R}$ and $\{a_n\}$ is a sequence of real or complex numbers, there exists a function $f \in C^\infty(\mathbb{R} \setminus \{x_0\})$ with

$$f(x) \sim a_0 + a_1(x - x_0) + 2!\, a_2(x - x_0)^2 + \cdots, \quad \text{as } x \to x_0. \quad (1.5.5)$$

The expansion (1.5.5) does not imply that f is smooth at $x = x_0$. Take $f(x) = e^{\frac{-1}{x^2}} \sin(e^{\frac{1}{x^2}})$ for instance: we have $f(x) = o(x^n)$ as $x \to 0$ for each n, but $f''(0)$ does not exist. However, a close examination of the method of proof of the theorem shows that if ρ_n is taken as an appropriate smooth function then $f(x) = \displaystyle\sum_{n=0}^{\infty} n!\, a_n \rho_n(x)(x - x_0)^n$ not only satisfies (1.5.5), but also $f^{(k)}(x)$ has an expansion that can be obtained by differentiation of (1.5.5). This yields the following *Borel theorem.*

Theorem 8. *Let $x_0 \in \mathbb{R}$ and let $\{a_n\}$ be a sequence of real or complex numbers. Then there exists a C^∞ function $f(x)$ with*

$$f^{(n)}(x_0) = a_n, \quad n = 0, 1, 2, \ldots. \quad (1.5.6)$$

The generalization of this result to n variables is also true. If we use the notation $\boldsymbol{D}^{\boldsymbol{k}}$ for the partial derivative operator $\frac{\partial^{k_1 + \cdots + k_n}}{\partial x_1^{k_1} \cdots \partial x_n^{k_n}}$, where $\boldsymbol{k} = (k_1, \ldots, k_n) \in \mathbb{N}^n$ is a multi-index, then we have

Theorem 9. *Let $\boldsymbol{x}_0 \in \mathbb{R}^n$ and let $\{a_{\boldsymbol{k}}\}$ be a sequence indexed by multi-indices $\boldsymbol{k} \in \mathbb{N}^n$. Then there exist $f \in C^\infty(\mathbb{R}^n)$ such that*

$$\boldsymbol{D}^{\boldsymbol{k}} f(\boldsymbol{x}_0) = a_{\boldsymbol{k}}, \quad \boldsymbol{k} \in \mathbb{N}^n. \quad (1.5.7)$$

1.6 Asymptotic Power Series in a Complex Variable

In the previous sections we have discussed general asymptotic expansions. Our aim in this section is to study the asymptotic power series expansion of analytic functions. As we shall see, the asymptotic power series in a complex variable has new and interesting features.

In order to appreciate the new ideas involved, we return to one of our previous examples, the integral

$$f(z) = \int_z^\infty \frac{e^{z-\xi}}{\xi} d\xi. \quad (1.6.1)$$

In formula (1.6.1) we take a path of integration that does not pass through the origin — where the integrand has a pole — and that ends on the positive real axis to ensure convergence. If z is a point with $\Re ez < 0$ then the value of the integral (1.6.1) depends on whether the path passes below or above the origin. Therefore (1.6.1) is single valued only in a split plane, where the branch line is the negative real axis. In this split plane we can take the path C_z as the arc of the circle $\mid \xi \mid = \mid z \mid$ to the positive real axis plus the interval $(\mid z \mid, \infty)$ of the positive real axis.

We develop the asymptotic series as $z \to \infty$ as before because the integration by parts is still valid. Thus,

$$f(z) = \sum_{k=1}^{n} (-1)^{k-1} (k-1)! \, z^{-k} + (-1)^n n! \int_z^\infty e^{z-\xi} \xi^{-n-1} d\xi. \quad (1.6.2)$$

To estimate the remainder we integrate once more by parts to obtain

$$R_n(z) = (-1)^n n! \int_z^\infty e^{z-\xi} \xi^{-n-1} d\xi$$

$$= (-1)^n n! \left[\frac{1}{z^{n+1}} - (n+1) \int_z^\infty e^{z-\xi} \xi^{-n-2} d\xi \right]. \quad (1.6.3)$$

On the arc, the real part of the exponent is non-positive and $\mid d\xi \mid = \mid z \mid d\phi$. Thus,

$$\mid R_n(z) \mid \leq n! \mid z \mid^{-n-1} \left[1 + (n+1)(\pi + \frac{1}{n+1}) \right],$$

and again we find that

$$R_n(z) = O\left(\frac{1}{\mid z \mid^{n+1}} \right), \quad \text{as } z \to \infty. \quad (1.6.4)$$

Accordingly, $f(z)$ has the asymptotic series expansion

$$f(z) \sim \sum_{n=1}^{\infty} (-1)^n (n-1)! \, z^{-n}, \quad \text{as } z \to \infty. \quad (1.6.5)$$

Observe that the right-hand side of (1.6.5), the series expansion, is a sum of *single valued* functions of z. Yet the function $f(z)$ is *multivalued* in the neighborhood of $z = \infty$. Hence it is possible to obtain an asymptotic series about a branch point while it is impossible to obtain a Taylor or Laurent series about a branch point. Incidentally, since the series is single valued but the function is not, we deduce at once that the series cannot converge to $f(z)$.

Now we describe an interesting phenomenon, the so-called *Stokes phenomenon*. Indeed, the analytic continuation of the asymptotic series might not be equal to the asymptotic series for the analytic continuation. For this purpose, we look at the multivalued function $f(z)$ given above. If $\pi < \arg z < 3\pi$, then the path of integration in (1.6.1) encircles the origin once and thus

$$f(z) = f(ze^{-2\pi i}) - 2\pi i\, e^z, \qquad (1.6.6)$$

as follows from the residue theorem. The expansion of $f(ze^{-2\pi i})$ has already been obtained because the argument of $ze^{-2\pi i}$ is now between $-\pi$ and π. As $(ze^{-2\pi i})^n = z^n$, we obtain

$$f(z) \sim -2\pi i e^z + \sum_{n=1}^{\infty} (-1)^{n-1}(n-1)!z^{-n}, \quad \text{as } z \to \infty, \ \pi < \arg z < 3\pi,$$

$$(1.6.7)$$

as the asymptotic expansion of the analytic continuation of $f(z)$ to the sector $\pi < \arg z < 3\pi$. But the analytic continuation of the series expansion for $f(z)$ given in (1.6.5) is single valued and its analytic continuation is the same series. This is the Stokes phenomenon.

The curves on which the above-mentioned change occurs are called *Stokes' lines*. In the present example they may be seen by inspection. Because the added term $-2\pi i e^{-z}$ is exponentially small in the range $\frac{\pi}{2} < \arg z < \frac{3\pi}{2}$, it is negligible in comparison with any of the terms of the series. Thus (1.6.5) actually holds in the extended range $-\pi < \arg z < \frac{3\pi}{2}$. The line $\arg z = \frac{3\pi}{2}$ is a Stokes line for $f(z)$. A similar analysis shows that (1.6.5) holds in the sector $\frac{-3\pi}{2} < \arg z < \frac{3\pi}{2}$, and that the ray $\arg z = \frac{-3\pi}{2}$ is also a Stokes line.

Let us give another example.

Example 14. Consider the function $g(z) = e^{\frac{-1}{z^2}}$ in the neighborhood of $z = 0$. Then $g(z) \sim 0$ to all orders with respect to the sequence $\{z^n\}$ in the sector $\frac{-\pi}{4} < \arg z < \frac{\pi}{4}$. On the other hand, $g(z)$ does not even have an expansion with respect to $\{z^n\}$ in the sector $\frac{\pi}{4} < \arg z < \frac{3\pi}{4}$. It is easy to see that the rays $\frac{\pi}{4} + \frac{k\pi}{2}$, $k = 0, \pm 1, \pm 2, \cdots$, are Stokes lines for the function $g(z)$.

We shall now study other properties of the asymptotic power series expansion of analytic functions. We shall work near $z = 0$, but using a conformal map we see that the results also hold around any other point $z \in \mathbb{C}$ or around $z = \infty$.

We have already studied the sum and product of asymptotic expansions; nothing new appears in this case. The integration of asymptotic power series in a complex variable does not present new features either. For the differentiation, however, we have the following result.

Theorem 10. *Let $f(z)$ be analytic in the region $S : 0 < |z| < r, \ \alpha < \arg z < \beta$ and have the asymptotic expansion*

$$f(z) \sim \sum_{n=0}^{\infty} a_n z^n, \quad as \ z \to 0. \tag{1.6.8}$$

Then, on any subsector $S_1 : \alpha_1 < \arg z < \beta_1$, with $\alpha < \alpha_1 < \beta_1 < \beta$ we have

$$f'(z) \sim \sum_{n=1}^{\infty} n \, a_n z^{n-1}, \quad as \ z \to 0. \tag{1.6.9}$$

Proof. Write

$$f(z) = \sum_{n=0}^{N} a_n z^n + z^N E(z), \tag{1.6.10}$$

where

$$\lim_{z \to 0} E(z) = 0, \quad \alpha < \arg z < \beta. \tag{1.6.11}$$

Differentiation yields

$$f'(z) = \sum_{n=1}^{N} n \, a_n z^{n-1} + N z^{N-1} E(z) + z^N E'(z). \tag{1.6.12}$$

Let ρ be a small positive number such that for each $z \in S_1, |z| < \dfrac{r}{2}$, the circle C_z with center z and radius $|z| \rho$ lies in S. If $M(z)$ denotes the maximum of $E(w)$ for w on the circle C_z, then on and inside C_z

$$|E'(w)| \le \frac{M(z)}{\rho |z|}, \tag{1.6.13}$$

by Cauchy's formula. Therefore

$$\left| f'(z) - \sum_{n=1}^{N} n \, a_n z^{n-1} \right| \le |z|^{N-1} \left[N \, |E(z)| + \frac{M(z)}{\rho} \right],$$

and the expression in brackets tends to zero as $z \to 0$ in S_1. ∎

We can rephrase Theorem 10 by saying that the asymptotic power series expansion of an analytic function is always strong. Therefore,

$$f^{(k)}(z) \sim \sum_{n=k}^{\infty} n(n-1)\cdots(n-k+1)\, a_n\, z^{n-k}, \quad \text{as } z \to 0, \qquad (1.6.14)$$

within any subsector of S.

Using (1.6.14) we immediately obtain the formula

$$a_n = \lim_{z \to 0} \frac{f^{(n)}(z)}{n!}, \qquad (1.6.15)$$

for the coefficients of the development. Actually we can say a little more.

Theorem 11.   *Let $f(z)$ be analytic in the sector $S : \alpha < \arg z < \beta$, $|z| < r$. Then $f(z)$ has an asymptotic power series expansion as $z \to 0$ if and only if the limits*

$$\lim_{\substack{z \to 0 \\ z \in S}} f^{(n)}(z) \qquad (1.6.16)$$

all exist.

Proof. We have already observed that if $f(z) \sim \sum_{n=0}^{\infty} a_n z^n$, as $z \to 0$ in S, then (1.6.15) holds. Conversely, suppose that all the limits in (1.6.16) exist. Then we can let a approach 0 in the Taylor formula

$$f(z) = f(a) + (z-a)f'(a) + \frac{(z-a)^2}{2!}\, f''(a) + \cdots + \frac{(z-a)^m}{m!}\, f^{(m)}(a)$$

$$+ \frac{1}{m!} \int_a^z (z-\xi)^m\, f^{(m+1)}(\xi)d\xi,$$

to obtain

$$f(z) = a_0 + a_1 z + a_2 z^2 + \cdots + a_m z^m + \frac{1}{m!} \int_0^z (z-\xi)^m f^{(m+1)}(\xi)\, d\xi. \quad (1.6.17)$$

But if M is a bound for $|f^{(m+1)}(\xi)|$ in a subsector $\alpha_1 < \arg z < \beta_1$, $|z| < \frac{r}{2}$, then

$$\left| \frac{1}{m!} \int_0^z (z-\xi)^m f^{(m+1)}(\xi)\, d\xi \right| \le \frac{M\, |z|^{m+1}}{m!} \int_0^1 (1-t)^m dt = O(\,|z|^{m+1}\,).$$

■

We would also like to consider the existence problem: if $\{a_n\}$ is a sequence of complex numbers, can we find an analytic function $f(z)$ in S with $f(z) \sim \sum_{n=0}^{\infty} a_n z^n$? Recall the proof of Theorem 7. There we constructed a function $f(z)$ as a sum of the form $\sum_{n=0}^{\infty} a_n \rho_n(z) z^n$, where the $\rho_n(z)$ were suitable cut-off functions. For our present problem, we have no analytic cut-off functions, but functions of the type

$$\rho_n(z) = 1 - e^{-b_n z^{-\delta}}, \qquad (1.6.18)$$

for suitable $b_n > 0$ and $0 < \delta < 1$ serve our purpose. Actually, if δ is small enough, the exponent will have a negative real part in any prescribed sector S and thus $\rho_n(z)$ will tend to 1 very fast as $z \to 0$ within S.

Theorem 12. *Let $\{a_n\}$ be a sequence of complex numbers. Then there exists an analytic function in the sector $S : \alpha < \arg z < \beta, \ | z | < r$ such that*

$$f(z) \sim \sum_{n=0}^{\infty} a_n z^n, \quad as \ z \to 0, \ z \in S. \qquad (1.6.19)$$

Proof. Let $\rho_n(z)$ be defined by (1.6.18) where $b_n = | a_n r^n |^{-1}$ if $a_n \neq 0$ and $b_n = 0$ if $a_n = 0$. Then if δ is small enough for $\Re e \, (z^\delta) > 0$ on S, we will have $\Re e(\omega) < 0$ if $\omega = -b_n z^{-\delta}$, thus $| -e^\omega | = | \int_0^\omega e^z dz | = | \int_0^1 e^{\omega t} \omega dt | \leq | \omega | \int_0^1 | e^{\omega t} | \, dt \leq | \omega |$,

$$| a_n \rho_n(z) z^n | \leq | a_n | \, | b_n | \, | z |^{n-\delta} \leq | \frac{z}{r} |^n \, | z |^{-\delta},$$

and the convergence on S of the series

$$f(z) = \sum_{n=0}^{\infty} a_n \, \rho_n(z) \, z^n \qquad (1.6.20)$$

follows.

To see that $f(z)$ has the asymptotic development (1.6.19) observe that

$$z^{-N} \left(f(z) - \sum_{n=0}^{N} a_n z^n \right) = - \sum_{n=0}^{N} a_n e^{-b_n z^{-\delta}} z^{-(N-n)}$$

$$+ \sum_{n=N+1}^{\infty} a_n \rho_n(z) \, z^{n-N}.$$

The first term tends to zero as $z \to 0$ in S and so does the second since

$$
\left| \sum_{n=N+1}^{\infty} a_n \rho_n(z) \, z^{n-N} \right| \le \sum_{n=N+1}^{\infty} \left| \frac{z}{r} \right|^n \left| z^{-\delta - N} \right| \le \frac{\left| \frac{z}{r} \right|^{N+1} |z|^{-\delta - N}}{1 - \left| \frac{z}{r} \right|} .
$$

∎

1.7 Asymptotic Approximations of Partial Sums

In this section we study some simple but rather useful methods for the asymptotic evaluation of indefinite integrals of the type:

$$
\int_a^x g(t) \, dt , \quad \text{as } x \to \infty, \tag{1.7.1}
$$

as well as sums of the type:

$$
\sum_{n=1}^{N} a_n , \quad \text{as } N \to \infty. \tag{1.7.2}
$$

As we will soon show, the methods for both situations are similar.

   Actually, the relation between indefinite integrals of the type (1.7.1) and sums of the type (1.7.2) is rather close since integrals can be approximated by sums and conversely. These ideas are studied in Section 8 when we consider the Euler-Maclaurin formula. Presently we give some direct methods. Let us start with the integral. From the elementary formula

$$
g(t) = (t \, g(t))' - t g'(t) , \tag{1.7.3}
$$

we observe that since the integral of $(t \, g(t))'$ is known explicitly, the integral of $g(t)$ could be approximated if we could approximate the integral of $t \, g'(t)$. Indeed, we have:

Theorem 13.   *Let $g(t)$ be differentiable and positive for $t > a$. Suppose that:*

$$
\frac{g'(t)}{g(t)} \sim \frac{\alpha}{t} , \quad \text{as } t \to \infty, \quad \alpha \ne 0, -1. \tag{1.7.4}
$$

(a) If $\alpha > -1$ then $\int_a^{\infty} g(t) \, dt$ diverges and we have

$$
\int_a^x g(t) \, dt \sim \frac{x g(x)}{\alpha + 1} \quad \text{as } x \to \infty. \tag{1.7.5}
$$

(b) If $\alpha < -1$ then $\int_a^\infty g(t)\,dt$ converges and

$$\int_x^\infty g(t)\,dt \sim \frac{-xg(x)}{\alpha+1}, \quad \text{as } x \to \infty. \tag{1.7.6}$$

Proof.

(a) The relation (1.7.4) can be integrated to obtain

$$\ln g(x) \sim \alpha \ln x, \quad \text{as } x \to \infty. \tag{1.7.7}$$

Thus if $\varepsilon > 0$ is arbitrary, we have

$$g(x) \geq x^{\alpha-\varepsilon}, \quad x \gg 1,$$

and it follows that $\int_a^\infty g(x)\,dx$ diverges. Using (1.7.3) we obtain

$$\int_a^x g(t)\,dt = xg(x) - ag(a) - \int_a^x tg'(t)\,dt,$$

or

$$\int_a^x g(t)\,dt + \int_a^x tg'(t)\,dt = xg(x) - ag(a),$$

but $tg'(t) \sim \alpha g(t)$ and thus

$$\int_a^x g(t)\,dt + \int_a^x tg'(t)\,dt \sim (1+\alpha) \int_a^x g(t)\,dt$$

and hence (1.7.5) follows.

(b) Using (1.7.7) we find that for each $\varepsilon > 0$

$$g(x) \leq x^{\alpha-\varepsilon}, \quad x \gg 1,$$

and thus $\int_a^\infty g(x)\,dx$ converges. Again the same argument applies, i.e.,

$$\int_x^\infty g(t)\,dt \sim \frac{1}{1+\alpha} \int_x^\infty (g(t) + tg'(t))\,dt = \frac{-xg(x)}{1+\alpha},$$

and (1.7.6) is established. ∎

Example 15. Let us consider the integral $\int_o^x \sqrt{t^4+1}\,dt$. Here $g(t) = \sqrt{t^4+1}$ and we have $\dfrac{g'(t)}{g(t)} = \dfrac{2t^3}{t^4+1} \sim \dfrac{2}{t}$, as $t \to \infty$. Therefore

$$\int_0^x \sqrt{t^4+1}\,dt \sim \frac{1}{3}x\sqrt{x^4+1} \quad \text{as} \quad x \to \infty.$$

The theorem requires that $\alpha \neq 0,\ 1$. The case $\alpha = 1$ is a boundary situation and might be somewhat complicated. For $\alpha = 0$ we have the following result: If $\dfrac{g'(t)}{g(t)} = o\left(\dfrac{1}{t}\right)$ as $t \to \infty$ then $\int_a^x g(t)dt \sim x\, g(x)$, as $x \to \infty$. The proof is the same as before, after observing the case $\int_a^x t\, g'(t)dt = o\left(\int_a^x g(t)\right)$, as $x \to \infty$.

For instance, for the integral $\int_a^x \dfrac{dt}{\ln t}$, $a > 1$, we have: $\dfrac{g'(t)}{g(t)} = \dfrac{-1}{t\ln t} = o\left(\dfrac{1}{t}\right)$ and therefore:

$$\int_a^x \frac{dt}{\ln t} \sim \frac{x}{\ln x}, \quad \text{as } x \to \infty.$$

The basic idea behind Theorem 13 is to find an appropriate function h such that $g(t) \sim h'(t)$ as $t \to \infty$. Indeed, the function h is given by $h(t) = \dfrac{tg(t)}{1+\alpha}$. Many times the theorem is not applicable, but the function $h(t)$ can be found by inspection. For instance, to evaluate

$$\int_0^x e^{t^2}\,dt$$

we observe that $e^{t^2} \sim \left(\dfrac{e^{t^2}}{2t}\right)'$ and thus we obtain

$$\int_0^x e^{t^2}\,dt \sim \frac{e^{x^2}}{2x}, \quad \text{as } x \to \infty.$$

Similarly, to approximate the integral $\int_x^\infty e^{-\sqrt{t^2+1}}\,dt$ we observe that $g(t) = e^{-\sqrt{t^2+1}}$ and by computing the derivative,

$$g'(t) = \frac{-t}{\sqrt{t^2+1}}\,e^{-\sqrt{t^2+1}} \sim g(t).$$

Thus,

$$\int_x^\infty e^{-\sqrt{t^2+1}} \sim e^{-\sqrt{x^2+1}}, \quad \text{as } x \to \infty.$$

Let us now consider the problem of the asymptotic evaluation of partial sums of the type:

$$S_N = \sum_{n=1}^N a_n.$$

Observe that the sum can be evaluated in closed form provided it is *telescopic*, i.e., $a_n = b_n - b_{n-1}$ for some *known* sequence b_n. In that case we clearly have

$$\sum_{n=1}^N (b_n - b_{n-1}) = b_N - b_0.$$

Proceeding in analogy with the analysis for integrals, we can try to examine the sequence $\dfrac{a_n - a_{n-1}}{a_n} = 1 - \dfrac{a_{n-1}}{a_n}$.

Theorem 14. *Let* $\displaystyle\sum_{n=1}^\infty a_n$ *be a series of positive terms. Suppose that*

$$\frac{a_{n-1}}{a_n} = 1 - \frac{\alpha}{n} + o\left(\frac{1}{n}\right) \qquad \text{as} \quad n \to \infty, \ \alpha \neq -1. \tag{1.7.8}$$

(a) If $\alpha > -1$ *the series diverges and we have*

$$\sum_{n=1}^N a_n \sim \frac{N a_N}{\alpha + 1}, \qquad \text{as } N \to \infty. \tag{1.7.9}$$

(b) If $\alpha < -1$ *then the series converges and we have*

$$\sum_{n=N}^\infty a_n \sim \frac{-N a_N}{\alpha + 1}, \qquad \text{as } N \to \infty. \tag{1.7.10}$$

Proof. Write, in analogy with (1.7.3),

$$a_n = [(n+1)a_n - n a_{n-1}] - n(a_n - a_{n-1}). \tag{1.7.11}$$

If $\alpha > -1$ the sequence $(n+1)a_n$ is increasing for large n and the divergence of $\sum_{n=1}^{\infty} a_n$ follows. If we now use (1.7.11) and (1.7.8),

$$(1+\alpha) \sum_{n=1}^{N} a_n \sim \sum_{n=1}^{N} [a_n + n(a_n - a_{n-1})] = (N+1)a_N,$$

and (1.7.9) is obtained.

When $\alpha < -1$, the series $\sum a_n$ is majorized by the convergent series $\dfrac{-1}{1+\alpha} \sum [(n+1)a_n - na_{n-1}]$. Using (1.7.11) and (1.7.8) then yields

$$(1+\alpha) \sum_{n=N}^{\infty} a_n \sim \sum_{n=N}^{\infty} [a_n + n(a_n - a_{n-1})]$$

$$= \sum_{n=N}^{\infty} [(n+1)a_n - na_{n-1}]$$

$$= -N\, a_{N-1}.$$

■

Let us consider some examples.

Example 16.   Let us consider the sum $\sum_{n=1}^{\infty} \dfrac{1}{n^p}$, where $p > 1$. The series is convergent and so as a first step we write

$$\sum_{n=1}^{N} \frac{1}{n^p} = \zeta(p) + o(1) \quad \text{as} \quad N \to \infty, \tag{1.7.12}$$

where ζ is the Riemann zeta function given by

$$\zeta(s) = \sum_{n=1}^{\infty} \frac{1}{n^s} \quad \text{for} \quad \Re e\, s > 1. \tag{1.7.13}$$

Next, we observe that $\sum_{n=1}^{N} \frac{1}{n^p} = \zeta(p) - \sum_{n=N+1}^{\infty} \frac{1}{n^p}$. Since $\dfrac{(n-1)^{-p}}{n^{-p}} = 1 + \dfrac{p}{n} + o\left(\dfrac{1}{n}\right)$, we obtain

$$\sum_{n=N+1}^{\infty} \frac{1}{n^p} \sim \frac{1}{(p-1)N^{p-1}},$$

and thus

$$\sum_{n=1}^{N} \frac{1}{n^p} = \zeta(p) - \frac{1}{(p-1)N^p} + o\left(\frac{1}{N^{p-1}}\right), \quad \text{as } N \to \infty. \qquad (1.7.14)$$

As the next example shows, the basic idea is to approximate a_n by an appropriate telescoping series $b_{n+1} - b_n$.

Example 17. Consider the series $\sum_{n=N}^{\infty} \frac{1}{n^2}$. We can approximate $\frac{1}{n^2}$ by the telescoping sequence $\frac{1}{n(n+1)} = \frac{1}{n} - \frac{1}{n+1}$. Indeed, $\frac{1}{n^2} = \frac{1}{n(n+1)} + \frac{1}{n^2(n+1)}$ and thus

$$\sum_{n=N}^{\infty} \frac{1}{n^2} = \sum_{n=N}^{\infty} \frac{1}{n(n+1)} + \sum_{n=N}^{\infty} \frac{1}{n^2(n+1)} = \frac{1}{N} + O\left(\frac{1}{N^2}\right) \text{ as } N \to \infty,$$

since $\sum_{n=N}^{\infty} \frac{1}{n^2(n+1)} \sim \sum_{n=N}^{\infty} \frac{1}{n^3} = O(\frac{1}{N^2})$.

We can continue this process. Write $\frac{1}{n^2(n+1)} = \frac{1}{n(n+1)(n+2)} + \frac{2}{n^2(n+1)(n+2)}$ and observe that

$$\sum_{n=N}^{\infty} \frac{1}{n(n+1)(n+2)} = \frac{1}{2} \sum_{n=N}^{\infty} \frac{1}{n(n+1)} - \frac{1}{(n+1)(n+2)} = \frac{1}{2N(N+1)}.$$

Therefore

$$\sum_{n=N}^{\infty} \frac{1}{n^2} = \frac{1}{N} + \frac{1}{2N(N+1)} + O\left(\frac{1}{N^3}\right) \quad \text{as} \quad N \to \infty.$$

Repeating this process, the infinite expansion

$$\sum_{n=N}^{\infty} \frac{1}{n^2} = \frac{1}{N} + \frac{1}{2N(N+1)} + \frac{2!}{3N(N+1)(N+2)}$$

$$+ \frac{3!}{4N(N+1)(N+2)(N+3)} + \cdots, \quad N \to \infty$$

$$(1.7.15)$$

is obtained.

Example 18. Let us study the partial sums of the harmonic series $\sum\limits_{n=1}^{N}\dfrac{1}{n}$.
Theorem 14 cannot be applied since $\alpha = -1$. However, we can approximate
$\dfrac{1}{n}$ by the telescoping series $\ln(1+\dfrac{1}{n}) = \ln(n+1) - \ln(n)$ since $\ln(1+\dfrac{1}{n}) \sim$
$\dfrac{1}{n} - \dfrac{1}{2n^2} + \dfrac{1}{3n^2} - ...,$ as $n \to \infty$. We find

$$\sum_{n=1}^{N}\frac{1}{n} = \sum_{n=1}^{N}\ln\left(1+\frac{1}{n}\right) + O(1) = \ln N + O(1), \quad \text{as } N \to \infty.$$

Since $\frac{1}{n} - \ln\left(1+\frac{1}{n}\right) \sim \frac{1}{2n^2}$, the series $\sum_{n=1}^{\infty}\left[\frac{1}{n} - \ln\left(1+\frac{1}{n}\right)\right]$ converges.
Its value is $\gamma = .5772157...$, the Euler constant, and

$$\gamma = \lim_{N\to\infty}\left(\sum_{n=1}^{N}\frac{1}{n} - \ln N\right). \tag{1.7.16}$$

Thus,

$$\sum_{n=1}^{N}\frac{1}{n} = \ln N + \gamma + O\left(\frac{1}{N}\right), \quad \text{as } N \to \infty.$$

The process can be continued. For instance, we obtain

$$\sum_{n=1}^{N}\frac{1}{n} = \sum_{n=1}^{N}\ln\left(1+\frac{1}{n}\right) + \sum_{n=1}^{\infty}\left(\frac{1}{n} - \ln\left(1+\frac{1}{n}\right)\right)$$
$$- \sum_{n=N+1}^{\infty}\left(\frac{1}{n} - \ln\left(1+\frac{1}{n}\right)\right)$$
$$= \ln(N+1) + \gamma - \frac{1}{2N} + O\left(\frac{1}{N^2}\right)$$
$$= \ln N + \gamma + \frac{1}{2N} + O\left(\frac{1}{N^2}\right), \quad \text{as } N \to \infty.$$

Other methods are required for rapidly divergent series.

Example 19. Consider the sum

$$S_N = \sum_{n=1}^{N} n!$$

Here the terms of the sum increase very fast. The maximum term is the last one and we could expect that its contribution is the most important. In fact,

$$\frac{S_N}{N!} = 1 + \frac{1}{N} + \frac{1}{N(N-1)} + \ldots + \frac{1}{N!},$$

and therefore

$$S_N = N! \left(1 + O\left(\frac{1}{N} \right) \right), \quad \text{as } N \to \infty.$$

Actually we can easily get more terms in the approximation:

$$S_N = N! \left(1 + \frac{1}{N} + O\left(\frac{1}{N^2} \right) \right) \quad \text{as } N \to \infty,$$

or

$$S_N = N! \left(1 + \frac{1}{N} + \frac{1}{N^2} + \frac{2}{N^3} + O\left(\frac{1}{N^4} \right) \right) \quad \text{as } N \to \infty.$$

1.8 The Euler–Maclaurin Summation Formula

In this section we give a precise formula for the approximation of sums by integrals, the celebrated Euler–Maclaurin formula.

We start by considering a very interesting sequence of polynomials, the so-called *Bernoulli polynomials*. Let us recall the elementary formulas

$$\sum_{n=1}^{N} n = \frac{N(N+1)}{2}, \tag{1.8.1a}$$

$$\sum_{n=1}^{N} n^2 = \frac{N(N+1)(2N+1)}{6}, \tag{1.8.1b}$$

$$\sum_{n=1}^{N} n^3 = \frac{N^2(N+1)^2}{4}. \tag{1.8.1c}$$

As should be clear, the sum $\sum_{n=1}^{N} n^k$ is a polynomial in N of degree $k + 1$. Except for some constants, these are the Bernoulli polynomials $B_k(x)$, constructed so as to satisfy

$$\sum_{n=M}^{N} n^k = \frac{1}{k+1} \left(B_k(N+1) - B_k(M) \right). \tag{1.8.2}$$

The Bernoulli polynomials are very important in many branches of mathematical analysis and number theory. The first few are

$$B_0(x) = 1,$$

$$B_1(x) = x - \frac{1}{2},$$

$$B_2(x) = x^2 - x + \frac{1}{6},$$

$$B_3(x) = x^3 - \frac{3}{2}x^2 + \frac{1}{2}x,$$

$$B_4(x) = x^4 - 2x^3 + x^2 + \frac{-1}{30}.$$

$$(1.8.3)$$

They satisfy rather interesting properties. Some of them are as follows:

$$\int_0^1 B_n(x)dx = 0, \quad n > 1, \tag{1.8.4}$$

$$B_n(1 - x) = (-1)^n B_n(x), \tag{1.8.5}$$

$$B_n'(x) = nB_{n-1}(x). \tag{1.8.6}$$

The Bernoulli polynomials can be written as

$$B_k(x) = \sum_{n=0}^{k} \binom{k}{n} B_k x^{k-n}, \tag{1.8.7}$$

where $B_k = B_k(0)$ are the *Bernoulli numbers* . These Bernoulli numbers are characterized by the recursion relation $B_0 = 1$ and

$$0 = \binom{k}{0} B_0 + \binom{k}{1} B_1 + \cdots + \binom{k}{k-1} B_{k-1}, \quad k \geq 2, \tag{1.8.8}$$

a formula that is often written in the symbolic form

$$B^k = (1 + B)^k, \quad k \geq 2, \tag{1.8.9}$$

where the right side is to be expanded according to the binomial theorem and the powers B^k are to be replaced by B_k. It is easy to show that the Bernoulli numbers of odd indices greater than 1 vanish:

$$B_{2k+1} = 0, \quad k = 1, 2, 3, \cdots . \tag{1.8.10}$$

The first few non-zero Bernoulli numbers are given by

$$B_1 = -\frac{1}{2}, \quad B_2 = \frac{1}{6}, \quad B_4 = -\frac{1}{30}, \quad B_6 = \frac{1}{42},$$

$$B_8 = -\frac{1}{30}, \quad B_{10} = \frac{5}{66}, \quad B_{12} = -\frac{691}{2730}, \quad B_{14} = \frac{7}{6}. \qquad (1.8.11)$$

We would also like to indicate that for $n \geq 1$, the only zeros of $B_{2n+1}(x)$ in the interval $[0, 1]$ are 0, $\frac{1}{2}$ and 1. Similarly, the only zeros of $B_{2n}(x) - B_{2n}$ in the same interval are 0 and 1.

After these preliminaries we give the Euler–Maclaurin summation formula.

Theorem 15. *Let f be k times continuously differentiable in $[N, M]$, where N and M are integers. Then*

$$\sum_{n=N+1}^{M} f(n) = \int_{N}^{M} f(x)dx + \sum_{j=1}^{k}(-1)^j \frac{B_j}{j!}(f^{(j-1)}(M) - f^{(j-1)}(N)) + R_k,$$

$$(1.8.12)$$

where the remainder R_k is given by

$$R_k(x) = \frac{(-1)^{k-1}}{k!} \int_{N}^{M} B_k(x - [\![\, x \,]\!])f^{(k)}(x)dx. \qquad (1.8.13)$$

Proof. Suppose first that f is defined in $[0, 1]$. Since $B_1(x) = x - \frac{1}{2}$, parts integration yields

$$\int_{0}^{1} f(x)dx = (B_1(x)f(x))\big|_0^1 - \int_{0}^{1} B_1(x)f'(x)dx.$$

Repeating this process and recalling (1.8.6) we obtain

$$\int_{0}^{1} f(x)dx = \sum_{j=1}^{k}(-1)^{j-1}\left(\frac{B_j(x)}{j!}f^{(j-1)}(x)\right)\bigg|_0^1 + (-1)^k \int_{0}^{1} \frac{B_k(x)}{k!}f^{(k)}(x)dx$$

and thus

$$f(1) = \int_{0}^{1} f(x)dx + \sum_{j=1}^{k}(-1)^j \frac{B_j}{j!}(f^{(j-1)}(1) - f^{(j-1)}(0))$$

$$+ (-1)^{k-1} \int_{0}^{1} \frac{B_k(x)}{k!}f^{(k)}(x)dx.$$

Replacing $f(x)$ by $f(n - 1 + x)$ yields

$$f(n) = \int_{0}^{1} f(n - 1 + x)dx + \sum_{j=1}^{k}(-1)^j \frac{B_j}{j!}(f^{(j-1)}(n) - f^{(j-1)}(n-1))$$

$$+(-1)^{k-1}\int_0^1 \frac{B_k(x)}{k!}f^{(k)}(n-1+x)dx.$$

summing from $n = N+1$ to $n = M$ then gives

$$\sum_{n=N+1}^{M} f(n) = \int_N^M f(x)dx + \sum_{j=1}^{k}(-1)^j \frac{B_j}{j!}(f^{(j-1)}(M) - f^{(j-1)}(N))$$

$$+\frac{(-1)^{k-1}}{k!}\int_N^M B_k(x - [\![\,x\,]\!])f^{(k)}(x)dx.$$

∎

Before we illustrate its use we would like to make several comments on this formula. Since $B_1 = -\dfrac{1}{2}$ and $0 = B_3 = B_5 = \cdots$, we can rewrite (1.8.12) in the following equivalent form

$$\sum_{n=N}^{M} f(n) = \int_N^M f(x)dx + \frac{1}{2}(f(N) + f(M)) + \sum_{j=1}^{q}\frac{B_{2j}}{2j!}(f^{(2j-1)}(M)$$

$$- f^{(2j-1)}(N)) + \tilde{R}_q,$$

$$(1.8.14)$$

where

$$\tilde{R}_q = -\frac{1}{(2q)!}\int_N^M B_{2q}(x - [\![\,x\,]\!])f^{(2q)}(x)dx$$

$$= \frac{1}{(2q+1)!}\int_N^M B_{2q+1}(x - [\![\,x\,]\!])f^{(2q+1)}(x)dx.$$

$$(1.8.15)$$

Next, we would like to say a little about the remainders. Observe that they are given in terms of integrals involving the functions $B_k(x - [\![\,x\,]\!])$, which are periodic of period 1. Their Fourier expansion is easily computed:

$$B_{2q}(x - [\![\,x\,]\!]) = 2(-1)^{q-1}(2q)!\sum_{n=1}^{\infty}\frac{\cos 2\pi nx}{(2\pi n)^{2q}}, \quad q \geq 1, \qquad (1.8.16a)$$

$$B_{2q-1}(x - [\![\,x\,]\!]) = 2(-1)^{q}(2q-1)!\sum_{n=1}^{\infty}\frac{\sin 2\pi nx}{(2\pi n)^{2q-1}}, \quad q \geq 1. \qquad (1.8.16b)$$

Setting $x = 0$ in (1.8.16a) yields the interesting relation

$$\sum_{n=1}^{\infty}\frac{1}{n^{2k}} = \frac{(-1)^{k-1}B_{2k}}{2(2k)!}(2\pi)^{2k}, \qquad (1.8.17)$$

of which the cases

$$\sum_{n=1}^{\infty} \frac{1}{n^2} = \frac{\pi^2}{6}, \quad \sum_{n=1}^{\infty} \frac{1}{n^4} = \frac{\pi^4}{90},$$

are readily obtained since $B_2 = \frac{1}{6}$, $B_4 = -\frac{1}{30}$.

The Fourier expansions (1.8.16a, b) can be used to estimate the function $B_k(x - [\![\, x \,]\!])$ for $k \geq 2$. In fact,

$$| B_k(x - [\![\, x \,]\!]) | \leq \frac{2k!}{(2\pi)^k} \sum_{n=1}^{\infty} \frac{1}{n^k} \leq \frac{k!}{12(2\pi)^{k-2}}, \qquad (1.8.18)$$

and using (1.8.17),

$$| B_{2q}(x - [\![\, x \,]\!]) | \leq | B_{2q} |. \qquad (1.8.19)$$

Let us now give some examples.

Example 20. Let us consider the function $f(x) = \frac{1}{x}$. We have

$$\sum_{n=1}^{N} \frac{1}{n} = \int_1^N \frac{dx}{x} + \frac{1}{2}\left(\frac{1}{N} + 1\right) - \int_1^N \frac{x - [\![\, x \,]\!] - \frac{1}{2}}{x^2}\, dx$$

$$= \ln N + \frac{1}{2N} + \frac{1}{2} - \int_1^{\infty} \frac{x - [\![\, x \,]\!] - \frac{1}{2}}{x^2}\, dx + \int_N^{\infty} \frac{x - [\![\, x \,]\!] - \frac{1}{2}}{x^2}\, dx$$

$$= \ln N + \left(\frac{1}{2} - \int_1^{\infty} \frac{x - [\![\, x \,]\!] - \frac{1}{2}}{x^2}\, dx\right) + O\left(\frac{1}{N}\right).$$

Comparison with (1.7.16) yields the formula

$$\gamma = \frac{1}{2} - \int_1^{\infty} \frac{x - [\![\, x \,]\!] - \frac{1}{2}}{x^2}\, dx. \qquad (1.8.20)$$

Using more terms in the Euler–Maclaurin formula we get

$$\sum_{n=1}^{N} \frac{1}{n} = \ln N + \gamma + \frac{1}{2N} - \sum_{j=1}^{k-1} \frac{B_{2j}}{2j N^{2j}} + R_k(N), \qquad (1.8.21)$$

where

$$| R_k(N) | = | \int_1^N \frac{B_{2k}(x - [\![\, x \,]\!])}{x^{2k+1}}\, dx | \leq \frac{B_{2k}}{(2k)N^{2k}}. \qquad (1.8.22)$$

Therefore,

$$\sum_{n=1}^{N} \frac{1}{n} \sim \ln N + \gamma + \frac{1}{2N} - \sum_{j=1}^{\infty} \frac{B_{2j}}{2jN^{2j}}. \tag{1.8.23}$$

Example 21. Let us apply the Euler–Maclaurin formula (1.8.12) to the function $f(x) = e^{xz}$, where z is a parameter. We have

$$\sum_{n=1}^{N} e^{nz} = \int_{0}^{N} e^{xz}\,dx + \sum_{j=1}^{k}(-1)^{j}\frac{B_{j}}{j!}z^{j-1}(e^{Nz}-1) + R_{k},$$

where

$$R_{k} = \frac{(-1)^{k-1}}{k!}\int_{0}^{1} B_{k}(x)z^{k}\sum_{n=1}^{N} e^{(n-1+x)z}\,dx.$$

But,

$$\sum_{n=1}^{N} e^{nz} = e^{z}\frac{e^{Nz}-1}{e^{z}-1}, \qquad \int_{0}^{N} e^{xz}\,dx = \frac{1}{z}(e^{Nz}-1),$$

$$\sum_{n=1}^{N} e^{(n-1+x)z} = e^{xz}\frac{e^{Nz}-1}{e^{z}-1},$$

and so

$$\frac{e^{z}}{e^{z}-1} = \frac{1}{z} + \sum_{j=1}^{k}(-1)^{j}\frac{B_{j}}{j!}z^{j-1} + \frac{(-1)^{k-1}}{k!}\frac{z^{k}}{e^{z}-1}\int_{0}^{1} B_{k}(x)e^{xz}\,dx.$$

The remainder is easy to estimate:

$$\left|\int_{0}^{1} B_{k}(x)e^{xz}\,dx\right| \leq \frac{e^{|z|}}{12(2\pi)^{q-2}},$$

and it follows that

$$\lim_{k\to\infty}\frac{z^{k}}{k!}\int_{0}^{1} B_{k}(x)\,e^{xz}\,dx = 0,$$

as long as $|z| < 2\pi$.

Therefore,

$$\frac{ze^{z}}{e^{z}-1} = \sum_{j=0}^{\infty}(-1)^{j}\frac{B_{j}}{j!}z^{j}, \qquad |z| < 2\pi. \tag{1.8.24}$$

If we now use the fact that $B_1 = -\dfrac{1}{2}$ and $0 = B_3 = B_5 = \cdots$, we obtain

$$\sum_{j=0}^{\infty} \frac{B_j}{j!} z^j = \sum_{j=0}^{\infty} (-1)^j \frac{B_j}{j!} z^j - z = \frac{ze^z}{e^z - 1} - z = \frac{z}{e^z - 1},$$

or

$$\frac{z}{e^z - 1} = \sum_{j=0}^{\infty} \frac{B_j}{j!} z^j. \tag{1.8.25}$$

The function $\dfrac{z}{e^z - 1}$ is the *generating function* for the Bernoulli numbers. A similar analysis yields the generating function for the Bernoulli polynomials, namely,

$$\frac{ze^{xz}}{e^z - 1} = \sum_{j=0}^{\infty} \frac{B_j(x)z^k}{k!}. \tag{1.8.26}$$

Our next illustration is the celebrated Stirling formula for the approximation of $n!$.

Example 22. Let us apply the Euler–Maclaurin formula to the function $f(x) = \ln x$. We obtain

$$\ln N! = \sum_{n=2}^{N} \ln n = \int_1^N \ln x\, dx + \frac{1}{2} \ln N + \sum_{j=1}^{k} \frac{B_{2j}}{(2j)!} (2j - 2)! \left(\frac{1}{N^{2j-1}} - 1 \right)$$

$$+ \int_1^N \frac{B_{2k}(x - [\![\, x\,]\!])}{2k} x^{-2k}\, dx,$$

or

$$\ln N! = \left(N + \frac{1}{2}\right) \ln N - N$$

$$+ A + \sum_{j=1}^{k} \frac{B_{2j}}{(2j - 1)2j} N^{-2j+1} + \frac{1}{2k} \int_N^{\infty} B_{2k}(x - [\![\, x\,]\!]) x^{-2k}\, dx,$$

where A is the constant given by

$$A = \lim_{N \to \infty} \left(\ln N! - \left(N + \frac{1}{2}\right) \ln N + N \right). \tag{1.8.27}$$

For the remainder we have the bound

$$\left| \frac{1}{2k} \int_N^{\infty} B_{2k}(x - [\![\, x\,]\!]) x^{-2k}\, dx \right| \leq \frac{|B_{2k}|}{(2k - 1)2k} N^{-2k+1}. \tag{1.8.28}$$

Therefore, we obtain the infinite asymptotic expansion

$$\ln N! \sim (N + \frac{1}{2}) \ln N - N + A + \sum_{j=1}^{\infty} \frac{B_{2j}}{(2j-1)2jN^{2j-1}}, \quad \text{as } N \to \infty. \quad (1.8.29)$$

In particular, the first approximation takes the form

$$\ln N! = (N + \frac{1}{2}) \ln N - N + A + R, \quad (1.8.30)$$

where

$$|R| < \frac{1}{12N},$$

because of (1.8.28) and the fact that $B_2 = \frac{1}{6}$. Finally we show that

$$A = \ln \sqrt{2\pi}. \quad (1.8.31)$$

Indeed, using (1.8.30) in the Wallis formula yields

$$\frac{\pi}{2} = \lim_{N \to \infty} \prod_{n=1}^{N} \frac{4n^2}{4n^2 - 1} = \lim_{N \to \infty} \frac{2^{4N}(N!)^4}{((2N)!)^2(2N+1)} = \frac{1}{4} e^{2A},$$

and (1.8.31) is easily deduced.

CHAPTER 2

Introduction to the Theory of Distributions

2.1 Introduction

The purpose of this chapter is to present the basic ideas of the theory of distributions. Distributions or generalized functions, as they are also known, have proved to be very useful in many branches of pure and applied mathematics. Many textbooks, monographs and articles have been written on their theory and their applications [12], [23], [53], [63], [64], [69], [71], [80], [97], [111]. Our present aim is to give a brief but solid introduction to the theory of distributions, particularly to those aspects that are important in the theory of asymptotic expansions.

The distributions were introduced by L. Schwartz [97] as a mathematically rigorous theory to justify many formal and heuristic, but quite successful, methods and procedures used in electricity, quantum mechanics and other branches of applied mathematics. Perhaps the most famous of these formal concepts is the celebrated *Dirac delta function* $\delta(x)$, which satisfies the apparently contradictory conditions of vanishing for non-zero values of x :

$$\delta(x) = 0, \quad x \neq 0, \tag{2.1.1}$$

while having a unit total mass,

$$\int_{-\infty}^{\infty} \delta(x)dx = 1. \tag{2.1.2}$$

Naturally, no classical function could satisfy (2.1.1) and (2.1.2) simultaneously, but still this "function" $\delta(x)$ has been shown to be tremendously useful in the study of many pure and applied problems. But much more paradoxical were its derivatives $\delta'(x), \delta''(x), \dots$, which were often needed in these formal manipulations. The reader is referred to Lützen's book [83] for an excellent account of the many developments that led to the theory of distributions.

There are several methods to introduce the theory of distributions and their generalizations. As is to be expected, some aspects of the theory become more transparent through the optics of one method as compared with others. For our present purposes, we choose to follow the functional approach, also called the Schwartz–Sobolev approach.

The basic idea of the functional approach is to study a function $f(x)$, not by looking at its point values, but rather by looking at its actions $< f, \phi > = \int f(x)\phi(x)dx$, $\phi \in \Phi$, on a set of "test" functions Φ. If the test functions are smooth enough then most ordinary functions will have well-defined actions $< f, \phi >$ for each $\phi \in \Phi$. But what is interesting is that many singular, non-classical "functions" also have well-defined actions. Thus, for instance, the action of the Dirac delta function $\delta(x - \xi)$ is given by its *sifting property*

$$< \delta(x - \xi), \phi(x) > = \int \delta(x - \xi)\phi(x)dx = \phi(\xi). \qquad (2.1.3)$$

One can think of Φ as a set of " instruments " to help us measure the point values of f. The values $< f, \phi >$ are then "observations" that depend not only on the values of f but also on the instruments used. Looking at Figure 1 and assuming that $\int \phi_i(x)dx = 1$, the values $< f, \phi_i >$ provide better and better approximations of $f(x_0)$ if f is continuous at $x = x_0$. Observe, however, that the values $< f, \phi >$ can be computed even if f is not continuous at $x = x_0$.

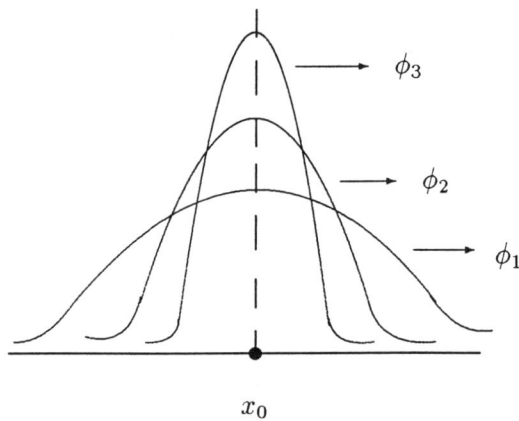

$$x_0$$

Figure 1

From a mathematical point of view the space of test functions Φ is chosen as a topological vector space. The distributions or generalized functions corresponding to Φ are the elements of the dual space Φ', that is, the continuous linear functionals defined on Φ. The selection of the appropriate space Φ depends on the problem under consideration, but we can remark that the smaller Φ and the stronger its topology, the larger Φ' will be.

We finish this section by commenting on the notation used for tensors. If $\boldsymbol{T} = (T_{i_1 \cdots i_N})$ and $\boldsymbol{S} = (S_{i_1 \cdots i_N})$ are two tensors of order N then their inner product is the complete contraction

$$T \parallel S = T_{i_1 \cdots i_N} S_{i_1 \cdots i_N}, \tag{2.1.4}$$

where we have used the summation convention. In the case where the tensors are symmetric an alternative description is available. If $k = (k_1, \ldots, k_n) \in \mathbb{N}^n$ is a multi-index with $|k| = \sum_{i=1}^n k_i = N$, then we can assign to k the number $T(k) = T_{i_1 \cdots i_N}$ where $(i_1, \ldots, i_N)$ is such that $|\{k : i_k = j\}| = k_j$. Observe that (2.1.4) takes the form

$$T \parallel S = \sum_{|k| = N} \frac{N!}{k!} T(k) S(k), \tag{2.1.5}$$

where $k! = k_1! \cdots k_n!$. We remark the use of the notation D^N for the tensor derivative operator $\left(\frac{\partial^N}{\partial x_{i_1} \cdots \partial x_{i_N}} \right)$ and D^k for the operator $\frac{\partial^{|k|}}{\partial x_i{}^{k_i} \cdots \partial x_n{}^{k_n}}$. It is clear that $D^N(k) = D^k$ if $|k| = N$. Also, x^N is the tensor with components $x_{i_1} \cdots x_{i_N}$ and $x^k = x^N(k) = x_1^{k_1} \cdots x_n^{k_n}$ if $|k| = N$.

2.2 The Space of Distributions $\mathcal{D}'$

In this section we define and study the distributions of the space $\mathcal{D}'$. As we mention in the introduction, distributions and generalized functions are defined through their action on some class of "test" function. Therefore, we start with the space of test functions.

Definition. *Let U be an open set in $\mathbb{R}^n$. The space $\mathcal{D}(U)$ consists of those smooth function defined in U whose support is compact. Convergence in $\mathcal{D}(U)$ is defined as follows: a net $\{\phi_\sigma\}$ of $\mathcal{D}(U)$ converges to $\phi \in \mathcal{D}(U)$ if*

(a) There is σ_0 and a fixed compact subset K of U with $\operatorname{supp} \phi_\sigma \subseteq K$ for $\sigma \geq \sigma_0$.
(b) For each multi-index $k \in \mathbb{N}^n$, $D^k \phi_\sigma$ converges uniformly to $D^k \phi$.

A typical element of $\mathcal{D}(\mathbb{R}^n)$ is the function

$$\phi(x) = \begin{cases} e^{(a^2 - |x|)^{-1}} & |x| \leq a, \\ 0, & |x| \geq a. \end{cases} \tag{2.2.1}$$

If $x_0 \in U$ and the ball $B(x_0, a) = \{x \in \mathbb{R}^n : |x - x_0| \leq a\}$ is contained in U then $\phi(x - x_0)$ belongs to $\mathcal{D}'(U)$.

We shall usually work in $\mathbb{R}^n$, but observe that the above definition can also be used to define the space $\mathcal{D}'(U)$ if U is a smooth manifold.

The notation $\mathcal{D}$ will be used for the space $\mathcal{D}(\mathbb{R}^n)$ whenever n is clear from the context.

Definition. *A distribution or generalized function $f(x)$ defined in U is a continuous linear functional on the space $\mathcal{D}(U)$, that is, an element of the dual space $\mathcal{D}'(U)$.*

The simplest distributions are given by the locally integrable functions. Indeed, if f is a function, defined in U, that is *locally integrable*, i.e., that satisfies

$$\int_K |f(x)|dx < \infty, \tag{2.2.2}$$

for each compact $K \subseteq U$, then we can construct a distribution in U by the formula

$$< f, \phi > = \int_U f(x)\phi(x)dx. \tag{2.2.3}$$

Observe that (2.2.3) is a well defined functional on $\mathcal{D}(U)$ since the integral is actually taken over supp ϕ, which is compact. The continuity of this functional is easy to see since, if $\{\phi_\sigma\}$ converges to 0 in $\mathcal{D}$, then there is a fixed compact K with supp $\phi_\sigma \subseteq K$ for $\sigma > \sigma_0$ and thus

$$\left|\int_U f(x)\phi_\sigma(x)d\,x\right| \leq \left(\int_K |f(x)|d\,x\right) \max\{|\phi_\sigma| : x \in U\} \to 0,$$

because of the uniform convergence of $\{\phi_\sigma\}$ to 0.

We use the same notation f or $f(x)$ for the locally integrable function $f(x)$ and for the distribution it defines. This is justified since two locally integrable functions define the same functional if and only if they are equal almost everywhere.

Distributions arising from locally integrable functions are called regular distributions. All other distributions are called singular.

We shall now give several examples of singular distributions.

Example 23. If $y \in U$ the Dirac delta function concentrated at y is the distribution $\delta(x - y)$ given by

$$< \delta(x - y), \phi(x) > = \phi(y). \tag{2.2.4}$$

Observe that the notation indicates clearly that the evaluation is with respect to x and that y is a fixed element of U.

Example 24. The function $f(x) = \dfrac{1}{x}, x \neq 0$, is not locally integrable in $\mathbb{R}$. Hence it does not define a regular distribution. However, we can construct a distribution out of $\frac{1}{x}$ by using principal value integrals. Indeed, define

$$< \mathcal{P}.v. \left(\frac{1}{x}\right), \phi(x) > = \mathcal{P}.v. \int_{-\infty}^{\infty} \frac{\phi(x)}{x}dx, \quad \phi \in \mathcal{D}, \tag{2.2.5}$$

where $\mathcal{P}.v.$ stands for the principal value of the integral, defined as

$$\mathcal{P}.v. \int_{-\infty}^{\infty} \frac{\phi(x)}{x} dx = \lim_{\varepsilon \to 0} \int_{|x| \geq \varepsilon} \frac{\phi(x)}{x} dx. \tag{2.2.6}$$

To see that $\mathcal{P}.v.\left(\frac{1}{x}\right)$ is a continuous linear functional on $\mathcal{D}(\mathbb{R})$ we proceed as follows. Let $\phi \in \mathcal{D}(\mathbb{R})$ and let A be large enough so that supp $\phi \subseteq [-A, A]$. Since ϕ is smooth at $x = 0$ we can find a smooth function ψ such that

$$\phi(x) = \phi(0) + x\psi(x). \tag{2.2.7}$$

Thus

$$\lim_{\varepsilon \to 0} \int_{|x| \geq \varepsilon} \frac{\phi(x)}{x} dx = \lim_{\varepsilon \to 0} \int_{\varepsilon \leq |x| \leq A} \frac{\phi(x)}{x} dx$$

$$= \lim_{\varepsilon \to 0} \int_{\varepsilon \leq |x| \leq A} \left(\frac{\phi(0)}{x} + \psi(x) \right) dx$$

$$= \int_{-A}^{A} \psi(x) dx,$$

and it follows that $\mathcal{P}.v. \int_{-\infty}^{\infty} \frac{\phi(x)}{x} dx = \int_{-A}^{A} \psi(x) dx$ exists for each $\phi \in \mathcal{D}$. The continuity is obtained by observing that if $\{\phi_\sigma\} \to 0$ in $\mathcal{D}$ then supp $\phi_\sigma \subseteq [-A, A]$ for $\sigma \geq \sigma_0$ and some large A and also $\psi_\sigma(x) = \frac{\phi_\sigma(x) - \phi_\sigma(0)}{x}$ converges to zero uniformly in $[-A, A]$. Hence for $\sigma \geq \sigma_0$

$$\mathcal{P}.v. \int_{-\infty}^{\infty} \frac{\phi_\sigma(x)}{x} dx = \int_{-A}^{A} \psi_\sigma(x) dx \to 0.$$

Examples of singular distributions in several variables include the following.

Example 25. Let Σ be a smooth oriented hypersurface in $\mathbb{R}^n$. Then the distribution $\delta(\Sigma)$ is defined as

$$< \delta(\Sigma), \phi > = \int_{\Sigma} \phi(y) d\sigma(y), \ \phi \in \mathcal{D}(\mathbb{R}^n), \tag{2.2.8}$$

where $d\sigma$ is the surface measure on Σ. More generally, if f is a distribution on Σ then the distribution $f\delta(\Sigma)$ is defined as

$$< f\delta(\Sigma), \phi > = < f, \phi|\Sigma >, \tag{2.2.9}$$

where $\phi|\Sigma$ is the restriction of ϕ to Σ and where the last operation takes place in $\mathcal{D}'(\Sigma) \times \mathcal{D}(\Sigma)$. Distributions of the form $f\delta(\Sigma)$ are called layers concentrated on Σ.

The convergence of the test functions is rather stringent. On the other hand, the convergence of distributions is defined as weak convergence.

Definition. *A net $\{f_\sigma\}$ of $\mathcal{D}'(U)$ converges to $f \in \mathcal{D}'(U)$ if $< f_\sigma, \phi > \rightarrow < f, \phi >$ for each $\phi \in \mathcal{D}(U)$.*

Example 26. If $\varepsilon > 0$ then the function $f_\varepsilon(x) = \frac{1}{x+i\varepsilon}$ is locally integrable and therefore defines a regular distribution in $\mathcal{D}(\mathbb{R})$. If we let ε approach 0, we obtain a singular distribution $\frac{1}{x+i0}$ defined as

$$< \frac{1}{x+i0}, \phi(x) >= \lim_{\varepsilon \to 0} \int_{-\infty}^{\infty} \frac{\phi(x)}{x + i\varepsilon} dx. \qquad (2.2.10)$$

The distribution $\frac{1}{x+i0}$ is related to the distribution $\mathcal{P}.v. \left(\frac{1}{x}\right)$ which we introduced before, but they do not coincide. Indeed, writing $\phi(x) = \phi(0) + x\psi(x)$ as in (2.2.7), we obtain

$$\lim_{\varepsilon \to 0} \int_{-\infty}^{\infty} \frac{\phi(x)}{x + i\varepsilon} dx = \lim_{\varepsilon \to 0} \int_{-A}^{A} \frac{\phi(0)}{x + i\varepsilon} dx$$
$$+ \int_{-A}^{A} \left(\frac{x}{x + i\varepsilon}\right) \psi(x) dx$$
$$= -i\pi\phi(0) + \int_{-A}^{A} \psi(x) dx,$$

or

$$\frac{1}{x + i0} = -i\pi\delta(x) + \mathcal{P}.v. \left(\frac{1}{x}\right). \qquad (2.2.11)$$

A similar analysis yields

$$\frac{1}{x - i0} = i\pi\delta(x) + \mathcal{P}.v. \left(\frac{1}{x}\right). \qquad (2.2.12)$$

Subtracting (2.2.11) from (2.2.12), we obtain

$$\delta(x) = \lim_{\varepsilon \to 0} \frac{\varepsilon}{\pi(x^2 + \varepsilon^2)}, \qquad (2.2.13)$$

which is the distributional form of the well-known limit

$$\phi(0) = \lim_{\varepsilon \to 0} \frac{\varepsilon}{\pi} \int_{-\infty}^{\infty} \frac{\phi(x)}{x^2 + \varepsilon^2} dx. \qquad (2.2.14)$$

2.3 Algebraic and Analytic Operations

We shall now discuss the basic algebraic and analytic operations that can be applied to the distributions. The basic idea behind the definitions of the operations is to use *duality*, which means that the operations in $\mathcal{D}'(U)$ are the adjoints of the corresponding operations in $\mathcal{D}(U)$.

Let us start with the linear changes of variables. Let A be a non-singular $n \times n$ matrix. We wish to define $f(A\boldsymbol{x})$ if $f \in \mathcal{D}'(\mathbb{R}^n)$. In order to do so we first suppose that f is a regular distribution. Then $f(A\boldsymbol{x})$ is also a well-defined locally integrable function, and changing variables, we obtain

$$< f(A\boldsymbol{x}), \, \phi(\boldsymbol{x}) > = \int_{\mathbb{R}^n} f(A\boldsymbol{x}) \, \phi(\boldsymbol{x}) d\boldsymbol{x}$$

$$= \frac{1}{|\det A|} \int_{\mathbb{R}^n} f(\boldsymbol{u}) \, \phi(A^{-1}\boldsymbol{u}) d\boldsymbol{u}$$

$$= \frac{1}{|\det A|} < f(\boldsymbol{x}), \, \phi(A^{-1}\boldsymbol{x}) > .$$

Therefore in the general case we *define* the distribution $f(A\boldsymbol{x})$ as

$$< f(A\boldsymbol{x}), \, \phi(\boldsymbol{x}) > = \frac{1}{|\det A|} < f(\boldsymbol{x}), \, \phi(A^{-1}\boldsymbol{x}) > . \qquad (2.3.1)$$

Observe that this definition makes sense because if $\phi \in \mathcal{D}$ then $\phi(A^{-1}\,\boldsymbol{x})$.

In particular, if $A = -I$, where I is the identity matrix, then we obtain the distribution $f(-\boldsymbol{x})$:

$$< f(-\boldsymbol{x}), \phi(\boldsymbol{x}) > = < f(\boldsymbol{x}), \phi(-\boldsymbol{x}) > . \qquad (2.3.2)$$

Example 27. Using (2.3.2) immediately gives the formula

$$\delta(-\boldsymbol{x}) = \delta(\boldsymbol{x}), \qquad (2.3.3)$$

so that the delta function is even. Actually $\delta(\boldsymbol{x})$ is spherically symmetric; i.e., if T is any rotation then

$$\delta(T\boldsymbol{x}) = \delta(\boldsymbol{x}), \qquad (2.3.4)$$

as follows from (2.3.1) since $\det T = 1$.

Considering $A = \lambda > 0$, we get

$$\delta(\lambda\boldsymbol{x}) = \lambda^{-n} \, \delta(\boldsymbol{x}), \qquad (2.3.5)$$

and thus $\delta(\boldsymbol{x})$ is homogeneous of degree $-n$.

Translations can be handled similarly.

Definition. *Let $f \in \mathcal{D}'(\mathbb{R}^n)$ and let $\boldsymbol{c} \in \mathbb{R}^n$. Then the distribution $f(\boldsymbol{x}+\boldsymbol{c})$ is defined as*

$$< f(\boldsymbol{x}+\boldsymbol{c}), \phi(\boldsymbol{x}) > \; = \; < f(\boldsymbol{x}), \phi(\boldsymbol{x}-\boldsymbol{c}) > . \qquad (2.3.6)$$

Observe that the notation $\delta(\boldsymbol{x}-\boldsymbol{y})$ that we have used for the Dirac delta function concentrated at the point $\boldsymbol{y}$ is consistent with this definition.

The next operation we want to study is differentiation. Suppose first that f is continuously differentiable in U. Then for each i, $\dfrac{\partial f}{\partial x_i}$ is also a distribution in $\mathcal{D}'(U)$ and we have for $\phi \in \mathcal{D}(U)$

$$< \frac{\partial f}{\partial x_i}, \phi > \; = \; \int_U \frac{\partial f}{\partial x_i} \phi \, d\boldsymbol{x} = - \int f \frac{\partial \phi}{\partial x_i} \, d\boldsymbol{x} = - < f, \frac{\partial \phi}{\partial x_i} > ,$$

where we have used the very important fact that ϕ vanishes outside a compact subset of U and thus the boundary terms in the integration by parts vanish. Motivated by this situation we give the following definition.

Definition. *Let $f \in \mathcal{D}'(U)$. The distribution $\dfrac{\partial f}{\partial x_i}$ is defined as*

$$< \frac{\partial f}{\partial x_i}, \phi > \; = \; - < f, \frac{\partial \phi}{\partial x_i} > . \qquad (2.3.7)$$

The operation of partial derivation is well defined since if $\phi \in \mathcal{D}(U)$ then so is $\dfrac{\partial \phi}{\partial x_i}$. More than that, the operator $\boldsymbol{D}_i : \mathcal{D}(U) \to \mathcal{D}(U)$ given by $\boldsymbol{D}_i(\phi) = \dfrac{\partial \phi}{\partial x_i}$ is clearly continuous and so from (2.3.7) we see that the operator $\boldsymbol{D}_i : \mathcal{D}'(U) \to \mathcal{D}'(U)$ is also continuous. This means that if $f_\sigma \to f$ then also $\dfrac{\partial f_\sigma}{\partial x_i} \to \dfrac{\partial f}{\partial x_i}$.

Observe that in general if $\boldsymbol{k} \in \mathbb{N}^n$ then we can define $\boldsymbol{D}^{\boldsymbol{k}} f$ by

$$< \boldsymbol{D}^{\boldsymbol{k}} f, \phi > \; = \; (-1)^{|\boldsymbol{k}|} < f, \boldsymbol{D}^{\boldsymbol{k}} \phi > . \qquad (2.3.8)$$

Notice that the partial derivatives of a distribution are always defined. Therefore, *in the distributional sense any distribution is infinitely differentiable.*

Example 28. Let $H(x)$ be the Heaviside function, defined as

$$H(x) = \begin{cases} 1, & x > 0, \\ 0, & x < 0. \end{cases}$$

Let us compute $H'(x)$:

$$< H'(x), \phi(x) > = - < H(x), \phi'(x) > = - \int_0^\infty \phi'(x)du = \phi(0).$$

Therefore

$$H'(x) = \delta(x). \qquad (2.3.9)$$

Example 29. We can differentiate the function $\delta(x)$ to any order. In one variable the distributions $\delta^{(k)}(x)$ are defined by

$$< \delta^{(k)}(x), \phi(x) > = (-1)^k \phi^{(k)}(0), \qquad (2.3.10)$$

while in n variables we have

$$< D^k \delta(x), \phi(x) > = (-1)^{|k|} D^k \phi(0). \qquad (2.3.11)$$

Example 30. Let us consider the locally integrable function $\ln |x|$. We have

$$< (\ln |x|)', \phi(x) > = - < \ln |x|, \phi'(x) >$$

$$= - \int_{-\infty}^\infty \ln |x| \phi'(x) dx$$

$$= - \lim_{\varepsilon \to 0} \int_{|x| \geq \varepsilon} \ln |x| \phi'(x) dx$$

$$= \lim_{\varepsilon \to 0} \left[\int_{|x| \geq \varepsilon} \frac{\phi(x)}{x} dx - (\phi(\varepsilon) - \phi(-\varepsilon)) \ln \varepsilon \right]$$

$$= \mathcal{P}.v. \int_{-\infty}^\infty \frac{\phi(x)}{x} dx,$$

since $(\phi(\varepsilon) - \phi(-\varepsilon)) \ln \varepsilon = O(\varepsilon \ln \varepsilon) = o(1)$ as $\varepsilon \to 0$. Thus

$$(\ln |x|)' = \mathcal{P}.v. \left(\frac{1}{x} \right). \qquad (2.3.12)$$

Example 31. Let $\varepsilon > 0$ and consider the distribution $\ln(x + i\varepsilon) = \ln |x + i\varepsilon| + i \arg(x + i\varepsilon)$, where $\arg(x + i\varepsilon)$ is chosen between 0 and π.

Observe that $\lim_{\varepsilon \to 0} \ln(x + i\varepsilon) = \ln|x| + i\pi H(-x)$ in the pointwise and distributional senses. Differentiating this relation and using the fact that $(H(-x))' = -\delta(x)$ we obtain

$$\frac{1}{x + i0} = \lim_{\varepsilon \to 0} \frac{1}{x + i\varepsilon} = \mathcal{P}.v. \left(\frac{1}{x}\right) - i\pi\delta(x),$$

in agreement with (2.2.11).

Next we consider multiplication. As it turns out, the multiplication of two arbitrary distributions cannot be defined in general. However, we can always multiply a distribution and a smooth function.

Definition. *Let $f \in \mathcal{D}'(U)$ and $\psi \in C^\infty(U)$. The distribution ψf is defined as*

$$< \psi f, \phi > = < f, \psi\phi > . \tag{2.3.13}$$

Observe that the definition makes sense since if $\psi \in C^\infty(U)$ and $\phi \in \mathcal{D}(U)$ then $\psi\phi \in \mathcal{D}(U)$.

Example 32. Let us compute $\psi(x)\delta(x)$:

$$< \psi(x)\delta(x), \phi(x) > = < \delta(x), \psi(x)\phi(x) > = \psi(0)\phi(0).$$

Thus

$$\psi(x)\delta(x) = \psi(0)\delta(x). \tag{2.3.14}$$

More generally,

$$\psi(x)\delta^{(k)}(x) = \sum_{j=0}^{k} \binom{k}{j} (-1)^j \psi^{(j)}(0)\delta^{(k-j)}(x). \tag{2.3.15}$$

Example 33. We have $x \ \mathcal{P}.v.(\frac{1}{x}) = 1$ because

$$< x \ \mathcal{P}.v.(\frac{1}{x}), \phi(x) > \ = < \mathcal{P}.v.(\frac{1}{x}), x\phi(x) > = \mathcal{P}.v. \int_{-\infty}^{\infty} \phi(x)dx$$

$$= \int_{-\infty}^{\infty} \phi(x)dx.$$

Since $x\delta(x) = 0$, we obtain

$$(\delta(x) \ x) \ \mathcal{P}.v.\frac{1}{x} = 0 \neq \delta(x)(x \ \mathcal{P}.v.(\frac{1}{x})).$$

Thus the multiplication is not associative.

Next let us consider non-linear changes of variable. Let U and V be open sets in $\mathbb{R}^n$ and let $\boldsymbol{\Psi} : U \to V$ be a smooth homeomorphism. If f is locally integrable in V then for each $\phi \in \mathcal{D}(U)$ we have

$$< f(\boldsymbol{\Psi}(\boldsymbol{u})), \phi(\boldsymbol{u}) > = \int_U f(\boldsymbol{\Psi}(\boldsymbol{u}))\phi(\boldsymbol{u})d\boldsymbol{u} = \int_V \frac{f(\boldsymbol{v})\phi(\boldsymbol{\Psi}^{-1}(\boldsymbol{v}))}{J(\boldsymbol{v})}\,d\boldsymbol{v},$$

where J is the Jacobian of the transformation. Therefore we have the following definition.

Definition. *If $f \in \mathcal{D}'(V)$ the distribution $f(\boldsymbol{\Psi}(\boldsymbol{u}))$ in $\mathcal{D}'(U)$ is defined as*

$$< f(\boldsymbol{\Psi}(\boldsymbol{u})), \phi(\boldsymbol{u}) > = < f(\boldsymbol{v}), \frac{\phi(\boldsymbol{\Psi}^{-1}(\boldsymbol{v}))}{J(\boldsymbol{v})} > . \tag{2.3.16}$$

Example 34. Let $\psi : \mathbb{R} \to \mathbb{R}$ be smooth and with non-vanishing derivative. Suppose $\psi(x_0) = 0$. Then

$$< \delta(\psi(x)), \phi(x) > = < \delta(y), \frac{\phi(\psi^{-1}(y))}{|\,\psi'(\psi^{-1}(y))\,|} > = \frac{\phi(x_0)}{|\,\psi'(x_0)\,|},$$

or

$$\delta(\psi(x)) = \frac{\delta(x - x_0)}{|\,\psi'(x_0)\,|}. \tag{2.3.17}$$

Similarly

$$\delta'(\psi(x)) = \frac{\delta'(x - x_0)}{|\,\psi'(x_0)^2\,|} + \frac{\psi''(x_0)\delta(x - x_0)}{\psi'(x_0)^3}. \tag{2.3.18}$$

In several dimensions for a change $\boldsymbol{y} = \boldsymbol{\Psi}(\boldsymbol{x})$ with $\boldsymbol{\Psi}(\boldsymbol{x}_0) = \boldsymbol{0}$, we have

$$\delta(\boldsymbol{\Psi}(\boldsymbol{x})) = \frac{\delta(\boldsymbol{x} - \boldsymbol{x}_0)}{J(\boldsymbol{x}_0)} = \frac{\delta(\boldsymbol{x} - \boldsymbol{x}_0)}{\det\left(\frac{\partial \psi_i}{\partial x_j}\right)|_{\boldsymbol{x}_0}}. \tag{2.3.19}$$

2.4 Regularization, Pseudofunction and Hadamard Finite Part

As we have seen, each locally integrable function defines a regular distribution. However, distributions can also be constructed from non-locally integrable functions. In this section we study the process of *regularization*, that is, the process of associating a distribution to a non-locally integrable function.

We have already seen some examples. The principal value distribution $\mathcal{P}.v.\left(\frac{1}{x}\right)$ is a regularization of $\frac{1}{x}$. The distributions $\frac{1}{x+i0}$ and $\frac{1}{x-i0}$ are also regularizations of $\frac{1}{x}$.

Suppose, to fix the ideas, that the function $f(x)$ is locally integrable in $U\setminus\{x_0\}$, where $x_0 \in U$. Then a *regularization* $F(x)$ of $f(x)$ is a distribution that satisfies

$$< F(x), \phi(x) > = \int_U f(x)\,\phi(x)dx, \qquad (2.4.1)$$

for each $\phi \in \mathcal{D}(U)$ for which the integral exists, in the Lebesgue sense. Observe that the integral in (2.4.1) exists for any $\phi \in \mathcal{D}'(U)$ such that $x_0 \notin \operatorname{supp} \phi$, but might exist for some $\phi \in \mathcal{D}(U)$ with $x_0 \in \operatorname{supp} \phi$. In the case where $f(x)$ is not locally integrable near $x = x_0$ then the integral will not converge if $\phi(x_0) \neq 0$. In general, there is a value k, $1 \leq k \leq +\infty$, such that the integral exists if $\mathcal{D}^j \phi(x_0) = 0, \mid j \mid < k$, but not otherwise.

Regularizations, if they exist, are not uniquely determined. For instance, $\mathcal{P}.v.\left(\frac{1}{x}\right)$, $\frac{1}{x+i0}$, and $\frac{1}{x-i0}$ are three different regularizations of $\frac{1}{x}$. Actually, if $F_0(x)$ is a regularization of $f(x)$ then so are $F_0(x) + \sum_{|j|<k} c_j \mathcal{D}^j \delta(x)$ for arbitrary constants c_j (if $k = \infty$ the c_j should vanish for $\mid j \mid$ large).

Example 35. The function $f(x) = e^{\frac{1}{x}} H(x)$ is locally integrable in $\mathbb{R}\setminus\{0\}$. However, it cannot be regularized. To see it, let $\phi \in \mathcal{D}(\mathbb{R})$ be a test function that satisfies $\phi(x) = 1$ for $\mid x \mid \leq 1$ and $\operatorname{supp} \phi \subseteq [-2, 2]$. Then, if $\phi_\varepsilon(x) = (x - \varepsilon)^{-1} H(x - \varepsilon)\, e^{\frac{-1}{(x-\varepsilon)}} \phi(x)$, we have $\phi_\varepsilon \in \mathcal{D}(\mathbb{R})$ and $\phi_\varepsilon \to \phi_0$ as $\varepsilon \to 0^+$. But $< f(x), \phi_\varepsilon > = \int_\varepsilon^2 \phi_\varepsilon(x) f(x) dx \to \infty$ as $\varepsilon \to 0^+$. Thus $< f, \phi_o >$ cannot be defined.

Example 36. The function $g(x) = H(x) e^{\frac{1}{x}} \sin e^{\frac{1}{x}}$ has a similar order of growth as the function $f(x) = H(x) e^{\frac{1}{x}}$ of the previous example. However, $g(x)$ can be regularized in $\mathcal{D}'(\mathbb{R})$: a regularization is provided by $x^2 \frac{d}{dx}(H(x) \cos e^{\frac{1}{x}})$, where $\frac{d}{dx}$ is the distributional derivative and where we observe that $H(x) \cos e^{\frac{1}{x}}$ is a regular distribution. Notice that in this case $k = \infty$.

The functions of the previous examples are difficult to handle because they become very large near $x = x_0$. However, if $f(x)$ has an *algebraic singularity*, in the sense that $f(x) \mid x - x_0 \mid^m$ is integrable near $x = x_0$ for some m, then $f(x)$ can be regularized. One such regularization is given by

$$< F(x), \phi(x) >= \int_{U \setminus B} f(x) \phi(x) dx$$

$$+ \int_B f(x) \left[\phi(x) - \sum_{|j| < m} \frac{D^j \phi(x_0)(x - x_0)^j}{j!} \right] dx, \qquad (2.4.2)$$

where B is a ball with center at x_0, small enough to be contained in U. Naturally this regularization depends on $B : F = F_B$.

Next we shall discuss several methods for the regularization of non-locally integrable functions. We already discussed the principal value integrals. Here is another example.

Example 37. Let $K(x)$ be a function defined in $\mathbb{R}^n \setminus \{0\}$ which is homogeneous of degree $-n$, that is, $K(\lambda x) = \lambda^{-n} K(x)$, $\lambda > 0$. Suppose also that on the unit sphere $S = \{x \in \mathbb{R}^n :| x |= 1\}$ we have

$$\int_S K(\omega) d\sigma(\omega) = 0. \qquad (2.4.3)$$

Then $K(x)$ will not be integrable near $x = 0$, but we can construct the principal value regularization $\mathcal{P}.v.(K(x))$, defined as

$$< \mathcal{P}.v.(K(x)), \phi(x) >= \lim_{\varepsilon \to 0} \int_{|x| \geq \varepsilon} K(x) \phi(x) dx. \qquad (2.4.4)$$

The existence of this principal value can be seen by using polar coordinates $x = r\omega$, with $r > 0$ and $\omega \in S$, so that

$$\int_{|x| \geq \varepsilon} K(x) \phi(x) dx = \int_\varepsilon^\infty r^{-1} \Phi_K(r) dr,$$

where

$$\Phi_K(r) = \int_S K(\omega) \phi(r\omega) d\omega. \qquad (2.4.5)$$

The function $\Phi_K(r)$ is smooth in $\mathbb{R}$ and, because of (2.4.3), $\Phi_K(0) = 0$. Thus $\int_0^\infty r^{-1} \Phi_K(r) dr$ converges.

The principal value method is very useful, but it does it not work in many situations. For instance, if $\int_S K(\omega) d\sigma(\omega) \neq 0$ in the previous example then the principal value of the integral $\int_{\mathbb{R}^n} K(x) \phi(x) dx$ does not exist if $\phi(0) \neq 0$. Similarly, the principal value of the integral $\int_{-\infty}^\infty \frac{\phi(x)}{x^2} dx$ does not exist if $\phi \in \mathcal{D}(\mathbb{R})$ and $\phi(0) \neq 0$.

The next method we are going to consider is that of *analytic continuation*. Suppose that $f_\lambda(x)$ are locally integrable functions that depend analytically on the parameter $\lambda \in \Lambda_0$. It is often the case that there is a larger region Λ_1 such that $f_\lambda(x)$ is analytic there, but not locally integrable. If for each $\phi \in \mathcal{D}(U)$ the function $\Phi(\lambda) = < f_\lambda(x), \phi(x) >$ initially defined for $\lambda \in \Lambda_0$ can be continued to Λ_1 then we obtain a regularization of f_λ for $\lambda \in \Lambda_1 \setminus \Lambda_0$.

The chief example is provided by the function $f_\lambda(x) = H(x)x^\lambda$. The function $f_\lambda(x)$ is locally integrable in $\mathbb{R}$ if $\Re \lambda > -1$, and in this case it defines a regular distribution, customarily denoted as x_+^λ. Let now $\phi \in \mathcal{D}(\mathbb{R})$ and set

$$\Phi(\lambda) =< x_+^\lambda, \phi(x) >= \int_0^\infty x^\lambda \phi(x)dx, \quad \Re \lambda > -1. \qquad (2.4.6)$$

As we shall presently show, $\Phi(\lambda)$ can be continued analytically to $\mathbb{C} \setminus \{-1, -2, -3, \cdots\}$. Indeed, the formula

$$\Phi(\lambda) = \int_0^1 x^\lambda (\phi(x) - \phi(0))dx + \int_1^\infty x^\lambda \phi(x)dx + \frac{\phi(0)}{\lambda + 1} \qquad (2.4.7)$$

gives the continuation to the region $\Re \lambda > -2$, $\lambda \neq -1$. More generally the formula

$$\Phi(\lambda) = \int_0^1 x^\lambda \left(\phi(x) - \sum_{j=0}^n \frac{\phi^{(j)}(0)}{j!}x^j\right)dx + \int_1^\infty x^\lambda \phi(x)dx + \sum_{j=0}^n \frac{\phi^{(j)}(0)}{j!(\lambda + j + 1)} \qquad (2.4.8)$$

gives the continuation to the region $\Re \lambda > -(n+2)$, $\lambda \neq -1, -2, \cdots,$ $-(n+1)$.

Therefore, the method of analytic continuation allows us to define the generalized function x_+^λ for $\lambda \neq -1, -2, -3, \cdots$. As can be seen from (2.4.8), the analytic function $\Phi(\lambda)$ has simple poles at $\lambda = -1, -2, -3,$ $\cdots$ with residues

$$\Re es_{\lambda=-k} \Phi(\lambda) = \frac{\phi^{(k-1)}(0)}{(k-1)!}.$$

This means that the generalized function x_+^λ has poles at $\lambda = -1, -2, -3, \cdots$ with residues

$$\Re es_{\lambda=-k} x_+^\lambda = \frac{(-1)^{k-1}\delta^{(k-1)}(x)}{(k-1)!}. \qquad (2.4.9)$$

The generalized function x_-^λ, regularization of $|x|^\lambda H(-x)$, can be defined as $x_-^\lambda = (-x)_+^\lambda$ so that it is also analytic in $\mathbb{C} \setminus \{-1, -2, -3, \cdots\}$ with simple poles at these singularities and with

$$\Re es_{\lambda=-k} x_-^\lambda = \frac{\delta^{(k-1)}(x)}{(k-1)!}. \qquad (2.4.10)$$

Using the functions x_+^λ and x_-^λ we can define the distributions $\mid x \mid^\lambda = x_+^\lambda + x_-^\lambda$ and $\mid x \mid^\lambda \operatorname{sgn} x = x_+^\lambda - x_-^\lambda$.

Observe that $\mid x \mid^\lambda$ is analytic in the larger region $\mathbb{C}\backslash\{-1,-3,-5,\cdots\}$ since the residues of x_+^λ and x_-^λ at the poles $\lambda = -2,-4,-6,\cdots$ cancel each other. The values of $\mid x \mid^\lambda$ for $\lambda = -2,-4,-6,\cdots$ are denoted as $x^{-2}, x^{-4}, x^{-6}\cdots$. Similarly, $\mid x \mid^\lambda \operatorname{sgn}(x)$ is analytic in the larger region $\mathbb{C}\backslash\{-2,-4,-6,\cdots\}$. The values of $\mid x \mid^\lambda \operatorname{sgn}(x)$ for $\lambda = -1,-3,-5,\cdots$ are denoted as $x^{-1}, x^{-3}, x^{-5}, \cdots$.

Observe that

$$< x^{-1}, \phi(x) > = \lim_{\lambda \to -1} < x_+^\lambda - x_-^\lambda, \phi(x) >$$

$$= \lim_{\lambda \to -1} \int_0^\infty x^\lambda(\phi(x) - \phi(-x))dx,$$

or

$$< x^{-1}, \phi(x) > = \int_0^\infty \frac{\phi(x) - \phi(-x)}{x} dx, \qquad (2.4.11)$$

and it follows that

$$x^{-1} = \mathcal{P}.v. \left(\frac{1}{x}\right). \qquad (2.4.12)$$

We also have

$$< x^{-2}, \phi(x) > = \int_0^\infty \frac{\phi(x) + \phi(-x) - 2\phi(0)}{x^2} dx. \qquad (2.4.13)$$

We would also like to point out the formulas

$$\frac{dx_+^\lambda}{dx} = \lambda x_+^{\lambda-1}, \quad \frac{dx_-^\lambda}{dx} = -\lambda x_-^{\lambda-1}, \qquad (2.4.14)$$

$$\frac{d\mid x \mid^\lambda}{dx} = \lambda \mid x \mid^\lambda \operatorname{sgn}(x), \quad \frac{d\mid x \mid^\lambda \operatorname{sgn}(x)}{dx} = \lambda \mid x \mid^\lambda, \qquad (2.4.15)$$

so that, in particular,

$$\frac{dx^{-n}}{dx} = -nx^{-n-1}, \quad n = 1,2,3,\cdots. \qquad (2.4.16)$$

The analytic continuation of the generalized functions $(x + i0)^\lambda$ and $(x - i0)^\lambda$ can be obtained from the formulas

$$(x + i0)^\lambda = x_+^\lambda + e^{i\lambda\pi} x_-^\lambda,$$

$$(x - i0)^\lambda = x_+^\lambda + e^{-i\lambda\pi} x_-^\lambda. \qquad (2.4.17)$$

It follows that $(x + i0)^\lambda$ and $(x - i0)^\lambda$ are entire functions of λ since the residues at all the poles $\lambda = -1, -2, -3, \cdots$ cancel. We recall the values

$$(x + i0)^{-1} = x^{-1} - i\,\pi\delta(x),$$

$$(x - i0)^{-1} = x^{-1} + i\,\pi\delta(x),$$

and more generally,

$$(x + i0)^{-n} = x^{-n} - \frac{i\,\pi(-1)^{n-1}}{(n-1)!}\delta^{(n-1)}(x), \qquad (2.4.18a)$$

$$(x - i0)^{-n} = x^{-n} + \frac{i\,\pi(-1)^{n-1}}{(n-1)!}\delta^{(n-1)}(x), \qquad (2.4.18b)$$

These can be written as

$$(x \pm i0)^{-n} = \pm\frac{2\pi i(-1)^n}{(n-1)!}\delta^{\pm(n-1)}, \qquad (2.4.19)$$

where

$$\delta^{\pm(m)} = \frac{1}{2}\delta^{(m)}(x) \pm \frac{(-1)^{m+1}m!}{2\pi\,i}x^{-m-1}, \qquad (2.4.20)$$

are the Heisenberg distributions, of importance in quantum mechanics.

The final regularization procedure that we would like to consider is the Hadamard finite part. This method allows us to obtain the regularization of functions like $\frac{H(x)}{x^n}$, $n = 1, 2, 3, \cdots$ for which the analytic continuation method fails. The basic idea is the following. Suppose $f(x)$ is locally integrable in $\mathbb{R}\backslash\{0\}$ and that it vanishes for $x < 0$. The Hadamard finite part of the integral $\int_0^a f(x)dx$ is constructed as follows. Let

$$F(\varepsilon) = \int_\varepsilon^a f(x)dx. \qquad (2.4.21)$$

Suppose we can split $F(\varepsilon)$ into two parts as

$$F(\varepsilon) = F_1(\varepsilon) + F_2(\varepsilon), \qquad (2.4.22)$$

where

$$F_1(\varepsilon) = a_1\phi_1(\varepsilon) + \cdots + a_n\phi_n(\varepsilon), \qquad (2.4.23)$$

the functions $\phi_1(\varepsilon), \cdots, \phi_n(\varepsilon)$ being taken from some fixed set of functions (usually inverse powers of ε and logarithms), while

$$\lim_{\varepsilon \to 0} F_2(\varepsilon) = A \qquad (2.4.24)$$

exists. Then $F_1(\varepsilon)$ and $F_2(\varepsilon)$ are called the infinite and finite parts. The Hadamard finite part is then defined as

$$F.p. \int_0^a f(x)dx = A. \qquad (2.4.25)$$

The Hadamard finite part provides a method of regularization by defining

$$< \mathcal{P}f\ (f(x)), \phi(x) > = F.p. \int_0^\infty f(x)\phi(x)dx, \qquad (2.4.26)$$

for $\phi \in \mathcal{D}(\mathbb{R})$ if the finite part exists. We call the distribution $\mathcal{P}f(f(x))$ a pseudofunction. Consider for instance the pseudofunction $\mathcal{P}f\left(\frac{H(x)}{x}\right)$. We have

$$< \mathcal{P}f\left(\frac{H(x)}{x}\right), \phi(x) > = F.p. \int_0^\infty \frac{\phi(x)}{x}dx,$$

but

$$\int_\varepsilon^\infty \frac{\phi(x)}{x}dx = -\phi(0)\ln\varepsilon + \int_\varepsilon^1 \frac{\phi(x) - \phi(0)}{x}dx + \int_1^\infty \frac{\phi(x)}{x}dx,$$

and so

$$F.p \int_0^\infty \frac{\phi(x)}{x}dx = \int_0^1 \frac{\phi(x) - \phi(0)}{x}dx + \int_1^\infty \frac{\phi(x)}{x}dx. \qquad (2.4.27)$$

More generally,

$$< \mathcal{P}f\left(\frac{H(x)}{x^n}\right), \phi(x) > = F.p \int_0^\infty \frac{\phi(x)}{x^n}dx$$

$$= \int_1^\infty \frac{\phi(x)}{x^n}dx + \int_0^1 \frac{1}{x^n}\left(\phi(x) - \sum_{j=0}^{n-1} \frac{\phi^{(j)}(0)}{j!}x^j\right)dx$$

$$- \sum_{j=0}^{n-2} \frac{\phi^{(j)}(0)}{j!(m - j - 1)}. \qquad (2.4.28)$$

Observe also that if $\lambda \neq -1, -2, -3, \cdots$

$$< x_+^\lambda, \phi(x) > = F.p. \int_0^\infty x^\lambda \phi(x)dx. \qquad (2.4.29)$$

We have

$$x^q \mathcal{P}f\left(\frac{H(x)}{x^k}\right) = \mathcal{P}f\left(\frac{H(x)}{x^{k-q}}\right), \qquad q = 1, 2, 3, \cdots, \qquad (2.4.30)$$

while for the derivatives the formulas take the form

$$\frac{d}{dx}(\ln x H(x)) = \mathcal{P}f\left(\frac{H(x)}{x}\right), \tag{2.4.31}$$

$$\frac{d}{dx}\left(\mathcal{P}f\left(\frac{H(x)}{x^k}\right)\right) = -k\,\mathcal{P}f\left(\frac{H(x)}{x^{k+1}}\right) + \frac{(-1)^k \delta^{(k)}(x)}{k!}. \tag{2.4.32}$$

Next we consider the regularization in $\mathcal{D}'(\mathbb{R}^n)$ of r^λ, where $r = \mid x \mid = \sqrt{x_1^2 + \cdots + x_n^2}$. This is achieved by using polar coordinates, so that

$$\int_{\mathbb{R}^n} \mid x \mid^\lambda \phi(x)dx = \int_0^\infty r^{\lambda+n-1}\Phi(r)dr, \tag{2.4.33}$$

where

$$\Phi(r) = \int_{|\omega|=1} \phi(r\omega)d\sigma(\omega). \tag{2.4.34}$$

Therefore, we define the finite part $\mathcal{P}f(\mid x \mid^\lambda)$ as

$$< \mathcal{P}f(\mid x \mid^\lambda), \phi(x) > = < \mathcal{P}f(r^{\lambda+n-1}), \Phi(r) >$$

$$= F.p. \int_0^\infty r^{\lambda+n-1}\Phi(r)dr. \tag{2.4.35}$$

Since $\Phi^{(2m+1)}(0) = 0$, $m = 0, 1, 2, \cdots$ it follows that we also have

$$< \mathcal{P}f(\mid x \mid^\lambda), \phi(x) > = < r_+^{\lambda+n-1}, \Phi(r) >$$
$$\lambda \neq -n, -n-2, -n-4, -n-6, \cdots. \tag{2.4.36}$$

2.5 Support and Order

We now extend the notion of support for generalized functions. Recall that if $f(x)$ is an ordinary function defined in U then its support is the closed subset of U

$$\text{supp } f = \{x \in U : f(x) \neq 0\}. \tag{2.5.1}$$

Alternatively, the support of f can be characterized in terms of its complement $V = U \setminus \text{supp } f$. Indeed, V is the largest open set on which f vanishes. Having in mind these ideas, we give the following definitions.

Definition. *Let $f \in \mathcal{D}'(U)$. Let V be an open subset of U. The restriction of f to V is the distribution $f \mid V$ given by*

$$< f \mid V, \phi > = < f, \tilde{\phi} >, \quad \phi \in \mathcal{D}(V),$$

where $\tilde{\phi}$ is the extension of ϕ to U obtained by setting $\tilde{\phi}(x) = 0$ if $x \in U \setminus V$.

We say that f vanishes on V if $f \mid V = 0$. This means that

$$< f, \phi > = 0, \quad \text{supp } \phi \subseteq V.$$

Definition. *Let $f \in \mathcal{D}'(U)$. The support of f is the closed set* supp f *characterized as the complement of the largest open set on which f vanishes.*

If $f(x)$ is a locally integrable function, the two notions coincide. Observe that the support of a regular distribution cannot be too small. For instance, there are no locally integrable functions whose support is exactly the one point set $\{y\}$. On the other hand, the Dirac delta funtion $\delta(x - y)$ and its derivatives have as support the set $\{y\}$. Similarly, the single layer $\delta(\Sigma)$ has as support the hypersurface Σ, but no regular distribution can have Σ as its support.

A very important class of distributions is the space $\mathcal{E}'(U)$ formed by those distributions of $\mathcal{D}'(U)$ whose support is compact. The notation $\mathcal{E}'(U)$ suggests that this space is the dual of a certain space $\mathcal{E}(U)$. Indeed, let $\mathcal{E}(U)$ be the space of all smooth functions defined in U with the topology generated by the family of seminorms

$$\| \phi \|_{K,j} = sup \{| D^j \phi(x) | : x \in K\}, \tag{2.5.2}$$

for $j \in \mathbb{N}^n$ and K a compact set in U. Then the space $\mathcal{D}(U)$ is a dense subspace of $\mathcal{E}(U)$ and the inclusion is continuous. It follows that the dual space $\mathcal{E}'(U)$ can be identified with a subspace of $\mathcal{D}'(U)$: precisely the distributions with compact support. The notation $\mathcal{E}'$ is usually employed for $\mathcal{E}'(\mathbb{R}^n)$.

It is interesting to observe that if U is an open set of $\mathbb{R}^n$ then $\mathcal{E}'(U)$ is a proper subspace of $\mathcal{D}'(U)$, however, the two spaces $\mathcal{E}'(U)$ and $\mathcal{D}'(U)$ coincide if U is a compact manifold. In particular $\mathcal{E}'(S) = \mathcal{D}'(S)$ if S is a sphere, $S = \{x \in \mathbb{R}^n : \mid x - a \mid = r\}$.

As we have already mentioned, any distribution of the type

$$f(x) = \sum_{|k| \le m} a_k D^k \delta(x - y), \tag{2.5.3}$$

for $m \in \mathbb{N}$ and arbitrary constants $\{a_k\}$ has as support the set $\{y\}$. That all distributions supported at $\{y\}$ have this form can be proved with the help of Borel's theorem, theorem 9 of Chapter 1.

Theorem 16. *Any distribution whose support is the set $\{y\}$ has form (2.5.3) for some $m \in \mathbb{N}$ and some constants $\{a_k\}$, $\mid k \mid \le m$.*

Proof. Let $b = \{b_k\}$ be an arbitrary sequence, indexed by $k \in \mathbb{N}^n$. Then there exists $\phi = \phi_b \in \mathcal{E}(\mathbb{R}^n)$ with $D^k \phi(y) = b_k$, $k \in \mathbb{N}^n$. The formula

$$< T, b > = < f(x), \phi_b(x) > \qquad (2.5.4)$$

defines a linear functional T on the space $\mathbb{R}^{\mathbb{N}^n}$ of such sequences b since the hypothesis that f is supported at $\{y\}$ implies that (2.5.4) does not depend on the extension ϕ_b but only on b.

It follows that there exists $m \in \mathbb{N}$ and constants a_k for $|k| \leq m$ such that

$$< T, b > = \sum_{|k| \leq m} (-1)^{|k|} a_k b_k, \qquad (2.5.5)$$

and so

$$f(x) = \sum_{|k| \leq m} a_k D^k \delta(x - y). \qquad \blacksquare \qquad (2.5.6)$$

Observe, in particular, that this theorem implies that a series of Dirac delta functions of the form

$$\sum_{k \in \mathbb{N}^n} a_k D^k \delta(x - y) \qquad (2.5.7)$$

cannot converge in $\mathcal{D}'$ unless only finitely many a_k do not vanish.

If V is an open subset of U, we say that a distribution $f \in \mathcal{D}'(U)$ is continuous in V if $f \mid V$ is a regular distribution generated by a continuous function. The statements "f is of class C^k in V ", "f is smooth in V" or "f is real-analytic in V" admit similar interpretations. The *singular support* of f, denoted by sing supp f, is the complement of the largest open set in which f is smooth.

For instance, the singular support of the distributions x_+^λ is $\{0\}$ while its support is the interval $[0, \infty)$. On the other hand, the singular support of a single layer $\delta(\Sigma)$ is Σ, its support. Two distributions whose singular supports do not meet can be multiplied in a natural way.

The last concept we would like to introduce now is the notion of order. To do so, let us first recall the definition of a Radon measure. A Radon (signed) measure μ in U can be defined in either of two equivalent ways. First, μ can be considered as a $\sigma-$additive finite set function on the class of Borel subsets of U. Alternatively μ can be considered as a continuous linear functional in the space $C_o(U)$ of continuous functions with compact support.

Clearly, any Radon measure in U defines a distribution of $\mathcal{D}'(U)$. Any locally integrable function is a Radon measure. A Dirac delta function $\delta(x - y)$ is also a Radon measure, but its derivatives are not.

A distribution $f \in \mathcal{D}'(U)$ which is a Radon measure has order 0. The distributions that can be written as $f = L\,g$, where g is a Radon measure and L is a differential operator of order k, have order less than or equal to k.

Any distribution of compact support has finite order, but there are distributions of infinite order.

Example 38. The function $\delta^{(n)}(x)$ is of order n. On the other hand, the generalized function $\sum_{n=0}^{\infty} \delta^{(n)}(x - n)$ has infinite order.

2.6 Homogeneous Distributions

We shall now define and study the basic properties of the homogeneous and associated homogeneous generalized functions. These types of generalized functions are the building blocks of many of the asymptotic expansions we study in this work.

Let us start with the homogeneous distributions.

Definition. *A distribution $f \in \mathcal{D}'(\mathbb{R}^n)$ is called homogeneous of degree ρ if*

$$f(\lambda x) = \lambda^{\rho} f(x) \quad , \lambda > 0. \tag{2.6.1}$$

An example of a homogeneous generalized function is provided by the Dirac delta function $\delta(x)$, which is homogeneous of degree $-n$: $\delta(\lambda x) = \lambda^{-n}\delta(x)$, $\lambda > 0$. More generally, if $k \in \mathbb{N}^n$ is a multi-index, then $D^k\delta(x)$ is homogeneous of degree $\rho = -n - |k|$.

Another example, perhaps the most natural one, is provided by the distributions x_+^{ρ} and x_-^{ρ} for $\rho \neq -1, -2, -3, \cdots$. They are homogeneous of degree ρ:

$$(\lambda x)_+^{\rho} = \lambda^{\rho} x_+^{\rho}, \quad (\lambda x)_-^{\rho} = \lambda^{\rho} x_-^{\rho}, \quad \lambda > 0.$$

These relations can be proved by analytic continuation considerations. In fact, they are clearly true if $\Re e\,\rho > -1$, and since both sides are analytic functions of ρ for $\rho \neq -1, -2, -3, \cdots$, it follows that they hold in the larger region.

The linear combinations $|x|^{\rho}$ and $|x|^{\rho}\,\mathrm{sgn}(x)$ are also homogeneous of degree ρ in their respective regions of definition. It follows, in particular, that the distribution x^{-n} is homogeneous of degree $-n$ for $n = 1, 2, 3, \cdots$.

On the other hand, the distributions $\mathcal{P}f\left(\frac{H(x)}{x^n}\right)$ are not homogeneous of degree $-n$. Rather, we have

$$\mathcal{P}f\left(\frac{H(\lambda x)}{(\lambda x)^n}\right) = \frac{1}{\lambda^n}\,\mathcal{P}f\left(\frac{H(x)}{x^n}\right) + \frac{\ln\lambda(-1)^{n-1}\delta^{(n-1)}(x)}{\lambda^n(n-1)!}. \tag{2.6.2}$$

No other examples of homogeneous generalized functions can be given in one dimension. Indeed, we have

Theorem 17. *Let $f \in \mathcal{D}'(\mathbb{R})$ be a homogeneous distribution of degree ρ. If $\rho \neq -1, -2, -3, \cdots$ there exist constants c_1 and c_2 such that*

$$f(x) = c_1 x_+^\rho + c_2 x_-^\rho. \tag{2.6.3}$$

If $\rho = -n$, $n = 1, 2, 3, \cdots$, then there exist constants c_1 and c_2 such that

$$f(x) = c_1 x^{-n} + c_2 \delta^{(n-1)}(x). \tag{2.6.4}$$

Proof. Differentiation of the relation $f(\lambda x) = \lambda^\rho f(x)$ with respect to λ and setting $\lambda = 1$ yields the differential equation

$$x f'(x) = \rho f(x). \tag{2.6.5}$$

This ordinary differential equation can be integrated for $x \neq 0$ to obtain $f(x) = c_1 x^\rho, x > 0$, $f(x) = c_2 |x|^\rho$, $x < 0$. When $\rho \neq -1, -2, -3, \cdots$ it follows that $f(x) = c_1 x_+^\rho + c_2 x_-^\rho + g(x)$, where g is supported at $\{0\}$. But then $g(x) = a_0 \delta(x) + \cdots + a_m \delta^{(m)}(x)$ and the relation $g(\lambda x) = \lambda^\rho g(x)$; i.e.,

$$\sum_{j=0}^{m} a_j \lambda^{-j} \delta^{(j)}(x) = \lambda^\rho \sum_{j=0}^{m} a_j \delta^{(j)}(x) \tag{2.6.6}$$

gives $a_0 = \cdots = a_m = 0$.

If $\rho = -n, n = 1, 2, 3, \cdots$, then $f(x) = c_1 x^{-n} + d \mathcal{P}f(H(x) x^{-n}) + g(x)$, with supp $g \subseteq \{0\}$. Since x^{-n} is homogeneous of degree n while $\mathcal{P}f(H(x) x^{-n})$ is not, according to (2.6.2), it follows that $d = 0$. Finally, from (2.6.6) it follows that $g(x) = a \delta^{(n-1)}(x)$ and (2.6.4) is obtained. ∎

Let us now study the situation in $\mathbb{R}^n$. Let $K(\omega)$ be a distribution defined on the unit sphere $S = \{ x \in \mathbb{R}^n : | x | = 1 \}$, i.e, $K \in \mathcal{D}'(S)$. Then we can associate with it the distribution $f(x) = \mathcal{P}f(| x |^\rho) K \left(\frac{x}{|x|} \right)$ by setting

$$< f(x), \phi(x) > = < \mathcal{P}f(H(r) r^{\rho+n-1}), \Phi_K(r) >, \tag{2.6.7}$$

where

$$\Phi_K(r) = < K(\omega), \phi(r\omega) >, \quad r \in \mathbb{R}. \tag{2.6.8}$$

We have

$$< f(\lambda \boldsymbol{x}), \phi(\boldsymbol{x}) > = \frac{1}{\lambda^n} < f(\boldsymbol{x}), \phi\left(\frac{\boldsymbol{x}}{\lambda}\right) >$$

$$= \frac{1}{\lambda^n} < \mathcal{P}f\left(H(r)r^{\rho+n-1}\right), \Phi_K\left(\frac{r}{\lambda}\right) >,$$

and since $\mathcal{P}f(H(r)\, r^{\rho+n-1}) = r_+^{\rho+n-1}$ is homogeneous of degree $\rho + n - 1$ if $\rho \neq -n, -n-1, -n-2, \cdots$ it follows that in this case

$$f(\lambda \boldsymbol{x}) = \lambda^\rho f(\boldsymbol{x}). \tag{2.6.9}$$

If, however, $\rho = -n - m$ for some $m = 0, 1, 2, \cdots$ then because of (2.6.2)

$$< f(\lambda \boldsymbol{x}), \phi(\boldsymbol{x}) > = \frac{1}{\lambda^n} < \mathcal{P}f\left(H(r)r^{-n-1}\right), \Phi_K\left(\frac{r}{\lambda}\right) >$$

$$= \frac{1}{\lambda^n} \left[\lambda^{-m-1} < \mathcal{P}f(H(r)r^{-m-1}), \Phi_K(r) > + \frac{\lambda^{-m-1} \ln \lambda \Phi_K^{(m)}(0)}{m!} \right],$$

but

$$\Phi_K^{(m)}(0) = \boldsymbol{D}^m \phi(0) \, \| < K(\boldsymbol{\omega}), \boldsymbol{\omega}^m >, \tag{2.6.10}$$

where $\boldsymbol{T} \parallel \boldsymbol{S}$ stands for the total contraction of the tensors $\boldsymbol{T}$ and $\boldsymbol{S}$. Hence

$$f(\lambda \boldsymbol{x}) = \lambda^\rho \, f(\boldsymbol{x}) + \frac{(-1)^m \lambda^\rho \ln \lambda}{m!} < K(\boldsymbol{\omega}), \boldsymbol{\omega}^m > \| \boldsymbol{D}^m \delta(\boldsymbol{x}). \tag{2.6.11}$$

Therefore, $f(\boldsymbol{x})$ is homogeous of degree $\rho = n - m$ if and only if the symmetric tensor of order m $< K(\boldsymbol{\omega}), \boldsymbol{\omega}^m >$ vanishes, that is, if and only if

$$< K(\omega_1, \cdots, \omega_n), \omega_1^{k_1} \cdots \omega_n^{k_n} > = 0, \quad k_1 + \cdots + k_n = m. \tag{2.6.12}$$

It can be shown that, aside from the Dirac deltas concentrated at the origin, this construction gives all the homogeneous generalized functions in $\mathbb{R}^n$.

Theorem 18. *Let $f(\boldsymbol{x})$ be a homogeneous distribution of degree ρ in $\mathbb{R}^n$. If $\rho \neq -n, -n-1, -n-2, \cdots$ then there exists $K \in \mathcal{D}'(S)$ such that*

$$f(\boldsymbol{x}) = \mathcal{P}f(|\boldsymbol{x}|^\rho) K\left(\frac{\boldsymbol{x}}{|\boldsymbol{x}|}\right). \tag{2.6.13}$$

If $\rho = -n - m$, with $m = 0, 1, 2, \cdots$ then there exist $K \in \mathcal{D}'(S)$ with $< K(\omega), \omega^m >= 0$ and constants a_k for each multiindex k with $\mid k \mid = m$ such that

$$f(x) = \mathcal{P}f(\mid x \mid^\rho)K\left(\frac{x}{\mid x \mid}\right)$$

$$+ \sum_{\mid k \mid = m} a_k D^k \delta(x). \qquad (2.6.14)$$

Homogeneous functions can also be studied from another perspective. Let us consider the dilatation operators T_λ, $\lambda > 0$ defined in the space $\mathcal{D}'(\mathbb{R}^n)$ by

$$T_\lambda(f)(x) = f(\lambda x). \qquad (2.6.15)$$

Then a distribution $f(x)$ is homogeneous of degree ρ if and only if it is a characteristic vector of each of the operators T_λ, associated to the characteristic value λ^ρ

$$T_\lambda(f) = \lambda^\rho f. \qquad (2.6.16)$$

Characteristic vectors play an important role in the diagonalization of matrices and operators since the matrix with respect to a basis formed by characteristic vectors is diagonal. When diagonalization is not possible, one considers the so-called *associated vectors*. Indeed if μ is a characteristic value of the operator T, then characteristic vectors of T associated to μ are called associated vectors of order 0. Associated vectors of order k are defined as those vectors V_k for which

$$TV_k = \mu V_k + V_{k-1} \qquad (2.6.17)$$

where V_{k-1} is associated vector of order $k - 1$. This means that we can find vectors $V_0, \cdots, V_{k-1}$ such that

$$TV_0 = \mu V_0 \ ,$$
$$TV_1 = \mu V_1 + V_0 \ ,$$

$$\vdots$$

$$TV_{k-1} = \mu V_{k-1} + V_{k-2} \ ,$$
$$TV_k = \mu V_k + V_{k-1} \ . \qquad (2.6.18)$$

Notice that the matrix of an operator with respect to a basis $V_0, \cdots, V_k$ that satisfies (2.6.18) has the Jordan canonical form

$$\begin{pmatrix} \mu & 1 & & 0 \\ & \ddots & \ddots & 1 \\ 0 & & & \mu \end{pmatrix}.$$

Returning to the dilatation operators T_λ, we could try to define recursively the associated homogeneous functions of order k and degree ρ as those distributions $f_k(x)$ for which

$$f_k(\lambda x) = \lambda^\rho f_k(x) + \lambda^\rho a(\lambda) f_{k-1}(x) \tag{2.6.19}$$

for some associated homogeneous distributions $f_{k-1}(x)$ of order $k-1$ and degree ρ and some function $a(\lambda)$. However (2.6.19) implies that

$$a(\lambda\mu) = a(\lambda) + a(\mu),$$

and it follows that $a(\lambda) = b\ln\lambda$ for some constant b, which can be absorbed in f_{k-1}. Therefore we define an associated homogeneous distribution of order k and degree ρ as a function that satisfies

$$f_k(\lambda x) = \lambda^\rho f_k(x) + \lambda^\rho \ln\lambda f_{k-1}(x), \tag{2.6.20}$$

for some associated homogeneous distribution $f_{k-1}(x)$ of order $k-1$ and degree ρ.

We have already seen the very important associated homogeneous distributions $\mathcal{P}f\left(\frac{H(x)}{x^n}\right)$, $n = 1, 2, 3, \cdots$, which according to (2.6.2) are associated homogeneous of order 1 and degree $-n$.

Also, the distributions $\mathcal{P}f(|x|^{-n-m})K(\frac{x}{|x|})$ for which $< K(\omega), \omega^m >$ does not necessarily vanish, are associated homogeneous of order 1 and degree ρ. In particular, if $K(\omega) = 1$, the distribution $\mathcal{P}f(|x|^{-n-2m})$, $m = 0, 1, 2, \cdots$ is associated homogeneous of order 1 and degree $\rho = -n - 2m$.

Associated homogeneous generalized functions also arise by parametric differentiation. Indeed, if $f_\rho(x)$ is a homogeneous generalized function of degree ρ that depends smoothly on ρ then differentiation with respect to ρ of the equation

$$f_\rho(\lambda x) = \lambda^\rho f_\rho(x)$$

yields

$$\frac{\partial f_\rho}{\partial \rho}(\lambda x) = \lambda^\rho \frac{\partial f_\rho}{\partial \rho}(x) + \lambda^\rho \ln\lambda f_\rho(x). \tag{2.6.21}$$

Thus $\dfrac{\partial f_\rho}{\partial \rho}$ is associated homogeneous of order 1. More generally, if f_ρ is associated homogeneous of order k and degree ρ then $\dfrac{\partial f_\rho}{\partial \rho}$ is associated homogeneous of order $k+1$ and degree ρ.

We also mention in this connection that if f is associated homogeneous of order k and degree ρ then $\dfrac{\partial f}{\partial x_i}$ is associated homogeneous of order k and degree $\rho - 1$.

Example 39. Since x_+^ρ is homogeneous of degree ρ it follows that the generalized function $\frac{\partial}{\partial \rho}\left(x_+^\rho\right) = x_+^\rho \ln x_+$ is associated homogeneous of order 1 and degree ρ. More generally, the generalized function $x_+^\rho \left(\ln x_+\right)^k$ is associated homogeneous of order k and degree ρ. Observe that $x_+^\rho \left(\ln x_+\right)^k$ is a regular distribution for $\Re e \, \rho > -1$ and extends by analytic continuation to $\mathbb{C} \setminus \{-1, -2, -3, \cdots\}$.

2.7 Distributional Derivatives of Discontinuous Functions

Let $F(x)$, $x \in \mathbb{R}$, have a jump discontinuity at $x = \xi$ of magnitude a, but be smooth everywhere else. Thus $F'(x)$ exists in both the intervals $x < \xi$ and $x > \xi$, but not at $x = \xi$. To find the distributional derivative of such a function we define the function $f(x)$ as

$$f(x) = F(x) - a \, H(x - \xi), \tag{2.7.1}$$

where $H(x)$ is the Heaviside function. This function is continuous everywhere and its derivatives coincide with those of $F(x)$ in both sides of ξ. Accordingly, the differentiation of both sides of (2.7.1) yields

$$f'(x) = \bar{F}'(x) - a\delta(x - \xi),$$

where the overbar stands for the distributional derivative. Writing $[F] = F_+ - F_- = a$, the jump of F across the discontinuity, the above result becomes

$$\bar{F}'(x) = F'(x) + [F] \, \delta(x - \xi). \tag{2.7.2}$$

If there are several points $\xi_1, \cdots, \xi_n$ of jump discontinuities of magnitudes $a_1, \cdots, a_n$, then we have

$$\bar{F}'(x) = F'(x) + \sum_{m=1}^{n} a_m \delta(x - \xi_m). \tag{2.7.3}$$

These concepts can be readily extended to functions of several variables. Let $F(\boldsymbol{x}, t)$ be a function that is defined in $\mathbb{R}^n \times \mathbb{R}$ that is everywhere smooth except for a jump discontinuity across the moving surface $\Sigma(t)$. We assume that Σ is smooth and that it divides the space in two regions S_+ and S_-. We denote by $\boldsymbol{n} = (n_i)$ the unit normal vector that points in the positive direction. Then the formula corresponding to (2.7.2) is [33], [35], [40]

$$\overline{grad} \, F(\boldsymbol{x}, t) = grad \, F(\boldsymbol{x}, t) + \boldsymbol{n} \, [F]\delta(\Sigma), \tag{2.7.4}$$

where $[F] = F_+ - F_-$ is the value of the jump of F across Σ and where $\delta(\Sigma)$ is a single layer concentrated on Σ, whose action on a test function $\phi(x, t)$ is given by

$$< \delta(\Sigma), \phi > = \int_{-\infty}^{\infty} \int_{\Sigma} \phi(x, t) d\sigma(x) dt. \qquad (2.7.5)$$

Here $d\sigma$ is the surface element on Σ.

The time derivative can also be computed. Indeed

$$\frac{\overline{\partial F}}{\partial t} = \frac{\partial F}{\partial t} - G[F]\delta(\Sigma), \qquad (2.7.6)$$

where G is the normal speed of the surface.

If $F(x, t)$ is a vector field with jump discontinuities across Σ, then the following formulas follow easily from (2.7.4):

$$\overline{\text{div }} F = \text{div } F + n \cdot [F] \, \delta(\Sigma), \qquad (2.7.7)$$

$$\overline{\text{curl }} F = \text{curl } F + (n \times [F]) \, \delta(\Sigma). \qquad (2.7.8)$$

These formulas are extremely useful in many fields of mathematical and physical science. We illustrate the use of formula (2.7.4) for deriving the distributional derivatives of $\frac{1}{r}$, where $r = (x_1^2 + x_2^2 + x_3^2)^{\frac{1}{2}}$ in $\mathbb{R}^3$. The function $\frac{1}{r}$ has an infinite, integrable singularity at $x = 0$, but we can write it as the limit of functions with jump discontinuities as

$$\frac{1}{r} = \lim_{\varepsilon \to 0} \frac{H(r - \varepsilon)}{r}. \qquad (2.7.9)$$

This arrangement will help us in obtaining the derivatives of $\frac{1}{r}$, since the distributional derivatives can be taken inside the limit. Now, the function $F(x) = \frac{H(r - \varepsilon)}{r}$ has a jump discontinuity across the sphere S_ε of radius ε and formula (2.7.4) is applicable. We rewrite (2.7.4) in the component form

$$\frac{\overline{\partial F}}{\partial x_j} = \frac{\partial F}{\partial x_j} + n_j[F]\delta(S_\varepsilon). \qquad (2.7.10)$$

In the present case it becomes

$$\frac{\overline{\partial}}{\partial x_j} \left(\frac{H(r - \varepsilon)}{r} \right) = \frac{-x_j}{r^3} H(r - \varepsilon) + \frac{n_j}{\varepsilon} \delta(S_\varepsilon), \qquad (2.7.11)$$

where $n_j = \dfrac{x_j}{r}$. Since

$$\lim_{\varepsilon \to 0} \int_{S_\varepsilon} \phi(\omega) \frac{1}{\varepsilon} n d\sigma(\omega) = \lim_{\varepsilon \to 0} \frac{1}{\varepsilon} \int_{S_1} \phi(\omega) n \varepsilon^2 d\sigma(\omega) = 0,$$

it follows from (2.7.11) that

$$\frac{\overline{\partial}}{\partial x_j} \left(\frac{1}{r} \right) = -\mathcal{P}f \left(\frac{x_i}{r^3} \right) = -x_j \, \mathcal{P}f \left(\frac{1}{r^3} \right), \qquad (2.7.12)$$

i.e, the first order distributional and classical derivatives are equivalent.

To compute the second order distributional derivatives we apply formula (2.7.4) to $\dfrac{\overline{\partial}}{\partial x_j} \left(\frac{1}{r} \right)$ and get

$$\frac{\overline{\partial}^2}{\partial x_i \partial x_j} \left(\frac{H(r - \varepsilon)}{r} \right) = \left(\frac{3x_i x_j - r^2 \delta_{ij}}{r^5} \right) H(r - \varepsilon) - \frac{x_i x_j}{r^4} \delta(S_\varepsilon), \quad (2.7.13)$$

where δ_{ij} is the Kronecker delta. Now,

$$\lim_{\varepsilon \to 0} \int_{S_\varepsilon} \frac{x_i x_j}{r^4} \phi(x) d\sigma(x) = \frac{4\pi}{3} \delta_{ij} \phi(0),$$

so that (2.7.13) yields

$$\frac{\overline{\partial}^2}{\partial x_i \partial x_j} \left(\frac{1}{r} \right) = (3x_i x_j - r^2 \delta_{ij}) \mathcal{P}f \left(\frac{1}{r^5} \right) - \frac{4\pi}{3} \delta_{ij} \delta(x). \qquad (2.7.14)$$

These ideas can be generalized in order to obtain the distributional derivatives of $\mathcal{P}f(\frac{1}{r^k})$ in the space $\mathcal{D}'(\mathbb{R}^n)$ [8], [37], [42] . We shall need these formulas in Chapter 4. In order to write these formulas we use the symmetric notation explained in the introduction to this chapter. In particular, $\boldsymbol{D}^N$ is the symmetric tensor with components $\dfrac{\partial^N}{\partial x_{i_1} \cdots \partial x_{i_N}}$, and $\boldsymbol{x}^N$ is the tensor with components $x_{i_1} \cdots x_{i_N}$. If $\boldsymbol{S}$ and $\boldsymbol{T}$ are symmetric tensors then $\boldsymbol{S}\,\boldsymbol{T}$ is their symmetric product, namely, the symmetrization of their tensor product $\boldsymbol{S} \otimes \boldsymbol{T}$. Similarly, $\boldsymbol{S}^q$ denotes the symmetric product $\boldsymbol{S} \cdots \boldsymbol{S}, q$ times . With this notation (2.7.14) takes the form

$$\boldsymbol{D}^2 \mathcal{P}f \left(\frac{1}{r} \right) = 3\boldsymbol{x}^2 \mathcal{P}f \left(\frac{1}{r^5} \right) - \boldsymbol{\Delta} \mathcal{P}f \left(\frac{1}{r^3} \right) - \frac{4\pi}{3} \boldsymbol{\Delta} \delta(\boldsymbol{x}).$$

Here $\boldsymbol{\Delta} = (\delta_{ij})$ is the second order identity symmetric tensor.

If $-\lambda - n$ is not an even integer (positive or negative), then the formulas for the derivatives of r_+^λ are equivalent to the ordinary ones in the sense that no extra delta terms arise. The formula is

$$D^N(r_+^\lambda) = \sum_{j=0}^{\left[\!\left[\frac{N}{2}\right]\!\right]} \frac{\lambda(\lambda - 2)\cdots(\lambda - 2N + 2 + 2j)N!}{2^j\, j!(N - 2j)!}\, x^{N-2j}\Delta^j r_+^{\lambda-2N+2j}.$$

$$(2.7.15)$$

When $\lambda = -k = -n - 2m$ for some $m \in \mathbb{Z}$, the formulas for $D^N \mathcal{P}f\left(r^{-k}\right)$ will contain deltas concentrated at the origin. We define

$$c_{m,n} = \frac{2\Gamma(m + \frac{1}{2})\pi^{n-\frac{1}{2}}}{\Gamma(m + \frac{n}{2})} \qquad (2.7.16)$$

and the constants $\beta_{q,p}$ by $\beta_{0,0} = 0$,

$$\beta_{q,0} = \frac{1}{k} + \frac{1}{k + 2} + \cdots + \frac{1}{k + 2q - 2}, \qquad q \geq 1, \qquad (2.7.17)$$

while if $p \geq 1$ we set $\beta_{0,p} = 0$ and

$$\beta_{q,p} = \beta_{q,p-1} - \beta_{q-1,p-1}, \quad q \geq 1. \qquad (2.7.18)$$

Then the formula takes the form

$$D^N \mathcal{P}f\left(\frac{1}{r^k}\right) =$$

$$\sum_{j=0}^{\left[\!\left[\frac{N}{2}\right]\!\right]} \frac{(-1)^{N-j}2^{N-2j}\Gamma(\frac{k}{2} + N - j)N!}{\Gamma\left(\frac{k}{2}\right)(N - 2j)!\, j!}\, \Delta^j x^{N-2j}\mathcal{P}f\left(\frac{1}{r^{k+2N-2j}}\right)$$

$$- \sum_{j=\frac{|m|-m}{2}}^{\left[\!\left[\frac{N}{2}\right]\!\right]} \frac{N!\Gamma\left(\frac{k}{2} + j\right)c_{m+j,n}\beta_{N,j}}{(N - 2j)!\,\Gamma\left(\frac{k}{2}\right)\, j!(2m + 2j)!}\, \Delta^j D^{N-2j}\nabla^{2m+2}\delta(x). \quad (2.7.19)$$

In addition to the distribution $\delta(\Sigma)$, there are various other basic surface distributions [33], [35], [39], [40]. One of them is the normal derivative operator, defined as

$$< d_n\delta(\Sigma), \phi > = -\int_{-\infty}^{\infty}\int_{\Sigma(t)} \frac{d\phi}{dn}\, d\sigma\, dt, \qquad (2.7.20)$$

where $\dfrac{d\phi}{dn} = \dfrac{\partial\phi}{\partial x_i}n_i$ is the derivative of ϕ in the normal direction. This is called the dipole layer. Another surface distribution is $\delta'(\Sigma)$, given as

$$\delta'(\Sigma) = n_i \frac{\overline{\partial}}{\partial x_i}\left(\delta(\Sigma)\right). \qquad (2.7.21)$$

These distributions are connected by the relation

$$d_n\delta(\Sigma) = \delta'(\Sigma) - 2\Omega\delta(\Sigma), \qquad (2.7.22)$$

where Ω is the mean curvature of $\Sigma(t)$.

We also have the derivative formulas

$$\frac{\overline{\partial}}{\partial x_i}\left(\delta(\Sigma)\right) = n_i\delta'(\Sigma), \qquad (2.7.23)$$

$$\frac{\overline{\partial}}{\partial t}\left(\delta(\Sigma)\right) = -G\delta'(\Sigma). \qquad (2.7.24)$$

Similarly, one can consider multilayer distributions of the form $f d_n^N\delta(\Sigma)$, where $f \in \mathcal{D}'(\Sigma)$; its action on a test function $\phi \in \mathcal{D}(\mathbb{R}^n \times \mathbb{R})$ is given by

$$< f d_n^N\delta(\Sigma), \phi > = (-1)^N < f, \frac{d^N\phi}{dn^N} >, \qquad (2.7.25)$$

where $\dfrac{d^N\phi}{dn^N}$ is considered as a test function defined only on Σ.

In order to find the derivatives of multilayers we need to introduce some geometrical notions. Let f be a quantity defined only on the moving surface $\Sigma(t)$. Then the $\delta-$ derivatives of f are defined as

$$\frac{\delta f}{\delta x_i} = \left(\frac{\partial\tilde{f}}{\partial x_i} - n_i\frac{d\tilde{f}}{dn}\right)\Big|_\Sigma, \qquad (2.7.26a)$$

$$\frac{\delta f}{\delta t} = \left(\frac{\partial\tilde{f}}{\partial t} + G\frac{d\tilde{f}}{dn}\right)\Big|_\Sigma, \qquad (2.7.26b)$$

where $\tilde{f}$ is any extension of f to a neighborhood of $\Sigma(t)$ in $\mathbb{R}^n \times \mathbb{R}$. If $f \in \mathcal{D}'(\Sigma)$, these derivatives can still be defined by an approximation procedure or by expressing them as linear combinations with smooth coefficients of derivatives with respect to a Gaussian coordinate system in $\Sigma(t)$.

We denote by μ_{ij} the quantity $\dfrac{\delta n_i}{\delta x_j}$. The matrix $\boldsymbol{\mu} = (\mu_{ij})$ is the second fundamental form of the surface expressed in the Cartesian coordinates of the surrounding space $\mathbb{R}^n$.

It is convenient to permit one or both indices in μ_{ij} to take the value t. This gives

$$\mu_{ti} = \frac{\delta(-G)}{\delta x_i} = \frac{\delta n_i}{\delta t}, \qquad (2.7.27a)$$

$$\mu_{ii} = \frac{\delta(-G)}{\delta t}. \qquad (2.7.27b)$$

The P-th power of the matrix μ is denoted by $\mu^{(P)}$, that is,

$$\mu_{ij}{}^{(P)} = \mu_{ik}{}^{(P-1)} \mu_{kj}, \quad P \geq 2, \qquad (2.7.28)$$

where the summation convention is used for the repeated index $k = 1, \cdots, n$.
Also

$$\mu_{it}{}^{(P)} = \mu_{ik}{}^{(P-1)}\mu_{kt}, \quad P \geq 2, \qquad (2.7.29a)$$

$$\mu_{tt}{}^{(P)} = \mu_{tk}{}^{(P-1)}\mu_{kt}, \quad P \geq 2. \qquad (2.7.29b)$$

It is also useful to introduce the 0th-order power

$$\mu_{ij}{}^{(0)} = \delta_{ij} - n_i n_j. \qquad (2.7.30)$$

The quantities $\mu_{it}{}^{(0)}$ and $\mu_{tt}{}^{(0)}$ are not defined.

Then the formula for the first order derivative of multilayers is the following

$$\frac{\overline{\partial}}{\partial x_i} \left(f d_n^N \delta(\Sigma) \right) = f n_i d_n^{N+1} \delta(\Sigma) + \sum_{M=0}^{N} \frac{N!}{M!} \frac{\delta}{\delta x_k} \left(\mu_{ik}{}^{(N-M)} f \right) d_n^M \delta(\Sigma). \tag{2.7.31}$$

Similarly, the formula for the time derivative can be written as

$$\frac{\overline{\partial}}{\partial t} \left(f d_n^N \delta(\Sigma) \right) = -G f \, d_n^{N+1} \delta(\Sigma) + \left(\frac{\delta f}{\delta t} + G \omega_1 f \right) d_n^N \delta(\Sigma)$$

$$+ \sum_{M=0}^{N-1} \frac{N!}{M!} \frac{\delta}{\delta x_k} \left(\mu_{tk}{}^{(N-M)} f \right) d_n^M \delta(\Sigma), \tag{2.7.32}$$

where $\omega_1 = \mu_{ii}$ is the trace of μ.

Formulas for higher order derivatives of multilayers are available [40]. They involve the tensors $\lambda^{(N,P)}$ which are the higher order analogs of the second fundamental form μ. Costen [20] was the first to give the formulas for $\dfrac{\overline{\partial}}{\partial x_i} \left(f \delta(\Sigma) \right)$ and $\dfrac{\overline{\partial}}{\partial t} \left(f \delta(\Sigma) \right)$; interestingly, he derived his formulas without the use of distributions.

2.8 Tempered Distributions and the Fourier Transform

In the previous sections we have defined the function spaces $\mathcal{D}$ and $\mathcal{E}$ as well as their duals $\mathcal{D}'$ and $\mathcal{E}'$. The third space we need for our study is the space $\mathcal{S} = \mathcal{S}(\mathbb{R}^n)$ which consists of the smooth functions $\phi(x)$ of *rapid decay* at infinity, that is,

$$D^j \phi(x) = o(|\, x\, |^{-\infty}), \quad \text{as} \quad |\, x\, | \to \infty, \tag{2.8.1}$$

for each $j \in \mathbb{N}^n$. The topology in $\mathcal{S}$ is defined in terms of the seminorms

$$\| \phi \|_{k,j} = \sup\{|\, x^k \, D^j \phi(x) \,|: x \in \mathbb{R}^n\}, \tag{2.8.2}$$

where $k, j \in \mathbb{N}^n$.

Clearly every function of $\mathcal{D}$ belongs to $\mathcal{S}$. Actually $\mathcal{D}$ is dense in $\mathcal{S}$.

A linear continuous functional over the space $\mathcal{S}$ is called a distribution of slow growth or a tempered distribution. This yields the space $\mathcal{S}'$ dual of $\mathcal{S}$; $\mathcal{S}'$ is a subspace of $\mathcal{D}'$.

Not every regular distribution of $\mathcal{D}'$ belongs to $\mathcal{S}'$. However, if f is a locally integrable function of slow growth, that is, if $f(x) = O(|\, x\, |^q)$ as $|\, x\, | \to \infty$ for some q, then $f \in \mathcal{S}'$. Notice also that $\mathcal{S}'$ contains certain locally integrable functions that are not of slow growth such as the function $f(x) = (\sin e^x)' = e^x \cos\, e^x$.

Most of the usual algebraic and analytic operations on distributions are closed in $\mathcal{S}'$. Observe however that if $f \in \mathcal{S}'$ and ψ is smooth then ψf belongs to $\mathcal{D}'$ but not necessarily to $\mathcal{S}'$. For ψf to belong to $\mathcal{S}'$ we need to require that ψ and all of its derivatives be of slow growth at infinity.

We shall now study the Fourier transform. We start with the transform of test functions. If $\phi \in \mathcal{S}(\mathbb{R}^n)$, its Fourier transform is given by

$$\mathcal{F}\{\phi(x); u\} = \widehat{\phi}(u) = \int_{\mathbb{R}^n} \phi(x) e^{i\, x \cdot u}\, dx. \tag{2.8.3}$$

It is easy to see that $\widehat{\phi}$ also belongs to $\mathcal{S}(\mathbb{R}^n)$. The inverse transform is

$$\mathcal{F}^{-1}\{\widehat{\phi}(u); x\} = \phi(x) = \frac{1}{(2\pi)^n} \int_{\mathbb{R}^n} \widehat{\phi}(u) \, e^{-i\, u \cdot x}\, du, \tag{2.8.4}$$

which is very closely related to $\mathcal{F}$ itself. We have

Theorem 19. *The Fourier transform is an isomorphism of $\mathcal{S}(\mathbb{R}^n)$ to itself.*

The definition of the Fourier transform of tempered distributions is motivated by the Parseval relation

$$\int_{\mathbb{R}^n} \widehat{\psi}(u)\phi(u)du = \int_{\mathbb{R}^n} \psi(x)\widehat{\phi}(x)dx, \tag{2.8.5}$$

which is valid if $\phi, \psi \in \mathcal{S}$.

Definition. *Let $f \in \mathcal{S}'(\mathbb{R}^n)$. Then its Fourier transform is the tempered distribution $\widehat{f}(u) = \mathcal{F}\{f(x); u\}$ given by*

$$< \widehat{f}(u), \phi(u) > = < f(x), \widehat{\phi}(x) > . \tag{2.8.6}$$

Since $\mathcal{F}$ is an isomorphism of $\mathcal{S}$, it follows by duality that the same is true in $\mathcal{S}'$.

Theorem 20. *The Fourier transform is an isomorphism of $\mathcal{S}'(\mathbb{R}^n)$ to itself.*

The basic properties of the Fourier transform are as follows:

$$\widehat{[\widehat{\phi}]}\ (x) = (2\pi)^n \phi(-x), \tag{2.8.7}$$

$$\widehat{(D^k\phi)}(u) = (-iu)^k\ \widehat{\phi}(u), \tag{2.8.8}$$

$$\widehat{(x^k\phi(x))}(u) = (-iD^k)\widehat{\phi}(u), \tag{2.8.9}$$

$$\widehat{\phi(x - a)}(u) = e^{i\,a\cdot u}\widehat{\phi}(u), \tag{2.8.10}$$

$$\widehat{(e^{i\,a\cdot x}\phi(x))}(u) = \widehat{\phi}(u + a), \tag{2.8.11}$$

$$\widehat{\phi(Ax)}(u) = |\det A\,|^{-1}\ \widehat{\phi}((A^t)^{-1}u), \tag{2.8.12}$$

if A is a non-singular matrix and A^t is its transpose. In particular, if T is a rotation then

$$\widehat{\phi(Tx)}(u) = \widehat{\phi}(Tu), \tag{2.8.13}$$

while if $\lambda \neq 0$,

$$\widehat{\phi}\ (\lambda x)\ (u) = |\lambda\,|^{-n}\ \widehat{\phi}(\lambda^{-1}u). \tag{2.8.14}$$

It is important to notice that the Fourier transform of a distribution f is given by the formula

$$\widehat{f}\ (u) = < f(x), e^{i\,x\cdot u} >, \tag{2.8.15}$$

if the evaluation of $f(x)$ on $e^{i\,x\cdot u}$ is possible.

In particular (2.8.15) gives the Fourier transform of a distribution $f \in \mathcal{E}'(\mathbb{R}^n)$. It follows that in this case $\widehat{f}$ is a smooth function. The space of Fourier transforms of distributions of compact support $\mathcal{F}(\mathcal{E}')$ is denoted as

$\mathcal{O}_{exp}$. Their elements are smooth functions $\psi(x)$ that can be extended as entire functions of exponential type in $\mathbb{C}^n$; namely, there exist constants m, M and a such that

$$| \psi(z) | \leq M(1+ | z |)^m \, e^{a \, |z|}, \quad z \in \mathbb{C}^n. \qquad (2.8.16)$$

Formula (2.8.15) can also be used if $f \in L^1(\mathbb{R}^n)$. In this case $\widehat{f}(u)$ is an uniformly continuous function in $\mathbb{R}^n$ that vanishes at infinity.

When $f \in L^2(\mathbb{R}^n)$ then (2.8.15) might be divergent and therefore the definition, (2.8.6), has to be used. Nevertheless, the Fourier transform is very well behaved in $L^2(\mathbb{R}^n)$ since, except for the constant factor $(2\pi)^n$, $\mathcal{F}$ is an isometry of this Hilbert space [108]:

$$\| \widehat{f} \|_{L^2} = (2\pi)^n \| f \|_{L^2} . \qquad (2.8.17)$$

We now give the Fourier transform of several generalized functions.

Example 40. Using (2.8.15) with $f(x) = \delta(x)$ we obtain

$$\widehat{\delta}(u) = 1. \qquad (2.8.18)$$

Use of (2.8.7)then yields

$$\widehat{1} = (2\pi)^n \, \delta(x). \qquad (2.8.19)$$

With the help of (2.8.8) and (2.8.9), we obtain

$$\widehat{D^k \delta(x)}(u) = (-iu)^k, \qquad (2.8.20)$$

$$\widehat{x^k}(u) = (2\pi)^n (-iD)^k \, \delta(u). \qquad (2.8.21)$$

Formulas (2.8.10) and (2.8.11), in turn yield

$$\widehat{\delta(x-a)}(u) = e^{ia \cdot u}, \qquad (2.8.22)$$

$$\widehat{e^{ia\cdot x}}(u) = (2\pi)^n \delta(u+a). \qquad (2.8.23)$$

Example 41. The Fourier transforms of the trigonometric functions $\sin wx$ and $\cos wx$, in one variable, are obtained by using (2.8.23) as

$$\widehat{(\sin wx)}(u) = i\pi(\delta(u-w) - \delta(u+w)), \qquad (2.8.24)$$

$$\widehat{(\cos wx)}(u) = \pi(\delta(u-w) + \delta(u+w)). \qquad (2.8.25)$$

Example 42. For the Heaviside function $H(x)$ we have

$$\widehat{H}(u) = \pi\delta(u) + i\,u^{-1}. \tag{2.8.26}$$

Similarly,

$$\widehat{H}(-x)(u) = \pi\delta(u) - iu^{-1}. \tag{2.8.27}$$

Upon inversion, the Fourier transform of the Heisenberg delta functions $\delta^{\pm}(x) = \frac{1}{2}\left(\delta(x) \mp (\pi i x)^{-1}\right)$ are obtained as

$$\widehat{\delta^{\pm}}(u) = H(\mp u). \tag{2.8.28}$$

Example 43. For the distributions x_{+}^{λ} and x_{-}^{λ} we have

$$\widehat{x_{+}^{\lambda}}(u) = \Gamma(\lambda+1)e^{i\pi(\lambda+1)\,\mathrm{sgn}\frac{u}{2}}\,|u|^{-\lambda-1}. \tag{2.8.29}$$

$$\widehat{x_{-}^{\lambda}}(u) = \Gamma(\lambda+1)e^{i\pi(\lambda+1)\mathrm{sgn}\frac{u}{2}}\,|u|^{-\lambda-1}\,\mathrm{sgn}(u). \tag{2.8.30}$$

Therefore,

$$\widehat{|x|^{\lambda}}(u) = -2\Gamma(\lambda+1)\sin\frac{\lambda\pi}{2}\,|u|^{-\lambda-1}, \tag{2.8.31}$$

$$\widehat{(|x|^{\lambda}\,\mathrm{sgn}\,x)}(u) = 2i\,\Gamma(\lambda+1)\cos\frac{\lambda\pi}{2}\,|u|^{-\lambda-1}\,\mathrm{sgn}\,u. \tag{2.8.32}$$

In particular,

$$\widehat{x^{-m}}(u) = \frac{i^{m}\pi}{(m-1)!}\,u^{m-1}\,\mathrm{sgn}\,u. \tag{2.8.33}$$

By differentiating the formulas (2.8.29) and (2.8.30), we can derive the values of the Fourier transforms of the functions $x_{+}^{\lambda}\ln x_{+}$ and $x_{-}^{\lambda}\ln x_{-}$. For the special case of $\lambda = 0$, we have after a slight computation,

$$\widehat{(\ln|x|)}(u) = \widehat{(\ln x_{+})}(u) + \widehat{(\ln x_{-})}(u)$$

$$= -(\pi\mathcal{P}f\frac{1}{|u|} + 2\pi\gamma\delta(u)).$$

By taking the Fourier transform of both sides of the above formula and using the relation (2.8.7) we derive the formula

$$(\widehat{\mathcal{P}f\frac{1}{|x|}})(u) = -2(\gamma + \ln|u|).$$

Example 44. For the generalized function $|\,x\,|^\lambda \in \mathcal{S}'(\mathbb{R}^n)$ we have

$$\widehat{|\,x\,|^\lambda}\;(u) = \frac{2^{\lambda+n}\,\pi^{\frac{n}{2}}\,\Gamma(\frac{\lambda+n}{2})}{\Gamma(\frac{-\lambda}{2})}\;|\,u\,|^{-\lambda-n}\,. \tag{2.8.34}$$

Example 45. For the Gaussian function $f(x) = e^{-|x|^2}$ we have

$$\widehat{e^{-|x|^2}}\;(u) = \pi^{\frac{n}{2}}\,e^{\frac{-|\,u\,|^2}{4}}\,. \tag{2.8.35}$$

Thus, if $a > 0$,

$$\widehat{e^{-a|\,x\,|^2}}(u) = \left(\frac{\pi}{a}\right)^{\frac{n}{2}}\,e^{\frac{-|\,u\,|^2}{4a}}\,. \tag{2.8.36}$$

The formula (2.8.36) remains valid if a is a complex number with $\mathfrak{Re}\,a \geq 0$ provided the square root in $\left(\frac{\pi}{a}\right)^{\frac{n}{2}}$ is chosen appropriately. In particular,

$$\widehat{e^{i\,b\,|\,x\,|^2}}(u) = \left(\frac{\pi}{-i\,b}\right)^{\frac{n}{2}}\,e^{\frac{-i\,|\,u\,|^2}{4b}}\,, \tag{2.8.37}$$

where $\arg\left(\frac{\pi}{-i\,b}\right)^{\frac{n}{2}}$ is chosen as $\frac{\pi\,n}{4}$ if $b > 0$ and $\frac{-\pi n}{4}$ if $b < 0$.

We would like to remark that, as these examples show, the Fourier transform of a generalized function that depends only on $|\,x\,|$ will depend only on $|\,u\,|$. Indeed if $f(x) = F(|\,x\,|)$ is radial, then $\widehat{f}(u) = G\,(|\,u\,|)$ is also radial, where

$$G(\rho) = (2\,\pi)^{\frac{n}{2}}\int_0^\infty \frac{F(r)r^{\frac{n}{2}}\,J_{\frac{n-2}{2}}\,(\rho r)dr}{\rho^{\frac{n-2}{2}}}\,, \tag{2.8.38}$$

and J stands for the Bessel function.

It is also important to point out that the Fourier transform of a homogeneous function of degree λ is homogeneous of degree $-\lambda - n$.

Next, we introduce another operation, *the convolution*, which is closely related to the Fourier transform. If f and g are integrable functions in $\mathbb{R}^n$ then their convolution is the function $f * g$ defined as

$$(f * g)(x) = \int_{\mathbb{R}^n} f(y)\,g(x - y)dy\,. \tag{2.8.39}$$

It is not hard to see that $f * g$ is likewise integrable over $\mathbb{R}^n$.

The convolution of two arbitrary generalized functions cannot be defined in general. The situation is similar to the definition of the product of two generalized functions. Actually, the two problems are equivalent since, Theorem 21 asserts, the Fourier transform sends convolutions to products and conversely. We now discuss several cases where the convolution exists.

The first case when the convolution can be defined is when $f \in \mathcal{A}'(\mathbb{R}^n)$ and $\phi \in \mathcal{A}(\mathbb{R}^n)$, where $\mathcal{A}$ is any of the spaces $\mathcal{D}$, $\mathcal{E}$ or $\mathcal{S}$. In this case the formula

$$(f * \psi)(x) = < f(y), \psi(x - y) >, \qquad (2.8.40)$$

defines $f * \psi$ as a smooth function in $\mathbb{R}^n$. Observe that if $f \in \mathcal{S}'(\mathbb{R}^n)$ and $\psi \in \mathcal{S}(\mathbb{R}^n)$ then $f * \psi$ does not necessarily belong to $\mathcal{S}(\mathbb{R}^n)$; however, $f * \psi$ belongs to $\mathcal{S}'(\mathbb{R}^n)$.

Another case when we can define the convolution $f * g$ is when one of the factors has compact support. To see it, observe that if we use (2.8.39), evaluate at a test function ϕ and simplify we obtain

$$< f * g(x), \phi(x) > = < f(x)\, g(y), \phi(x + y) > . \qquad (2.8.41)$$

Formula (2.8.41) cannot be used to define the convolution of two distributions $f, g \in \mathcal{D}'(\mathbb{R}^n)$ since even if $\phi \in \mathcal{D}(\mathbb{R}^n)$, the function of two variables $\phi(x + y)$ has as support the set $X = \{(x, y) \in \mathbb{R}^n \times \mathbb{R}^n : x + y \in \text{supp } \phi\}$, which is not compact. But if $f \in \mathcal{E}'(\mathbb{R}^n)$ then supp $f(x)g(y) = \text{supp } f \times$ supp g meets X on a compact set and (2.8.41) can be used as a definition of $f * g$.

Similarly in the case of one variable (2.8.41) can be used to define the convolution $f * g$ of two distributions whose supports are both bounded on the left or both bounded on the right. In this case supp $f \times$ supp g is contained in a quadrant, which intersects X on a compact set.

Example 46. The Dirac delta function $\delta(x)$ is the identity for the convolution

$$\delta * f = f. \qquad (2.8.42)$$

More generally,

$$D^k\, \delta * f = D^k f, \qquad (2.8.43)$$

while

$$\delta(x - a) * f(x) = f(x - a). \qquad (2.8.44)$$

The formula

$$D^k(f * g) = D^k f * g = f * D^k g \qquad (2.8.45)$$

is also useful, especially in the study of differential equations.

The connection between the convolution and the Fourier transform is the following.

Theorem 21. *Let* $f, g \in \mathcal{S}'(\mathbb{R}^n)$ *such that their convolution* $f * g$ *is defined. Then*

$$\widehat{f * g} \quad = \hat{f}\,\hat{g}. \tag{2.8.46}$$

2.9 Distributions of Rapid Decay

The distributions of rapid decay at infinity are of fundamental importance in the study of the asymptotic expansion of generalized functions. We devote this section to study some of the more useful spaces of distributions of rapid decay.

In the previous section we introduced the space $\mathcal{S}$ of test functions of rapid decay and its dual $\mathcal{S}'$ of tempered distributions. In order to obtain a space of distributions of rapid decay we need a space of "tempered" test functions. However, unlike the case of $\mathcal{S}'$, which is the natural space of tempered distributions, there are many useful spaces of distributions of rapid decay.

Let us start with the space $\mathcal{P} = \mathcal{P}(\mathbb{R}^n)$ of test functions of "less than exponential growth." A smooth function $\phi(x)$ defined in $\mathbb{R}^n$ belongs to $\mathcal{P}$ if

$$D^k \phi(x) = o(e^{\gamma|x|}), \quad \text{as } |x| \to \infty, \tag{2.9.1}$$

for each $\gamma > 0$ and each $k \in \mathbb{N}^n$. The topology of $\mathcal{P}$ is generated by the family of seminorms

$$\| \phi \|_{\gamma, k} = \sup\{e^{-\gamma|x|} \mid D^k \phi(x) \mid : x \in \mathbb{R}^n\}. \tag{2.9.2}$$

This space was introduced for $n = 1$ to study distributional weight functions for orthogonal polynomials [89], (See also [71], Chapter 15). The space $\mathcal{P}$ contains the polynomials but it also contains functions of faster growth at infinity such as $\phi(x) = \exp(|x|^2 + 1)^{\frac{1}{4}}$.

The dual space $\mathcal{P}'(\mathbb{R}^n)$ consists of distributions of exponential decay at infinity. A typical element of $\mathcal{P}'(\mathbb{R}^n)$ is the distribution $e^{-|x|}$.

Next we consider the space $\mathcal{O}_q = \mathcal{O}_q(\mathbb{R}^n)$, where $q \in \mathbb{R}$ [12]. It consists of those smooth functions $\phi(x)$ that satisfy

$$D^k \phi(x) = O(|x|^q), \quad \text{as } |x| \to \infty, \tag{2.9.3}$$

for each $k \in \mathbb{N}^n$. A useful topology for $\mathcal{O}_q$ is obtained by considering the family of seminorms

$$\| \phi \|_{q, k} = \sup\{\rho_q(|x|) \mid D^k \phi(x) \mid : x \in \mathbb{R}^n\}, \tag{2.9.4}$$

where

$$\rho(r) = \begin{cases} r^{-q}, & r \geq 1, \\ 1, & 0 \leq r \leq 1. \end{cases} \tag{2.9.5}$$

This topology is stronger than the one considered by Bremmerman [12].

The space $\mathcal{O}_C(\mathbb{R}^n)$ is the inductive limit $\bigcup_{q \in \mathbb{R}} \mathcal{O}_q(\mathbb{R}^n)$. The elements of $\mathcal{O}_C(\mathbb{R}^n)$ are "tempered" test functions that satisfy (2.9.3) for some $q \in \mathbb{R}$. A net $\{\phi_\sigma\}$ of $\mathcal{O}_C$ converges to 0 if there is a q and a σ_o such that $\phi_\sigma \in \mathcal{O}_q$ for $\sigma \geq \sigma_o$ and $\| \phi_\sigma \|_{q,k} \to 0$ for each $k \in \mathbb{N}^n$.

Let us now consider the dual space $\mathcal{O}'_C(\mathbb{R}^n)$. Since the elements of $\mathcal{O}_C(\mathbb{R}^n)$ are tempered test functions, it could be expected that the distributions of $\mathcal{O}'_C(\mathbb{R}^n)$ decay very rapidly at infinity. However, it is not hard to see that functions that oscillate fast enough at infinity also belong to $\mathcal{O}'_C$. For instance, the function $e^{i|x|^2}$ belongs to $\mathcal{O}'_C(\mathbb{R}^n)$. As it turns out, the fact that many oscillatory kernels belong to $\mathcal{O}'_C$ is a rather fortunate circumstance in the study of asymptotic expansions.

Another useful space of tempered test functions is the space $\mathcal{O}_M(\mathbb{R}^n)$, which consists of smooth functions whose derivatives are bounded by polynomials, of possibly different degrees. Then the space $\mathcal{O}_C$ defined above is the subspace of $\mathcal{O}_M$ for which the bounding polynomials are of the same degree. An interesting feature of the spaces $\mathcal{O}_C$ and $\mathcal{O}_M$ is that the Fourier transform takes these spaces to their duals [64], [97], i.e.,

$$\mathcal{F}(\mathcal{O}_C) = \mathcal{O}'_M, \quad \mathcal{F}(\mathcal{O}_M) = \mathcal{O}'_C. \tag{2.9.6}$$

The space $\mathcal{O}_M$ is the space of multipliers of $\mathcal{S}$, that is, if ψ is smooth and $\psi\phi \in \mathcal{S}$ for every $\phi \in \mathcal{S}$ then $\psi \in \mathcal{O}_M$. It also follows that if $\psi \in \mathcal{O}_M$ and $f \in \mathcal{S}'$ then $\psi f \in \mathcal{S}'$. The space $\mathcal{O}_M$ is *the Moyal algebra* of $\mathcal{S}$.

Similarly, $\mathcal{O}'_C$ is the space of *convolutors* of $\mathcal{S}$: if g is a distribution and $g * \phi \in \mathcal{S}$ for every $\phi \in \mathcal{S}$ then $g \in \mathcal{O}'_C$. Again, this implies that if $g \in \mathcal{O}'_C$ and $f \in \mathcal{S}'$ then $g * f \in \mathcal{S}'$ [64], [97].

We shall also have use for the space $\mathcal{K}(\mathbb{R}^n)$, which is the subspace of $\mathcal{O}_M$ formed by those smooth functions ϕ for which there is a $q \in \mathbb{R}$ such that

$$D^k\phi(x) = O(|x|^{q-|k|}), \quad \text{as } |x| \to \infty, \tag{2.9.7}$$

for each $k \in \mathbb{N}^n$. A net $\{\phi_\sigma\}$ of $\mathcal{K}$ converges to 0 if there exists $q \in \mathbb{R}$ and σ_o such that ϕ_σ satisfies (2.9.7) for this q for all $\sigma \geq \sigma_o$ and $\rho_{q-|k|}(x) D^k\phi(x)$ converges to 0 uniformly. A typical element of $\mathcal{K}'$ is the kernel $e^{i|x|}$.

Observe that all the spaces $\mathcal{P}, \mathcal{O}_C, \mathcal{O}_M$ and $\mathcal{K}$ contain the polynomials. Therefore, if f is a distribution of any of their dual spaces, it has well-defined moments $\mu_k = \mu_k(f)$ given by

$$\mu_k = < f(x), x^k > \tag{2.9.8}$$

for each multi-index $k \in \mathbb{N}^n$. The existence of moments of all orders is an indication of the fact that it decays rapidly at infinity.

We also remark that all the dual spaces $\mathcal{P}', \mathcal{O}'_C, \mathcal{O}'_M$ and $\mathcal{K}'$ contain $\mathcal{E}'$, the space of distributions of compact support. This is very natural since the distributions of compact support are not only of rapid decay at infinity; they vanish near infinity.

As we mentioned earlier, oscillatory generalized functions occur many times in the space of functions of rapid decay, in the distributional sense. Therefore, it is worthwhile to give some basic facts about the simplest oscillatory distributions, the periodic distributions.

A distribution $f \in \mathcal{D}'(\mathbb{R})$ is called periodic of period p if $f(x+p) = f(x)$. It is not hard to see that any periodic distribution is tempered, i.e., $f \in \mathcal{S}'$. Using (2.8.9) it follows that its Fourier transform satisfies $e^{iup} \, \widehat{f}(u) = \widehat{f}(u)$ or $(e^{iup} - 1)\widehat{f}(u) = 0$. Since $e^{iup} - 1$ has simple zeros at $u = \frac{2\pi k}{p}$, $k \in \mathbb{Z}$ it follows that there are constants $c_k, k \in \mathbb{Z}$, such that

$$\widehat{f}(u) = \sum_{k=-\infty}^{\infty} c_k \, \delta(u - \frac{2\pi k}{p}). \tag{2.9.9}$$

Inverse Fourier transformation yields

$$f(x) = \sum_{k=-\infty}^{\infty} a_k \, e^{\frac{2\pi k i x}{p}}, \tag{2.9.10}$$

the Fourier series representation of f. Here $a_k = c_k/2\pi$.

When f is locally integrable, the Fourier coefficients a_k can be computed by integrating $f(x)e^{\frac{-2k\pi ix}{p}}$ over any interval of length p:

$$a_k = \frac{1}{p} \int_{\alpha}^{\alpha+p} f(x) \, e^{\frac{-2k\pi ix}{p}} \, dx \tag{2.9.11}$$

or

$$a_k = \frac{1}{p} < f(x), e^{\frac{-2k\pi ix}{p}} \, H(x - \alpha) \, H(\alpha + p - x) > . \tag{2.9.12}$$

Formula (2.9.12) remains valid if f is any periodic generalized function of period p. Actually, an evaluation of the type $< f(x), \psi(x)H(x - \alpha)H(\alpha + p - x) >$ cannot be defined in general if $f \in \mathcal{D}'$ and $\psi \in \mathcal{D}$. However, if f and ψ are *both* periodic of period p, such an evaluation makes sense and it is in fact independent of α. If ψ has a Fourier series of the type

$$\psi(x) = \sum_{k=-\infty}^{\infty} b_k \, e^{\frac{2k\pi ix}{p}}, \tag{2.9.13}$$

then

$$< f(x), \psi(x) \, H(x - \alpha) H(\alpha + p - x) >= p \sum_{k=-\infty}^{\infty} a_k b_{-k}. \qquad (2.9.14)$$

If $\psi \in S(\mathbb{R})$ is periodic of period p with Fourier series (2.9.13) then $b_k = O(|k|^{-\infty})$ as $|k| \to \infty$. If $f \in \mathcal{D}'(\mathbb{R})$ is periodic of period p with Fourier series (2.9.10) then $a_k = O(|k|^q)$ as $|k| \to \infty$ for some q.

The primitives of a periodic distribution are not periodic, in general. A necessary and sufficient condition for this to be the case is that the constant in the Fourier series expansion vanishes :$a_0 = 0$. Actually, any periodic distributions with $< f(x), H(x)H(p - x) >= 0$ can be written as $f = g^{(n)}$ for some periodic distribution g, where

$$g(x) = \sum_{k=-\infty}^{\infty}{}' \left(\frac{p}{2k\pi i}\right)^n a_k e^{\frac{2k\pi i x}{p}}, \qquad (2.9.15)$$

where the prime indicates that the term $k = 0$ is omitted. If n is large enough, the Fourier coefficients of g will be small and thus g will be bounded.

It follows that any periodic distribution without a constant term in its Fourier series expansion belongs to $\mathcal{K}'$. Indeed, the formula

$$< f, \phi >= (-1)^n \int_{-\infty}^{\infty} g(x) \, \phi^{(n)}(x) dx \qquad (2.9.16)$$

defines $< f, \phi >$ as a convergent integral if n is large enough.

We also remark that all the moments of such a periodic function vanish.

2.10 Spaces of Distributions Associated with an Asymptotic Sequence

We shall now consider some natural procedures to associate spaces of test and generalized functions to an asymptotic sequence [43], [44]. These spaces will play a very important role in the distributional theory of asymptotic expansions developed in the following chapters.

Let $\{\phi_n(x)\}$ be an asymptotic sequence as $x \to 0^+$. We suppose that all the functions $\phi_n(x)$ are defined in the interval $(0, b]$ and are continuous and positive there. The space $C = C(\{\phi_n(x)\}, (0, b])$ is the space of continuous functions $\psi(x)$ defined in $(0, b]$ that admit the asymptotic expansion

$$\psi(x) \sim a_1 \phi_1(x) + a_2 \phi_2(x) + a_3 \phi_3(x) + \cdots, \text{ as } x \to 0^+, \qquad (2.10.1)$$

for some constants $a_1, a_2, a_3, \cdots$. The space C becomes a Fréchet topological vector space when endowed with the seminorms

$$
\| \psi \|_m = \sup \left\{ \frac{\left| \psi(x) - \displaystyle\sum_{j=1}^{m-1} a_j \phi_j(x) \right|}{\phi_m(x)} : 0 < x \le b \right\}, \qquad (2.10.2)
$$

for $n = 1, 2, 3, \cdots$. Observe that $\| \psi \|_1$ is the uniform norm of ψ/ϕ_1 and thus the convergence of $\{\psi_k\}$ to ψ in C implies the uniform convergence of $\{\psi_k/\phi_1\}$ to ψ/ϕ_1.

Let us now consider the dual space C'. Some of the simplest elements of C' are the delta functions $\delta_m(x) = \phi_m^{-1}(x)\delta(x)$, defined as

$$
< \delta_m(x), \psi(x) > = a_m, \qquad (2.10.3)
$$

where the a_m are the coefficients in the asymptotic expansion (2.10.1). The continuity of $\delta_m(x)$ follows since $|< \delta_m(x), \psi(x) >| \le \| \psi \|_m$.

If $\mu(x)$ is a Radon measure in $(0, b]$ that satisfies

$$
\int_0^b \phi_1(x)d \mid \mu \mid (x) < \infty, \qquad (2.10.4)
$$

then it gives rise to an element of C' by setting

$$
< \mu(x), \psi(x) > = \int_0^b \psi(x)\, d\mu(x), \quad \psi \in C. \qquad (2.10.5)
$$

More generally, if $\mu(x)$ is a Radon measure in $(0, b]$ that satisfies

$$
\int_0^b \phi_m(x)d \mid \mu \mid (x) < \infty, \qquad (2.10.6)
$$

for some m, then the integral

$$
\int_0^b \psi(x)d\mu(x), \quad \psi \in C,
$$

can be divergent sometimes, but it can be regularized to give an element of C. One such regularization is given by

$$
< \hat{\mu}, \psi > = \int_0^b (\psi(x) - \sum_{j=1}^{m-1} a_j \phi_j(x))d\mu(x). \qquad (2.10.7)
$$

The regularization is not uniquely determined since if $c_1, \cdots, c_n$ are any constants then $\widehat{\mu} + \sum_{j=1}^{m-1} c_j \delta_j$ is also a regularization of μ.

It can be shown that the space C consists precisely of the regularizations of the Radon measures that satisfy (2.10.6) for some m.

A very important example is provided by the asymptotic sequence $\phi_j(x) = x^{\lambda_j}$, where $\lambda_1 < \lambda_2 < \lambda_3 < \cdots$ and $\lim_{m\to\infty} \lambda_m = \infty$. In this case we can regularize the integral

$$\int_0^b x^\lambda \psi(x) dx, \qquad (2.10.8)$$

for any $\lambda \in \mathbb{C}$ and $\psi \in C = C(\{x^{\lambda_m}\}, (0, b])$ by taking the Hadamard finite part

$$< \mathcal{P}f(x^\lambda), \psi(x) > = F.p. \int_0^b x^\lambda \psi(x) dx. \qquad (2.10.9)$$

Observe that if $\lambda_{n+1} + \lambda > -1$ and $\lambda + \lambda_j \neq -1$, $j = 1, \cdots n$, this regularization can be written as

$$< \mathcal{P}f(x^\lambda), \psi(x) > = \int_0^b x^\lambda (\psi(x) - \sum_{j=1}^n a_j x^{\lambda_j}) dx + \sum_{j=1}^n \frac{a_j b^{\lambda+\lambda_j+1}}{\lambda + \lambda_j + 1},$$

$$(2.10.10a)$$

while

$$< \mathcal{P}f(x^\lambda), \psi(x) > = \int_0^b x^\lambda (\psi(x) - \sum_{j=1}^n a_j x^{\lambda_j}) dx + \sum_{\substack{j=1 \\ j \neq i}} \frac{a_j b^{\lambda+\lambda_j+1}}{\lambda + \lambda_j + 1} + a_i \ln b,$$

$$(2.10.10b)$$

if $\lambda = -\lambda_i - 1$.

In the preceding discussion we just imposed continuity at the endpoint $x = b$. It ought to be clear, however, that other conditions (such as $\psi(b) = 0$ or $\psi(x) = 0$ for $x \geq b - \xi$, etc.) can also be imposed. In particular, if $b = \infty$ we consider the space $\mathcal{E}_o\{\phi_n\}$ obtained as the projective limit of the spaces $C(\{\phi_n\}, (0, R])$ as $R \to \infty$. That is, a function $\psi(x)$ defined in $(0, \infty)$ belongs to $\mathcal{E}\{\phi_n\}$ if and only if it is continuous in $(0, \infty)$ and if near $x = 0$ it admits the asymptotic development

$$\psi(x) = a_1 \phi_1(x) + \cdots + a_m \phi_m(x) + o(\phi_m(x)). \qquad (2.10.11)$$

The topology of $\mathcal{E}_o\{\phi_n\}$ is generated by the family of seminorms

$$\| \psi \|_{m,R} = \sup\{\phi_m^{-1}(x) \mid \psi(x) - \sum_{j=0}^{m-1} a_j \phi_j(x) \mid : 0 < x \leq R\}, \qquad (2.10.12)$$

for $m = 1, 2, 3, \cdots$ and $R > 0$. The dual $\mathcal{E}'_\circ\{\phi_n\}$ is formed by regularizations of Radon measures with bounded support that satisfy

$$\int_0^\infty \phi_m(x)d \mid \mu \mid (x) < \infty$$

for some m.

The elements of $\mathcal{E}'_\circ\{\phi_n\}$ are regularizations of measures. We can also consider spaces of regularizations of generalized functions in the following way. Suppose the system $\{\phi_m(x)\}$ is such that the sequence of derivatives $\{\phi_n^{(k)}(x)\}$ is also an asymptotic sequence for $k \leq p$. Then a function ψ belongs to $\mathcal{E}'_p\{\phi_n\}$ if it is of class C^p in $(0, \infty)$ and it admits a strong development for order p in terms of the ϕ_n's. This last condition means that

$$\psi^{(k)}(x) \sim a_1\phi_1^{(k)}(x) + a_2\phi_2^{(k)}(x) + a_3\phi_3^{(k)}(x) + \cdots, \text{ as } x \to 0^+, \quad (2.10.13)$$

for $0 \leq k \leq p$.

The space $\mathcal{E}_\infty\{\phi_n\}$ is denoted as $\mathcal{E}\{\phi_n\}$ since if $\phi_n(x) = x^n$ it reduces to the standard space $\mathcal{E}'[0, \infty) = \{f \in \mathcal{E}'(\mathbb{R}) : \text{supp } f \subseteq [0, \infty)\}$. Observe that $\delta_n(x) = \dfrac{(-1)^n \delta^{(n)}(x)}{n!}$, $n = 0, 1, 2, \cdots$, in this case.

In a completely analogous way we can define the spaces $\mathcal{D}\{\phi_n\}$, $\mathcal{S}\{\phi_n\}$, $\mathcal{P}\{\phi_n\}$, $\mathcal{O}_C\{\phi_n\}$, etc. For instance a function $\psi \in \mathcal{D}\{\phi_n\}$ is a smooth function defined in $(0, \infty)$ that vanishes outside $(0, A)$ for some A and that near $x = 0$ admits the strong expansion $\psi(x) \sim a_1\phi_1(x) + a_2\phi_2(x) + \cdots$.

In general, if $\psi \in \mathcal{D}\{\phi_n\}$ then $\psi(\lambda x)$ might not belong to $\mathcal{D}\{\phi_n\}$. However, if $\phi_n(x) = x^{\alpha_n}$ then the operator $T_\lambda(\psi) = \psi(\lambda x)$, $\lambda > 0$, is a well-defined isomorphism of $\mathcal{A}\{x^{\alpha_n}\}$ to itself and of $\mathcal{A}'\{x^{\alpha_n}\}$ to itself for any space $\mathcal{A} = \mathcal{E}, \mathcal{D}, \mathcal{S}$, etc. The notions of homogeneous and associated homogeneous generalized functions are meaningful in this context.

We have

$$\delta_n(\lambda x) = \lambda^{-1-\alpha_n}\delta_n(x), \quad (2.10.14)$$

since $< \delta_n(\lambda x), \psi(x) > = \lambda^{-1} < \delta_n(x), \psi(\lambda^{-1}x) >$ and the expansion of $\psi(\lambda^{-1}x)$ is $\psi(\lambda^{-1}x) \sim a_1\lambda^{-\alpha_1}x^{\alpha_1} + a_2\lambda^{-\alpha_2}x^{\alpha_2} + \cdots$, as $x \to 0^+$. Thus $\delta_n(x)$ is homogeneous of degree $-1 - \alpha_n$.

If $\beta \neq -1 - \alpha_n$ for every n then

$$\mathcal{P}f((\lambda x)^\beta) = \lambda^\beta \mathcal{P}f(x^\beta) \quad (2.10.15)$$

so that $\mathcal{P}f(x^\beta)$ is homogeneous of degree β if $\beta \neq -1 - \alpha_n$, $\forall n$.

The distributions $\delta_n(x)$ and $\mathcal{P}f(x^\beta)$, $\beta \neq -1 - \alpha_n$ are the only homogeneous distributions in $\mathcal{D}'\{x^{\alpha_n}\}$. The distributions $\mathcal{P}f(x^{-1-\alpha_n})$ are associated homogeneous distributions of order 1 and degree $-1 - \alpha_n$, since

$$\mathcal{P}f((\lambda x)^{-1-\alpha_n}) = \lambda^{-1-\alpha_n}\mathcal{P}f(x^{-1-\alpha_n}) + \lambda^{-1-\alpha_n}\ln\lambda\,\delta_n(x). \quad (2.10.16)$$

Most of these ideas can be generalized to functions of several variables defined in a cone. We discuss such spaces in Chapter 4 by using tools from the theory of topological tensor products.

CHAPTER 3

A Distributional Theory of Asymptotic Expansions

3.1 Introduction

The purpose of this chapter is to present the distributional theory of asymptotic expansions for functions of one variable. This chapter and the next, where the multidimensional expansions are studied, are the central part of the book.

The close ties between the theory of distributions and the theory of asymptotic expansions have become clear during recent years. Researchers in the field of asymptotic analysis have found distributions very convenient for performing the analytical operations involved and for assigning values to the divergent integrals that often arise [13], [26], [56], [57], [58], [68], [84], [85], [100], [103], [104], [117], [118], [119].

Workers in the theory of generalized functions have studied the asymptotic development of distributions to understand their local and global behavior [15], [43], [44], [92], [105], [112]. The applications of these theories have also provided results in this connection [14], [22], [30], [32], [51].

We start by considering the Taylor expansion and some of its generalizations in Section 3.2. Interesting classical results such as the Lagrange theorem on the inversion of power series [16] are closely related to these distributional Taylor expansions.

In Section 3.3 and 3.4, we introduce a simple but powerful technique for the asymptotic development of distributions: *the moment asymptotic expansion* [41], [43]. The moment asymptotic expansion holds in a large variety of situations, including the expansions of distributions of fast decay and of distributions of rapid oscillation. These expansions, in turn, immediately give the classical development of several integrals and series.

The moment asymptotic expansion is concerned with the expansion of distributional kernels of the type $f(\lambda x)$ as $\lambda \to \infty$. By using the notion of change of variables in distributions, our analysis can be extended to the expansion of some more general distributional kernels of the form $F(\lambda, x)$. Some of the best known methods for the asymptotic approximation of integrals containing a large parameter, such as the Laplace formula and the stationary phase formula, are obtained by using this procedure while the steepest descent method follows by considering the expansion of functionals in spaces of analytic functions. These ideas are discussed in Sections 3.5, 3.6 and 3.7.

The expansion of $f(\lambda x)$ for distributions that do not decay at infinity is considered in Section 3.8. While the moment asymptotic expansion is given in terms of Dirac delta functions concentrated at the origin, the general expansions are given in terms of homogeneous generalized functions. This analysis is complemented by the results of Section 3.9, where we show that the only possible way to have an asymptotic separation of variables $f(\lambda x) \sim \rho_1(\lambda) h_1(x) + \rho_2(\lambda) h_2(x) + \rho_3(\lambda) h_3(x) + \cdots$ is if the terms are homogeneous and associate homogeneous functions.

3.2 The Taylor Expansion of Distributions

We shall start by considering the Taylor expansion of distributions. As is well known, any smooth function near $x \in \mathbb{R}$ admits the Taylor expansion [45]

$$\phi(x + \varepsilon) \sim \sum_{n=0}^{\infty} \frac{\phi^{(n)}(x)}{n!} \varepsilon^n, \quad \text{as} \quad \varepsilon \to 0. \tag{3.2.1}$$

By duality, if $f(x)$ is a generalized function of any of the spaces $\mathcal{D}', \mathcal{E}', \mathcal{S}', \mathcal{K}'$, etc., then we have the following asymptotic Taylor expansion

$$f(x + \varepsilon) \sim \sum_{n=0}^{\infty} \frac{f^{(n)}(x)}{n!} \varepsilon^n, \quad \text{as} \quad \varepsilon \to 0. \tag{3.2.2}$$

The interpretation of (3.2.2) is in the weak or distributional sense: it means that for any test function $\phi(x)$ we have

$$< f(x + \varepsilon), \phi(x) > = \sum_{n=0}^{N} \frac{< f^{(n)}(x), \phi(x) >}{n!} \varepsilon^n + O(\varepsilon^{N+1}), \quad \text{as} \quad \varepsilon \to 0. \tag{3.2.3}$$

For instance, if $f(x) = \delta(x)$, relation (3.2.2) becomes

$$\delta(x + \varepsilon) \sim \sum_{n=0}^{\infty} \frac{\delta^{(n)}(x)\varepsilon^n}{n!}, \quad \text{as} \quad \varepsilon \to 0. \tag{3.2.4}$$

Evaluation of (3.2.4) at a test function yields (3.2.1) again.

The proof of (3.2.2) is straightforward. We present variuous examples to illustrate these concepts.

Example 47. If $\alpha \notin \mathbb{Z}$, let us take $f(x) = x_+^\alpha$. Then $f^{(k)}(x) = \alpha(\alpha - 1) \cdots (\alpha - k + 1) x_+^{\alpha - k}$. Thus,

$$(x - \varepsilon)_+^\alpha \sim \sum_{k=0}^{\infty} \binom{\alpha}{k} (-1)^k x_+^{\alpha - k} \varepsilon^k, \quad \text{as} \quad \varepsilon \to 0, \tag{3.2.5}$$

where $\binom{\alpha}{k} = \frac{\alpha(\alpha-1)\cdots(\alpha-k+1)}{k!}$. This means that if $\phi \in \mathcal{D}(\mathbb{R})$ then

$$F.p. \int_{\varepsilon}^{\infty} (x-\varepsilon)^{\alpha}\phi(x)dx \sim \sum_{k=0}^{\infty}\left[\binom{\alpha}{k}(-1)^{k} F.p. \int_{0}^{\infty} x^{\alpha-k}\phi(x)dx\right]\varepsilon^{k}$$

$$\text{as } \varepsilon \to 0.$$

$$(3.2.6)$$

Example 48. Alternatively, if $f(x) = \mathcal{P}f\left(\frac{H(x)}{x^{k}}\right)$, where $\mathcal{P}f$ stands for pseudofunction and $H(x)$ is the Heaviside function we obtain

$$\mathcal{P}f\left(\frac{H(x-\varepsilon)}{(x-\varepsilon)^{k}}\right) \sim \sum_{m=0}^{\infty}\left[\binom{m+k-1}{m}\mathcal{P}f\left(\frac{H(x)}{x^{m+k}}\right)\right.$$

$$+ \frac{(-1)^{m+k}}{(k-1)!m!}(\psi(m+k)$$

$$\left. - \psi(k))\delta^{(m+k-1)}(x)\right]\varepsilon^{m}.$$

$$(3.2.7)$$

Here $\psi(n) = 1 + 1/2 + \cdots + 1/(n-1) - \gamma$ is the digamma function and γ is Euler's constant [116]. If we now make the change of variables $\varepsilon = x_0/\lambda$ and use the relation

$$\mathcal{P}f\left(\frac{H(\mu x)}{(\mu x)^{k}}\right) = \frac{1}{\mu^{k}}\mathcal{P}f\left(\frac{H(x)}{x^{k}}\right) + \frac{(-1)^{k}\ln\mu}{(k-1)!\mu^{k}}\delta^{(k-1)}(x) \qquad (3.2.8)$$

we obtain

$$\mathcal{P}f\left(\frac{H(\lambda x - x_0)}{(\lambda x - x_0)^{k}}\right) \sim \sum_{m=0}^{\infty}\left\{\binom{m+k-1}{m}\mathcal{P}f\left(\frac{H(x)}{x^{(m+k)}}\right) + \right.$$

$$\frac{(-1)^{m+k}}{(k-1)!m!}(\psi(m+k) - \psi(k))\delta^{(m+k-1)}(x)$$

$$\left. + \frac{(-1)^{m+k}}{(k-1)!m!}\delta^{(m+k-1)}(x)\ln\lambda\right\}\frac{x_0^{n}}{\lambda^{n+1}}, \quad \text{as } \lambda \to \infty. \qquad (3.2.9)$$

Example 49. In Example 48 we used the derivative formula

$$\frac{d}{dx}\left(\mathcal{P}f\left(\frac{H(x)}{x^{k}}\right)\right) = -k\mathcal{P}f\left(\frac{H(x)}{x^{k+1}}\right) + \frac{(-1)^{k}\delta^{(k)}(x)}{k!}. \qquad (3.2.10)$$

We now show how this formula can be obtained from the results of this section. Indeed, if in (3.2.4) we replace ε by $-\varepsilon$ and divide by ε^k we obtain

$$\frac{1}{\varepsilon^k}\delta(x-\varepsilon) = \sum_{j=0}^{k} \frac{(-1)^j \delta^{(j)}(x)}{j!\varepsilon^{k-j}} + o(1), \quad \text{as} \quad \varepsilon \to 0, \qquad (3.2.11)$$

while the definition of the finite part $\mathcal{P}f\left(\frac{H(x)}{x^k}\right)$ gives

$$\mathcal{P}f\left(\frac{H(x)}{x^k}\right) = \frac{H(x-\varepsilon)}{x^k} - \sum_{j=0}^{k-2} \frac{(-1)^j \delta^{(j)}(x)}{j!(k-j-1)\varepsilon^{k-j-1}}$$
$$+ \frac{(-1)^{k-1}}{(k-1)!} \ln\varepsilon\, \delta^{(k-1)}(x) + o(1), \quad \text{as } \varepsilon \to 0.$$

$$(3.2.12)$$

Differentiation of this last relation yields

$$\frac{d}{dx}\mathcal{P}f\left(\frac{H(x)}{x^k}\right) = -k\frac{H(x-\varepsilon)}{x^{k+1}} + \frac{\delta(x-\varepsilon)}{\varepsilon^k} - \sum_{j=0}^{k-2}\frac{(-1)^j\delta^{(j+1)}(x)}{j!(k-j-1)\varepsilon^{k-j-1}}$$
$$+ \frac{(-1)^{k-1}}{(k-1)!}\ln\varepsilon\delta^{(k)}(x) + o(1)$$
$$= -k\frac{H(x-\varepsilon)}{x^{k+1}}$$
$$+ \sum_{j=0}^{k}\frac{(-1)^j\delta^{(j)}(x)}{j!\varepsilon^{k-j}} + \sum_{j=1}^{k-1}\frac{(-1)^j\delta^{(j)}(x)}{(j-1)!(k-j)\varepsilon^{k-j}}$$
$$+ \frac{(-1)^{k-1}}{(k-1)!}\ln\varepsilon\delta^{(k)}(x) + o(1)$$
$$= -k\,\mathcal{P}f\left(\frac{H(x)}{x^{k+1}}\right) + \frac{(-1)^k\delta^{(k)}(x)}{k!} + o(1),$$

and thus (3.2.10) follows.

The Taylor expansion of a smooth function is not convergent but asymptotic. However, if ϕ is real analytic then the expansion becomes a convergent series:

$$\phi(x+\varepsilon) = \sum_{n=0}^{\infty}\frac{\phi^{(n)}(x)}{n!}\varepsilon^n, \qquad (3.2.13)$$

whenever ϕ can be extended to an analytic function in the disc $D = \{z \in \mathbb{C} :| \ z - x \ |< r_x\}$. It follows that if $f \in \mathcal{E}'$ then the Taylor expansion becomes a convergent series

$$< f(x+\varepsilon), \phi(x) > = \sum_{n=0}^{\infty} \frac{< f^{(n)}(x), \phi(x) >}{n!} \varepsilon^n, \qquad (3.2.14)$$

if ε is small enough, say $| \ \varepsilon \ |< r = inf\{r_x : x \in supp \ f\}$. Since $supp \ f$ is compact then $r > 0$.

Convergence results cannot be considered in $\mathcal{D}$ since no function of $\mathcal{D}$ is real analytic. In spaces like $\mathcal{S}$, and $\mathcal{K}$, we need ϕ not only to be real-analytic but to admit an analytic extension to the strip $| \ Im \ \omega \ |< r$.

There is an interesting extension of the Taylor expansion of distributions. Let $\rho(x)$ be a smooth function defined in $\mathbb{R}$. Suppose that if $| \ \varepsilon \ |<< 1$ then the "perturbation" $x + \varepsilon\rho(x)$ is increasing. Then we have

$$f(x + \varepsilon\rho(x)) \sim f(x) + \varepsilon f'(x)\rho(x) + \frac{\varepsilon^2 f''(x)(\rho(x))^2}{2!} + \cdots , \quad as \ \varepsilon \to 0,$$
$$(3.2.15)$$

for each $f \in \mathcal{D}'$.

To derive (3.2.15), let $\phi \in \mathcal{D}$. Then making the change of variables

$$y = x + \varepsilon\rho(x), \qquad (3.2.16)$$

we obtain

$$< f(x + \varepsilon\rho(x)), \phi(x) > = < f(y), \psi(y) >, \qquad (3.2.17)$$

where

$$\psi(y) = \frac{\phi(x)}{1 + \varepsilon\rho'(x)}. \qquad (3.2.18)$$

The relation (3.2.18) gives $\psi(y)$ as an implicit function. However, when $| \ \varepsilon \ |<< 1$ the function $\psi(y)$ admits an asymptotic expansion in terms of explicit functions. To obtain such an expansion, let us start by considering the inversion of (3.2.16). Under our hypothesis, (3.2.16) defines x as an implicit function of y. As a first approximation from (3.2.16) itself we obtain

$$x = y + O(\varepsilon), \quad as \ | \ \varepsilon \ |\to 0. \qquad (3.2.19)$$

This approximation in turn can be used in conjunction with (3.2.16) to obtain a further approximation

$$x = y - \varepsilon\rho(x) = y - \varepsilon\rho(y + O(\varepsilon)) = y - \varepsilon\rho(y) + O(\varepsilon^2). \qquad (3.2.20)$$

We can repeat this process again:

$$x = y - \varepsilon\rho(x) = y - \varepsilon\rho\big(y - \varepsilon\rho(y) + O(\varepsilon^2)\big),$$

or

$$x = y - \varepsilon\rho(x) + \varepsilon^2\rho'(y)\rho(y) + O(\varepsilon^3). \tag{3.2.21}$$

The next iteration takes the form

$$x = y - \varepsilon\rho(y) + \varepsilon^2\rho'(y)\rho(y) - \varepsilon^3(\rho(y)(\rho'(y))^2 + \frac{(\rho(y))^2\rho''(y)}{2}) + O(\varepsilon^4). \tag{3.2.22}$$

As is clear, this iterative process can be continued to obtain as many terms of the asymptotic expansion as desired. But it is not obvious that the coefficients of ε^n follow a simple rule. However, the nice formula

$$x \sim y + \sum_{n=1}^{\infty} \frac{(-1)^n}{n!} \frac{d^{n-1}(\rho(y))^n}{dy^{n-1}} \varepsilon^n, \quad \text{as } \varepsilon \to 0, \tag{3.2.23}$$

was obtained by Lagrange over two hundred years ago.

The expansion of $\phi(x)$ as a function of y can be derived by using the development (3.2.23). The result is

$$\phi(x) \sim \phi(y) + \sum_{n=1}^{\infty} \frac{(-1)^n}{n!} \frac{d^{n-1}}{dy^{n-1}} (\phi'(y)(\rho(y))^n)\varepsilon^n, \quad \text{as } \varepsilon \to 0, \tag{3.2.24}$$

another formula of Lagrange. Observe that if $\phi(x) = x$ then (3.2.24) reduces to (3.2.23).

Finally, the asymptotic formula for $\psi(y)$ is obtained with the help of (3.2.24), since

$$\psi(y) = \frac{\phi(x)}{1 + \varepsilon\rho'(x)} \sim \phi(x)\,(1 - \varepsilon\rho'(x) + \varepsilon^2(\rho'(x))^2 - \cdots).$$

Therefore,

$$\psi(y) \sim \sum_{n=0}^{\infty} \frac{(-1)^n}{n!} \frac{d^n}{dy^n} (\phi(y)(\rho(y))^n)\varepsilon^n \quad \text{as } \varepsilon \to 0. \tag{3.2.25}$$

This expansion is valid uniformly on compacts. If $\phi \in \mathcal{D}$ then so is ψ and then (3.2.25) holds in $\mathcal{D}$. Therefore

$$< f(x + \varepsilon\rho(x)), \phi(x) > = < f(y), \psi(y) >$$

$$\sim < f(y), \sum_{n=0}^{\infty} \frac{(-1)^n}{n!} \frac{d^n}{dy^n} (\phi(y)(\rho(y))^n)\varepsilon^n >$$

$$\sim \sum_{n=0}^{\infty} < f^{(n)}(y)(\rho(y))^n, \phi(y) > \frac{\varepsilon^n}{n!}$$

and (3.2.15) follows.

The expansion (3.2.15) also holds in $\mathcal{E}'$. In this case it is not necessary to impose any conditions on $\rho(x)$, aside from its smoothness. In fact, if $f \in \mathcal{E}'$ then *supp f* is compact and thus $1 + \varepsilon\rho'(x) > 0$ for all $x \in$ *supp f* if $|\varepsilon|$ is small enough.

Example 50. Let $f(x) = \delta(x - a)$. Then we have

$$\delta(x - a + \varepsilon\rho(x)) \sim \sum_{n=0}^{\infty} \frac{\delta^{(n)}(x - a)}{n!}(\rho(x))^n \varepsilon^n. \qquad (3.2.26)$$

In particular, if $\rho(x) = -e^x$ then

$$\delta(x - a - \varepsilon e^x) \sim \sum_{n=0}^{\infty} \frac{(-1)^n \delta^{(n)}(x - a)}{n!} e^{-nx} \varepsilon^n. \qquad (3.2.27)$$

Thus, if $\phi \in \mathcal{E}$ and b is the solution of the equation

$$a = b - \varepsilon e^b, \qquad (3.2.28)$$

then

$$\frac{\phi(b)}{1 + a - b} \sim \sum_{n=0}^{\infty} \frac{d^n}{da^n}(\phi(a)e^{-na})\frac{\varepsilon}{n!}. \qquad (3.2.29)$$

Convergence results for (3.2.15) are valid if ϕ is real-analytic in a neighborhood of *supp f*.

3.3 The Moment Asymptotic Expansion

In this section we consider the asymptotic behavior of $f(\lambda x)$ as $\lambda \to \infty$. In a certain sense, this amounts to the study of the behavior of the distribution $f(x)$ at infinity. Several results on the asymptotic behavior of integrals and series can be written in this way.

The simplest result in this direction is the *moment asymptotic expansion* [41], [43], which can be written as

$$f(\lambda x) = \sum_{n=0}^{N} \frac{(-1)^n \mu_n \delta^{(n)}(x)}{n! \lambda^{n+1}} + O\left(\frac{1}{\lambda^{N+2}}\right), \qquad \text{as } \lambda \to \infty, \qquad (3.3.1)$$

where μ_k are the moments of the generalized function $f(x)$, given by,

$$\mu_k = < f(x), x^k > . \qquad (3.3.2)$$

The asymptotic expansion (3.3.1) will be valid in several important spaces of distributions. In fact, it holds for distributions of compact support, distributions of rapid decay at infinity as those of $\mathcal{P}'(\mathbb{R})$ and for rapidly oscillating distributions as those found in the space $\mathcal{O}'_C(\mathbb{R})$. It does not hold in other spaces such as $\mathcal{D}'(\mathbb{R})$ or $\mathcal{S}'(\mathbb{R})$.

The formula (3.3.1) gives a precise meaning to the series of delta functions used in the theory of orthogonal polynomials [74], [75], [89] as well as to the series of delta functions found in the solutions of differential equations [18], [62], [81], [99], [114]. In fact, if the moment sequence $\{\mu_n\}$ of a distribution $f(x)$ is known, then one is tempted to write

$$f(x) \sim \sum_{n=0}^{\infty} \frac{(-1)^n \mu_n \, \delta^{(n)}(x)}{n!} \tag{3.3.3}$$

since for every polynomial $P(x)$ one formally has

$$< f(x), P(x) > = \sum_{n=0}^{\infty} \frac{\mu_n \, P^{(n)}(0)}{n!}. \tag{3.3.4}$$

However, a series of delta functions, as given in (3.3.3), cannot converge unless $\mu_n = 0$ for $n \geq N$. Actually, if the polynomial $P(x)$ is replaced by a general test function in the relation (3.3.4), then the series on the right side would probably diverge.

The moment asymptotic expansion (3.3.1) is then a rigorous interpretation of such series of delta functions.

We shall now provide a proof of (3.3.1) in the space $\mathcal{E}'(\mathbb{R})$. The analyses in the spaces $\mathcal{P}'(\mathbb{R})$ and $\mathcal{O}'_\gamma(\mathbb{R})$ are given in the following sections, but the basic arguments can be seen in the simpler context of $\mathcal{E}'(\mathbb{R})$.

In the space $\mathcal{E}'(\mathbb{R})$ we consider the seminorms

$$\| \phi \|_{n,R} = \max\{| \phi^{(n)}(x) | : | x | \leq R \}, \tag{3.3.5}$$

for $n \in \mathbb{N}$, $R > 0$. These seminorms generate the topology of $\mathcal{E}(\mathbb{R})$. If $q = 0, 1, 2, \cdots$ we set

$$X_q = \{\phi \in \mathcal{E}(\mathbb{R}) : \phi^{(n)}(0) = 0 \ \text{for} \ n < q\}. \tag{3.3.6}$$

Lemma 1. *Let $\phi \in X_q$. Then for every $n \in \mathbb{N}$ and $R > 0$,*

$$\left\| \phi\left(\frac{x}{\lambda}\right) \right\|_{n,R} = O\left(\frac{1}{\lambda^q}\right), \qquad as \ \lambda \to \infty. \tag{3.3.7}$$

Proof. If $\phi \in X_q$, we can find a constant K such that

$$| \phi(x) | \leq K \mid x \mid^q, \qquad \mid x \mid \leq 1. \tag{3.3.8}$$

Therefore, if $\lambda > R$ we obtain

$$\| \phi\left(\frac{x}{\lambda}\right) \|_{0,R} = max\{| \phi\left(\frac{x}{\lambda}\right) | : | x | \leq R\} \leq \frac{K}{\lambda^q}.$$

If $n \leq q$ and $\phi \in X_q$ then $\phi^{(n)} \in X_{q-n}$ and thus

$$\| \phi\left(\frac{x}{\lambda}\right) \|_{n,R} = \| \frac{1}{\lambda^n} \phi^{(n)}\left(\frac{x}{\lambda}\right) \|_{0,R} = \frac{1}{\lambda^n} O\left(\frac{1}{\lambda^{q-n}}\right) = O\left(\frac{1}{\lambda^q}\right),$$

while if $n \geq q$, $\phi^{(n)} \in X_0$ and hence

$$\| \phi\left(\frac{x}{\lambda}\right) \|_{n,R} = O\left(\frac{1}{\lambda^n}\right).$$

■

Actually, we can reformulate the result of Lemma 1 in a slightly different way.

Lemma 2. *For every* $q = 0, 1, 2, \cdots$ *and for every continuous seminorm* $\| \ \|_1$ *in the space* $\mathcal{E}(\mathbb{R})$ *we can find another continuous seminorm* $\| \ \|_2$ *such that*

$$\| \phi\left(\frac{x}{\lambda}\right) \|_1 \leq \lambda^{-q} \| \phi(x) \|_2, \qquad \lambda > \lambda_0, \tag{3.3.9}$$

for each $\phi \in X_q$.

Use of estimate (3.3.7) or (3.3.9) permits us to obtain

Theorem 22. *Let* $f \in \mathcal{E}'(\mathbb{R})$ *and let* $\{\mu_n\}$ *be its moment sequence. Then*

$$f(\lambda x) \sim \sum_{n=0}^{\infty} \frac{(-1)^n \mu_n \ \delta^{(n)}(x)}{n! \lambda^{n+1}}, \qquad as \ \lambda \to \infty, \tag{3.3.10}$$

in the sense that for any $\phi \in \mathcal{E}(\mathbb{R})$ *we have*

$$< f(\lambda x), \phi(x) > = \sum_{n=0}^{N} \frac{\mu_n \phi^{(n)}(0)}{n! \lambda^{n+1}} + O\left(\frac{1}{\lambda^{N+2}}\right) \qquad as \ \lambda \to \infty. \tag{3.3.11}$$

Proof. Let $P_N(x) = \sum_{n=0}^{N} \frac{\phi^{(n)}(0)}{n!} x^n$ be the Taylor polynomial of order N of the function ϕ. Then we have

$$< f(\lambda x), \phi(x) > \, = \, < f(\lambda x), P_N(x) > \, + \, < f(\lambda x), \phi(x) - P_N(x) >$$

$$= \sum_{n=0}^{N} \frac{\mu_n \phi^{(n)}(0)}{n! \lambda^{n+1}} + R_N(\lambda),$$

where the remainder $R_N(\lambda)$ is given as $R_N(\lambda) = \, < f(\lambda x), \phi(x) - P_N(x) >$. Since $\phi_N = \phi - P_N \in X_{N+1}$, we obtain

$$| \, R_N(\lambda) \, | = | < f(\lambda x), \phi_N(x) > | = \frac{1}{\lambda} | < f(x), \phi_N(\frac{x}{\lambda}) > |$$

$$\leq \frac{M}{\lambda} \sum_{j=0}^{q} \| \, \phi_N(\frac{x}{\lambda}) \, \|_{j,R} = O \left(\frac{1}{\lambda^{N+2}} \right),$$

where the existence of M, q and R is guaranteed by the continuity of f. ∎

The proof of Theorem 22 actually provides a method for obtaining error bounds for the remainder $R_N(\lambda)$. Indeed, we have

$$| \, R(\lambda) \, | \leq \frac{M \, \| \, \phi_N(x) \, \|_2}{\lambda^{N+2}}, \qquad (3.3.12)$$

where M is the norm of f with respect to a seminorm $\| \; \|_1$,

$$M = \, \sup \{ | < f, \phi > | : \| \, \phi \, \|_1 = 1 \},$$

and $\| \; \|_2$ is the seminorm given by Lemma 2.

Suppose in particular that $f(x)$ is a Radon measure with support in $[-R, R]$, so that

$$| < f(x), \phi(x) > | \leq M \, \| \, \phi \, \|_{0,R},$$

where M is the total variation of f. Then (3.3.12) takes the form

$$| \, R(\lambda) \, | \leq \frac{MK}{\lambda^{N+2}}, \qquad \lambda > R,$$

where

$$K = \, \sup \left\{ \frac{| \, \phi(x) - P_N(x) \, |}{| \, x \, |^{N+1}} : 0 < | \, x \, | \leq 1 \right\}.$$

Examples of the moment asymptotic expansion in the space $\mathcal{E}'(\mathbb{R})$ include the following:

$$\delta^{(k)}(\lambda x - x_0) \sim \sum_{n=0}^{\infty} \frac{(-1)^n x_0^n \delta^{(n+k)}(x)}{n! \lambda^{n+k+1}}, \qquad \text{as } \lambda \to \infty, \qquad (3.3.13)$$

$$H(\lambda x - a)\, H(b - \lambda x) \sim \sum_{n=0}^{\infty} \frac{(-1)^n (b^{n+1} - a^{n+1}) \delta^{(n)}(x)}{(n+1)! \lambda^{n+1}}, \qquad \text{as } \lambda \to \infty.$$

$$(3.3.14)$$

A more interesting example is the following.

Example 51. Let $f(x)$ be the positive measure defined by

$$f(x) = \sum_{n=-\infty}^{\infty} 2^n \delta(x - 2^n). \tag{3.3.15}$$

Then $f \in \mathcal{S}'$. Also, $f(2x) = f(x)$ as follows directly from (3.3.15). Thus, $f(x) = F(\ln x)$, where F is periodic of period $\ln 2$. The mean of F, its average over an interval of length $\ln 2$, is $\dfrac{1}{\ln 2}$:

$$\frac{1}{\ln 2} \int_{\alpha}^{\alpha + \ln 2} F(u)\,du = \frac{1}{\ln 2} \int_{e^{\alpha}}^{2e^{\alpha}} f(x) \frac{dx}{x} = \frac{1}{\ln 2}.$$

Therefore we can write

$$f(x) = \frac{H(x)}{\ln 2} + g(x), \tag{3.3.16}$$

where $H(x)$ is the Heaviside function and g is a periodic function of $\ln x$ of period $\ln 2$ and zero mean. It follows that if $\phi \in \mathcal{S}$ then

$$\varepsilon \sum_{n=-\infty}^{\infty} 2^n \phi(\varepsilon 2^n) = \frac{1}{\ln 2} \int_{0}^{\infty} \phi(x)\,dx + \psi(\varepsilon), \tag{3.3.17}$$

where the oscillatory component $\psi(\varepsilon) = \psi(\phi; \varepsilon)$ is a periodic function of $\ln \varepsilon$ of period $\ln 2$ and zero mean.

Let us now write

$$f(x) = f_+(x) + f_-(x), \tag{3.3.18}$$

where

$$f_+(x) = \sum_{n=0}^{\infty} 2^n\, \delta(x - 2^n), \tag{3.3.19a}$$

$$f_-(x) = \sum_{n=1}^{\infty} 2^{-n}\, \delta(x - 2^{-n}). \tag{3.3.19b}$$

The distribution $f_-(x)$ has compact support and thus $f_-(\lambda x)$ admits the moment asymptotic expansion as $\lambda \to \infty$. The moments are

$$\mu_k = \ <f_-(x), x^k> \ = \sum_{n=1}^{\infty} 2^{-n} \, 2^{-nk} = \frac{1}{2^{k+1}-1},$$

hence

$$f_-(\lambda x) \sim \sum_{k=0}^{\infty} \frac{(-1)^k \delta^{(k)}(x)}{k!(2^{k+1}-1)\lambda^{k+1}}, \qquad \text{as } \lambda \to \infty. \qquad (3.3.20)$$

Therefore, the development of $f_+(\lambda x)$ takes the form

$$f_+(\lambda x) \sim \frac{H(x)}{\ln 2} + g(\lambda x) - \sum_{k=0}^{\infty} \frac{(-1)^k \delta^{(k)}(x)}{k!(2^{k+1}-1)\lambda^{k+1}}, \qquad \text{as } \lambda \to \infty. \quad (3.3.21)$$

Evaluating (3.3.21) at $\phi \in \mathcal{S}$ and setting $\lambda = \frac{1}{\varepsilon}$ we find

$$\sum_{n=0}^{\infty} 2^n \phi(\varepsilon 2^n) \sim \frac{1}{(\ln 2)\varepsilon} \int_0^{\infty} \phi(x)dx + \frac{\psi(\varepsilon)}{\varepsilon} - \sum_{k=0}^{\infty} \frac{\phi^{(k)}(0)\varepsilon^k}{k!(2^{k+1}-1)}, \qquad \text{as } \varepsilon \to 0,$$

$$(3.3.22)$$

where $\psi(\varepsilon) = \psi(\phi; \varepsilon)$ is a periodic function of $\ln \varepsilon$ of period $\ln 2$ and zero mean.

Formula (3.3.22) and its generalizations considered in Example 52 below play an important role in the counting algorithms used in data base systems [47].

The moment asymptotic expansion can be generalized to the spaces $\mathcal{E}_p\{\phi_n\}$, introduced in Section 2.10, for certain asymptotic sequences $\{\phi_n\}$. In particular, generalized moment asymptotic expansions are obtained for the sequence $\{x^{\alpha_n}\}$ if $\Re e \; \alpha_n \nearrow \infty$ as well as for sequences of the form $\{(\ln x)^r \, x^{\alpha_n}\}$ with $0 \leq r \leq k_n$.

Recall that $\phi_n^{-1}\delta(x)$ is the functional defined in $\mathcal{E}_p\{\phi_n\}$ by

$$<\phi_n^{-1}\delta(x), \phi(x)> \ = a_n, \qquad (3.3.23)$$

where $\phi(x)$ has the development

$$\phi(x) \sim \sum_{k=1}^{\infty} a_k \phi_k(x), \qquad \text{as } x \to 0^+. \qquad (3.3.24)$$

We have the following generalized moment asymptotic expansion.

Theorem 23. *Let α_n be a sequence with $\Re\, \alpha_n \nearrow \infty$ and let $f \in \mathcal{E}'_0\{x^{\alpha_n}\}$. Then*

$$f(\lambda x) \sim \sum_{n=1}^{\infty} \frac{\mu(\alpha_n)\,(x^{-\alpha_n}\,\delta(x))}{\lambda^{\alpha_n+1}}, \qquad as\ \lambda \to \infty, \qquad (3.3.25)$$

in the weak sense, where $\mu(\alpha_n)$ are the generalized moments

$$\mu(\alpha_n) = \,<f(x), x^{\alpha_n}> . \qquad (3.3.26)$$

Proof. Let $X_q = \{\psi \in \mathcal{E}\{x^{\alpha_n}\} : \,<\delta_j, \psi> = 0, j\ < q\}$. If $\psi \in X_q$, then $|\psi(x)| \le K\,|\,x\,|^{\alpha_q}$, $0 \le |\,x\,| \le 1$, where $K = \|\,\psi\,\|_{q,1}$.

It follows that $\|\,\psi(x/\lambda)\,\|_{m,R} = O\left(\frac{1}{\lambda^{\alpha_q}}\right)$ as $\lambda \to \infty$ if $1 \le m \le q$. But if $m > q$, then $\psi(x) - \sum_{j=q}^{m-1} a_j x^{\alpha_j} \in X_m$ and $\|\,\psi(x/\lambda)\,\|_{m,R} = \|\,\psi(x/\lambda) - \sum_{j=q}^{m-1} a_j\,x^{\alpha_j}/\lambda^{\alpha_j}\,\|_{m,R}$, and it follows that $\|\,\psi(x/\lambda)\,\|_{m,R} = O\left(\frac{1}{\lambda^{\alpha_m}}\right)$ as $\lambda \to \infty$.

Therefore, if $\psi \in X_q$,

$$\|\,\psi(x/\lambda)\,\| = O\,(\lambda^{-\alpha_q}) \qquad as\ \lambda \to \infty \qquad (3.3.27)$$

for every continuous seminorm $\|\quad\|$ in $\mathcal{E}_0\{x^{\alpha_n}\}$.

If $f \in \mathcal{E}'_0\{x^{\alpha_n}\}$ then $\|\,\psi\,\| = |\,<f, \psi>\,|$ is a continuous seminorm. Thus if $\psi \sim a_1 x^{\alpha_1} + a_2 x^{\alpha_2} + ...$, we obtain

$$<f(\lambda x), \psi(x)> = \frac{1}{\lambda} \,<f(x), \psi\left(\frac{x}{\lambda}\right)>$$

$$= \frac{1}{\lambda} \,<f(x), a_1\left(\frac{x}{\lambda}\right)^{\alpha_1} + ... + a_{q-1}\left(\frac{x}{\lambda}\right)^{\alpha_{q-1}} + \psi_1\left(\frac{x}{\lambda}\right)>$$

$$= \frac{\mu(\alpha_1)a_1}{\lambda^{\alpha_1+1}} + ... + \frac{\mu(\alpha_{q-1})a_{q-1}}{\lambda^{\alpha_{q-1}+1}} + R_q(\lambda),$$

where $\psi_1(x) = \psi(x) - \sum_{j=1}^{q-1} a_j x^{\alpha_j}$. But the remainder $R_q(\lambda)$ is bounded as

$$|\,R_q(\lambda)\,| = \frac{1}{\lambda}\,|<f(x), \psi_1\left(\frac{x}{\lambda}\right)>| = \frac{1}{\lambda}\,\|\,\psi_1(\frac{x}{\lambda})\,\| = O(\lambda^{-\alpha_q-1})$$

as $\lambda \to \infty$ since $\psi_1 \in X_q$. $\qquad\blacksquare$

A completely analogous analysis yields

Theorem 24. *Let $\{\alpha_n\}$ be a sequence with $\Re e \; \alpha_n \nearrow \infty$, let k_j be a sequence of non-negative integers and let $0 \leq p \leq \infty$. Every functional $f \in \mathcal{E}'_p \{(\ln x)^r x^{\alpha_j}, 0 \leq r \leq k_j\}$ has the asymptotic expansion*

$$f(\lambda x) \sim \sum_{j=1}^{\infty} \sum_{t=0}^{k_j} (-1)^{k_j - t} \left(\sum_{r=0}^{t} \binom{k_j - r}{t - r} \mu(t - r, \alpha_j) \delta_{k_j - r, j}(x) \right)$$

$$(\ln \lambda)^{k_j - t} \lambda^{-\alpha_j - 1}, \tag{3.3.28}$$

where $\delta_{r,j}(x) = (\ln x)^{-r} x^{-\alpha_j} \delta(x)$ and where

$$\mu(r, \alpha_j) = \; < f(x), (\ln x)^r x^{\alpha_j} > . \tag{3.3.29}$$

Example 52. We now generalize the expansion (3.3.22) to the case when $\phi \in \mathcal{S}\{x^{\alpha_n}\}$. In order to do so we need to regularize

$$f(x) = \sum_{n=-\infty}^{\infty} 2^n \delta(x - 2^n)$$

in $\mathcal{S}\{x^{\alpha_n}\}$. Equivalently, we need to regularize

$$f_-(x) = \sum_{n=1}^{\infty} 2^{-n} \delta(x - 2^{-n}).$$

We construct the regularization by using the finite part ideas of Chapter 2.

Let $\phi \in \mathcal{S}\{x^{\alpha_n}\}$ with expansion $\phi(x) \sim a_1 x^{\alpha_1} + a_2 x^{\alpha_2} + a_3 x^{\alpha_3} + \cdots$, as $x \to 0^+$. Suppose $\alpha_m = -1$. Then the series giving the value of $< f_-(x), \phi(x) - \sum_{j=1}^{m} a_j x^{\alpha_j} >$ converges. Thus it is enough to give the finite part of sum $\sum_{n=1}^{\infty} 2^n \phi(2^{-n})$ if $\phi(x) = x^{\alpha}$, $\alpha \leq -1$. If $\alpha < -1$ we have

$$\sum_{n=1}^{N} 2^{-n} 2^{-n\alpha} = \frac{2^{-1-\alpha} - (2^{-1-\alpha})^N}{1 - 2^{-1-\alpha}}.$$

Thus the finite part of the series is

$$< f_-(x), x^\alpha > = F.p. \sum_{n=1}^{\infty} 2^{-n} 2^{-n\alpha} = \frac{2^{-1-\alpha}}{1 - 2^{-1-\alpha}} = \frac{1}{2^{1+\alpha} - 1}.$$

When $\alpha = -1$ we have

$$< f_-(x), x^{-1} > = F.p. \sum_{n=1}^{\infty} 1 = F.p. \lim_{N \to \infty} \sum_{m=1}^{N} 1 = F.p. \lim_{N \to \infty} N = 0.$$

Therefore the regularization is given by

$$< f_-(x), \phi(x) >= \sum_{j=1}^{m-1} \frac{a_j}{2^{1+\alpha_j} - 1} + \sum_{n=1}^{\infty} 2^{-n} \left[\phi(2^{-n}) - \sum_{j=1}^{m} a_j 2^{-n\alpha_j} \right].$$

$$(3.3.30)$$

The moment asymptotic expansion for $f_-(\lambda x)$ in $\mathcal{S}\{x^{\alpha_n}\}$ takes the form

$$f_-(\lambda x) \sim \sum_{j=1}^{m-1} \frac{x^{-\alpha_j} \delta(x)}{(2^{1+\alpha_j} - 1)\lambda^{\alpha_j}} + \sum_{j=m+1}^{\infty} \frac{x^{-\alpha_j} \delta(x)}{(2^{1+\alpha_j} - 1)\lambda^{\alpha_j}}, \quad \lambda \to \infty,$$

$$(3.3.31)$$

where $\alpha_m = -1$.

Using (3.3.31) we obtain the expansion of $f_+(\lambda x)$ and setting $\lambda = \frac{1}{\varepsilon}$ we get the development of $< f_+(x), \phi(\varepsilon x) >$ as $\varepsilon \to 0$:

$$\sum_{n=0}^{\infty} 2^n \phi(\varepsilon 2^n) \sim \sum_{j=1}^{m-1} \frac{a_j \varepsilon^{\alpha_j}}{1 - 2^{1+\alpha_j}} + \left(\frac{F.p. \int_0^{\infty} \phi(x) dx}{\ln 2} - \frac{a_m \ln \varepsilon}{\ln 2} + \psi(\varepsilon) \right) \frac{1}{\varepsilon}$$

$$+ \sum_{j=m+1}^{\infty} \frac{a_j \varepsilon^{\alpha_j}}{1 - 2^{1+\alpha_j}}, \quad \text{as } \varepsilon \to 0, \qquad (3.3.32)$$

if $\phi(x) \sim a_1 x^{\alpha_1} + a_2 x^{\alpha_2} + a_3 x^{\alpha_3} + \cdots$, as $x \to 0^+$, and $\alpha_m = -1$. Here the oscillatory component $\psi(\varepsilon) = \psi(\phi; \varepsilon)$ is a periodic function of $\ln \varepsilon$ of period $\ln 2$ and zero mean.

Observe that the expansion (3.3.32) is similar but different from the moment asymptotic expansion. The difference arises in the coefficient of ε^{-1}. Also observe that the integral $\int_0^{\infty} \phi(x) dx$ is generally divergent if $\phi \in \mathcal{S}\{x^{\alpha_n}\}$ and thus it is necesary to consider its finite part $F.p. \int_0^{\infty} \phi(x) dx$. Notice that we have used the formula

$$F.p. \int_0^{\infty} \phi(\varepsilon x) dx = \frac{1}{\varepsilon} \left[F.p. \int_0^{\infty} \phi(x) dx - a_m \ln \varepsilon \right].$$

Replacing $\phi(x)$ by $\dfrac{\phi(x)}{x}$ in (3.3.32) and observing that $\dfrac{\phi(x)}{x}$ has the expansion $a_1 x^{\alpha_1 - 1} + a_2 x^{\alpha_2 - 1} + \cdots$ as $x \to 0$, we obtain the formula

$$\sum_{n=0}^{\infty} \phi(\varepsilon 2^n) \sim \sum_{j=1}^{k-1} \frac{a_j \varepsilon^{\alpha_j}}{1 - 2^{\alpha_j}} + \frac{F.p. \int_0^{\infty} \frac{\phi(x)}{x} dx}{\ln 2} - \frac{a_k \ln \varepsilon}{\ln 2} + \psi_1(\varepsilon)$$

$$+ \sum_{j=k+1}^{\infty} \frac{a_j \varepsilon^{\alpha_j}}{1 - 2^{\alpha_j}},$$

$$(3.3.33)$$

where $\alpha_k = 0$ and where $\psi_1(\varepsilon)$ is a periodic function of $\ln \varepsilon$ of period $\ln 2$ and zero mean.

Formulas (3.3.32) and (3.3.33) are easily generalized to the space $\mathcal{S}\{x^{\alpha_n} \ln x, x^{\alpha_n}\}$, where $\alpha_n \nearrow \infty$. Indeed, if $\phi(x) \sim \sum_{j=1}^{\infty} (a'_j \ln x + a_j) x^{\alpha_j}$, as $x \to 0^+$, with $\alpha_k = 0$ then

$$\sum_{n=0}^{\infty} \phi(\varepsilon 2^n) \sim \sum_{j=1}^{k-1} \left[\frac{a'_j \ln \varepsilon + a_j}{1 - 2^{\alpha_j}} - \frac{a'_j \ln 2}{(1 - 2^{\alpha_j})^2} \right] \varepsilon^{\alpha_j}$$

$$+ \frac{F.p. \int_0^{\infty} \frac{\phi(x)}{x} dx}{\ln 2} - \frac{a_k \ln \varepsilon}{\ln 2} - \frac{a'_k (\ln \varepsilon)^2}{2 \ln 2} + \psi(\varepsilon) \qquad (3.3.34)$$

$$+ \sum_{j=k+1}^{\infty} \left[\frac{a'_j \ln \varepsilon + a_j}{1 - 2^{\alpha_j}} - \frac{a'_j \ln 2}{(1 - 2^{\alpha_j})^2} \right] \varepsilon^{\alpha_j}, \quad \text{as } \varepsilon \to 0.$$

3.4 Expansions in the Space $\mathcal{P}'$

We shall now consider the moment asymptotic expansion in the space $\mathcal{P}'(\mathbb{R})$ of distributions of "less than exponential growth."

In the space $\mathcal{P}(\mathbb{R})$ we consider the seminorms

$$\| \phi \|_{\gamma, j} = \sup \{ |e^{-\gamma |x|} \phi^{(j)}(x)| : x \in \mathbb{R} \} \qquad (3.4.1)$$

for $\gamma > 0$ and $j \in \mathbb{N}$. They generate the topology of the space.

If $\phi \in X_q = \{\psi \in \mathcal{P}(\mathbb{R}) : \psi^{(j)}(0) = 0, j < q\}$, then for any $\gamma > 0$ we can find a constant K such that

$$|\phi(x)| \leq K|x|^q e^{\frac{\gamma |x|}{2}}.$$

Therefore if $\lambda > 1$,

$$e^{-\gamma |x|} |\phi(x/\lambda)| \leq \frac{K}{\lambda^q} e^{\frac{-\gamma |x|}{2}} |x|^q \leq \frac{K'}{\lambda^q},$$

and thus

$$\| \phi(x/\lambda) \|_{\gamma, 0} = O \left(\frac{1}{\lambda^q} \right), \quad \text{as } \lambda \longrightarrow \infty, \quad \phi \in X_q. \qquad (3.4.2)$$

Proceeding in a similar fashion we thus obtain

Theorem 25. *If $\| \ \|$ is any continuous seminorm in $\mathcal{P}(\mathbb{R})$ then*

$$\|\phi(x/\lambda)\| = O \left(\frac{1}{\lambda^q} \right), \quad \text{as } \lambda \longrightarrow \infty, \quad \phi \in X_q. \qquad (3.4.3)$$

If $f(x)$ is any distribution of the space $\mathcal{P}'(\mathbb{R})$ then

$$f(\lambda x) \sim \sum_{n=0}^{\infty} \frac{(-1)^n \mu_n \delta^{(n)}(x)}{n!\, \lambda^{n+1}}, \quad \text{as } \lambda \longrightarrow \infty \qquad (3.4.4)$$

in the space $\mathcal{P}(\mathbb{R})$, where $\mu_n = <f(x), x^n>$ are the moments of f.

Let us now give some examples.

Example 53. Let $f(x) = H(x)e^{-x}$, then the moments are $\mu_k = \int_0^{\infty} x^k e^{-x}\, dx = k!$ and thus we obtain

$$H(x)e^{-\lambda x} \sim \sum_{k=0}^{\infty} \frac{(-1)^k \delta^{(k)}(x)}{\lambda^{k+1}}, \quad \text{as } \lambda \longrightarrow \infty, \qquad (3.4.5)$$

in the space $\mathcal{P}'(\mathbb{R})$. This means that if $\phi \in \mathcal{P}$, the following expansion

$$\Phi(\lambda) = \int_0^{\infty} e^{-\lambda x}\phi(x)dx \sim \frac{\phi(0)}{\lambda} + \frac{\phi'(0)}{\lambda^2} + \frac{\phi''(0)}{\lambda^3} + \cdots, \quad \text{as } \lambda \longrightarrow \infty$$
$$(3.4.6)$$

of the Laplace transform $\Phi(\lambda)$ holds. This is the celebrated Watson's lemma. Expansion (3.4.5) is thus the distributional version of Watson's lemma.

More generally, if $f_z(x) = H(x)e^{-zx}$, where z is a complex number with $\mid z \mid = 1$, $\Re e\, z > 0$, we have $\mu_k = \int_o^{\infty} x^k e^{-zx}dx = \frac{k!}{z^{k+1}}$ and thus

$$H(x)e^{-\lambda zx} \sim \sum_{k=0}^{\infty} \frac{(-1)^k \delta^{(k)}(x)}{\lambda^{k+1} z^{k+1}}, \quad \text{as } \lambda \longrightarrow \infty. \qquad (3.4.7)$$

If we now take $\lambda \in \mathbb{C}$, with $\Re e\, \lambda > 0$ and set $z = \frac{\lambda}{|\lambda|}$, then (3.4.7) shows that (3.4.5) and (3.4.6) actually hold as $\lambda \longrightarrow \infty$ in the half plane $\Re e\, \lambda > 0$.

If z varies on the sector $\mid \arg \lambda \mid \leq \frac{\pi}{2} - \delta$, where $\delta > 0$ is fixed, then it is easy to see that $\mid <f_z, \phi> \mid \leq M\|\phi\|_{\gamma,0}$ for some $M > 0$, $\gamma > 0$ for all z in the sector and it follows that (3.4.5) and (3.4.6) are uniform as $\lambda \longrightarrow \infty$ in the sector $\mid \arg \lambda \mid \leq \frac{\pi}{2} - \delta$.

By considering the generalized functions $e^{-x}x_+^{\alpha}$, $\alpha \notin \mathbb{Z}$ and $e^{-x}Pf\left(\frac{H(x)}{x^q}\right)$, $q = 1, 2, \cdots$ we readily obtain

$$x_+^{\alpha} e^{-\lambda x} \sim \sum_{k=0}^{\infty} \frac{(-1)^k \Gamma(k+\alpha+1)\delta^{(k)}(x)}{k!\, \lambda^{k+\alpha+1}}, \quad \text{as } \lambda \to \infty, \qquad (3.4.8)$$

$$\mathcal{P}f\left(\frac{H(x)}{x^q}\right)e^{-\lambda x} \sim \sum_{k=0}^{q-1} \frac{(-1)^{q-1}}{k!(q-k-1)!}(\psi(q-k)+\ln\lambda)\lambda^{q-k-1}\delta^{(k)}(x)$$

$$+\sum_{k=q}^{\infty} \frac{(-1)^k(k-q)!\delta^{(k)}(x)}{k!\,\lambda^{q-k+1}}, \quad \text{as } \lambda \to \infty, \tag{3.4.9}$$

in the half plane $\Re\,\lambda > 0$. Thus if $\phi \in \mathcal{P}$ we obtain the following expansion of the Laplace transforms

$$F.p. \int_0^\infty x^\alpha e^{-\lambda x}\phi(x)dx \sim \sum_{k=0}^{\infty} \frac{\Gamma(k+\alpha+1)\phi^{(k)}(0)}{k!\,\lambda^{k+\lambda+1}}, \quad \text{as } \lambda \to \infty,$$

$$\tag{3.4.10}$$

$$F.p. \int_0^\infty x^{-q} e^{-\lambda x}\phi(x)dx \sim \sum_{k=0}^{q-1} \frac{(-1)^{k-q-1}}{k!\,(q-k-1)!}(\psi(q-k)+\ln\lambda)\lambda^{q-k-1}\phi^{(k)}(0)$$

$$+\sum_{k=q}^{\infty} \frac{(k-q)!\phi^{(k)}(0)}{k!\,\lambda^{q-k+1}}, \quad \text{as } \lambda \to \infty. \tag{3.4.11}$$

In Chapter 1 we used the Euler–Maclaurin summation formula to obtain Stirling's formula for the approximation of $n!$. We now present another derivation of Stirling's formula, based on the moment asymptotic expansion.

Example 54. We start with the representation

$$\ln\Gamma(\lambda) = (\lambda - \frac{1}{2})\ln\lambda - \lambda + \frac{1}{2}\ln 2\pi + \int_0^\infty \phi(t)e^{-\lambda t}dt, \tag{3.4.12}$$

where

$$\phi(t) = \frac{1}{t}\left(\frac{1}{2} - \frac{1}{t} + \frac{1}{e^t - 1}\right), \quad t > 0. \tag{3.4.13}$$

The values of $\phi(t)$ for t negative are irrelevant; we just assume that $\phi \in \mathcal{P}$. Observe that

$$\phi^{(k)}(0) = \frac{k!B_{k+2}}{(k+2)!}, \quad k = 0, 1, 2, \cdots, \tag{3.4.14}$$

where B_k are the Bernoulli numbers. Therefore (3.4.6) yields the expansion

$$\ln\Gamma(\lambda) \sim (\lambda - \frac{1}{2})\ln\lambda - \lambda + \frac{1}{2}\ln 2\pi + \sum_{k=0}^{\infty} \frac{B_{k+2}}{(k+1)(k+2)\lambda^{k+1}}. \tag{3.4.15}$$

Taking exponentials we find

$$\Gamma(\lambda) \sim \sqrt{2\pi}\lambda^{\lambda-\frac{1}{2}}e^{-\lambda}\left(1 + \frac{A_1}{\lambda} + \frac{A_2}{\lambda^2} + \frac{A_3}{\lambda^3} + \cdots\right), \qquad (3.4.16)$$

where

$$A_k = \sum_{q=1}^{k}\frac{1}{q!}\sum_{j_1+\cdots+j_q=k}\frac{B_{j_1+1}\cdots B_{j_q+1}}{j_1(j_1+1)\cdots j_q(j_q+1)}. \qquad (3.4.17)$$

The first few values of A_k are given by

$$A_1 = \frac{1}{12}, \quad A_2 = \frac{1}{288}, \quad A_3 = \frac{-139}{51840}, \quad A_4 = \frac{-31}{155520}. \qquad (3.4.18)$$

Hence

$$\Gamma(\lambda) \sim \sqrt{2\pi}\lambda^{\lambda-\frac{1}{2}}e^{-\lambda}\left(1 + \frac{1}{12\lambda} + \frac{1}{288\lambda^2} - \frac{139}{51840\lambda^3} - \frac{31}{155520\lambda^4} + \cdots\right). \qquad (3.4.19)$$

Example 55. Let us consider the digamma function $\psi(z) = \frac{\Gamma'(z)}{\Gamma(z)}$, the logarithmic derivative of the gamma function.

The function $\psi(z)$ has many interesting properties related to the properties of the gamma function. Indeed, since

$$\Gamma(z+1) = z\Gamma(z), \qquad (3.4.20)$$

it follows that

$$\psi(z+1) = \frac{1}{z} + \psi(z). \qquad (3.4.21)$$

Similarly, the formulas

$$\psi(1-z) - \psi(z) = \pi\cot\pi z, \qquad (3.4.22)$$

$$\psi(z) + \psi(z + \frac{1}{2}) + 2\ln 2 = 2 + \psi(2z), \qquad (3.4.23)$$

are obtained from the formulas

$$\Gamma(z)\Gamma(1-z) = \frac{\pi}{\sin\pi z}, \qquad (3.4.24)$$

and

$$2^{2z-1}\Gamma(z)\Gamma(z+\frac{1}{2}) = \sqrt{\pi}\Gamma(2z). \qquad (3.4.25)$$

The function $\psi(z)$ is meromorphic, with simple poles at the points $z = 0, -1, -2, \cdots$. The residues at all the poles are equal to -1.

To study the asymptotic behavior of $\psi(z)$ as $z \longrightarrow \infty$ we use the representation

$$\psi(z) = \ln z - \frac{1}{2z} - \int_0^\infty \left(\frac{1}{2} - \frac{1}{t} + \frac{1}{e^t - 1} \right) e^{-tz} dt. \qquad (3.4.26)$$

It follows that

$$\psi(z) \sim \ln z - \frac{1}{2z} - \sum_{k=2}^\infty \frac{B_k}{kz^k}, \quad \text{as } z \longrightarrow \infty, \qquad (3.4.27)$$

in the sector $|\arg z| < \frac{\pi}{2}$. Observe that (3.4.27) can be simplified by recalling that $0 = B_3 = B_5 = \cdots$.

In particular, since for $N = 1, 2, 3, \cdots$,

$$\psi(N) = \sum_{n=1}^{N-1} \frac{1}{n} - \gamma, \qquad (3.4.28)$$

where $\gamma = -\psi(1) = -\Gamma'(1)$ is Euler's constant, we recover the expansion

$$\sum_{n=1}^N \frac{1}{N} \sim \ln N + \gamma + \frac{1}{2N} - \sum_{j=1}^\infty \frac{B_{2j}}{2jN^{2j}}, \quad \text{as } N \to \infty. \qquad (3.4.29)$$

Similarly (3.4.23) yields

$$\psi(N + \frac{1}{2}) = -\gamma - 2\ln 2 + 2 \sum_{n=1}^N \frac{1}{2n - 1}, \qquad (3.4.30)$$

and the expansion

$$\sum_{n=1}^N \frac{1}{2n - 1} \sim \frac{1}{2} \ln(N + \frac{1}{2}) + \frac{\gamma}{2} + \ln 2 - \frac{1}{2N + 1} - \sum_{j=1}^\infty \frac{B_{2j}}{4j(N + \frac{1}{2})^{2j}}. \qquad (3.4.31)$$

Example 56. The function e^{-x^2} belongs to $\mathcal{P}'(\mathbb{R})$. Its moments are $\mu_n = \Gamma\left(\frac{n+1}{2}\right)$, n even; $\mu_n = 0$, n odd. Therefore

$$e^{-\lambda x^2} = e^{-(\lambda^{\frac{1}{2}}x)^2} \sim \sum_{n=0}^\infty \frac{\Gamma\left(\frac{2n+1}{2}\right) \delta^{(2n)}(x)}{(2n)! \, \lambda^{\frac{2n+1}{2}}}, \quad \text{as } \lambda \to \infty. \qquad (3.4.32)$$

This formula will play a very important role in the Section 3.5 where we discuss Laplace's asymptotic formula. Actually, Laplace's formula follows by a simple change of variables in (3.4.32).

We also have the generalized moment asymptotic expansion in the spaces $\mathcal{P}_p\{\phi_n\}$, $0 \le p \le \infty$.

Theorem 26. *Let $\{\alpha_n\}$ be a sequence with $\Re\, \alpha_n \nearrow \infty$ and let k_j be a sequence of non-negative integers. Every functional $f \in \mathcal{P}'_p\{(\ln x)^r x^{\alpha_j}\,,\ 0 \le r \le k_j\}$ has the asymptotic expansion*

$$f(\lambda x) \sim$$

$$\sum_{j=0}^{\infty} \sum_{t=0}^{k_j} (-1)^{k_j - t} \left(\sum_{r=0}^{t} \binom{k_j - r}{t - r} \mu(t - r, \alpha_j) \delta_{k_j - r, j}(x) \right) (\ln \lambda)^{k_j - t} \lambda^{-\alpha_j - 1},$$

$$\text{as } \lambda \to \infty, \tag{3.4.33}$$

where $\delta_{r,j}(x) = (\ln x)^{-r} x^{-\alpha_j} \delta(x)$ and where

$$\mu(r, \alpha_j) = \ <f(x), (\ln x)^r x^{\alpha_j} >\ . \tag{3.4.34}$$

Example 57. If $-1 < \alpha_1 < \alpha_2 < \alpha_3 < \cdots$ we readily obtain the expansion

$$H(x) e^{-\lambda x} \sim \sum_{j=1}^{\infty} \frac{\Gamma(\alpha_j + 1)(x^{-\alpha_j} \delta(x))}{\lambda^{\alpha_j + 1}}, \quad \text{as } \lambda \to \infty, \tag{3.4.35}$$

in the space $\mathcal{P}\{x^{\alpha_n}\}$. Thus if $\phi \in \mathcal{P}_0\{x^{\alpha_n}\}$ with expansion $\phi(x) \sim \sum_{j=1}^{\infty} a_j x^{\alpha_j}$, as $x \to 0^+$, then

$$\int_0^{\infty} e^{-\lambda x} \phi(x)\, dx \sim \sum_{j=1}^{\infty} \frac{\Gamma(\alpha_j + 1) a_j}{\lambda^{\alpha_j + 1}}, \quad \text{as } \lambda \to \infty. \tag{3.4.36}$$

3.5 Laplace's Asymptotic Formula

We shall now consider the asymptotic approximation of integrals of the form

$$\int_a^b e^{-\lambda h(x)} \phi(x) dx, \tag{3.5.1}$$

as $\lambda \to \infty$, where $h(x)$ is real.

It was already observed by Laplace that if $\lambda \gg 1$ the main contribution to (3.5.1) comes from the neighborhood of the minima of $h(x)$. This idea led him to the approximation

$$\int_a^b e^{-\lambda h(x)}\phi(x)dx \sim e^{-\lambda h(x_0)}\sqrt{\frac{2\pi}{h''(x_0)\lambda}}\phi(x_0), \qquad (3.5.2)$$

in the case where the minimum occurs at an interior point x_0, where $h''(x_0) > 0$.

We discuss (3.5.2) and its several variants in this section. But first, we would like to remark that (3.5.2) can be written in the distributional form

$$e^{-\lambda h(x)} \sim e^{-\lambda h(x_0)}\sqrt{\frac{2\pi}{h''(x_0)\lambda}}\delta(x - x_0). \qquad (3.5.3)$$

We now proceed to the derivation of Laplace's formula.

Suppose that the smooth function $h(x)$ has a minimum at $x = x_0$, where $h''(x_0) > 0$. Let

$$I(\lambda) = \int_{-\infty}^{\infty} e^{-\lambda h(x)}\phi(x)dx,$$

where the support of ϕ is a small enough neighborhood of x_0 so that it contains no other critical points of h.

Under these assumptions we can find an increasing smooth function $\psi(x)$ that satisfies $\psi(0) = 0$, $\psi'(x) > 0, \forall x \in \mathbb{R}$, such that $h(x) = h(x_0)+\psi(x)^2$ in supp ϕ. Therefore the change of variables $u = \psi(x)$ yields

$$I(\lambda) = \int_{-\infty}^{\infty} e^{-\lambda(h(x_0)+u^2)}\phi_1(u)du \sim e^{-\lambda h(x_0)}\sum_{n=0}^{\infty}\frac{\Gamma(\frac{2n+1}{2})\phi_1^{(2n)}(0)}{(2n)!\,\lambda^{\frac{2n+1}{2}}},$$

$$\text{as } \lambda \to \infty, \qquad (3.5.4)$$

where $\phi_1(u) = \dfrac{\phi(x)}{\psi'(x)}$ and where expansion (3.4.32), i.e.,

$$e^{-\lambda x^2} \sim \sum_{n=0}^{\infty}\frac{\Gamma(\frac{2n+1}{2})\delta^{(2n)}(x)}{(2n)!\,\lambda^{\frac{2n+1}{2}}}, \qquad \text{as } \lambda \to \infty, \qquad (3.5.5)$$

is used.

If we now recall the notion of change of variables in distributions, we readily obtain the expansion

$$e^{-\lambda h(x)} = e^{-\lambda(h(x_0)+\psi(x)^2)} \sim e^{-\lambda h(x_0)}\sum_{n=0}^{\infty}\frac{\Gamma(\frac{2n+1}{2})\delta^{(2n)}(\psi(x))}{(2n)!\,\lambda^{\frac{2n+1}{2}}},$$

$$\text{as } \lambda \to \infty, \qquad (3.5.6)$$

in the space $\mathcal{D}(U)$, U being a small enough neighborhood of x_0. Observe that (3.5.6) follows by direct substitution of $\psi(x)$ by x in (3.5.5). Observe also that

$$\delta(\psi(x)) = \frac{\delta(x - x_0)}{\psi'(x_0)} = \sqrt{\frac{2}{h''(x_0)}} \delta(x - x_0),$$

and thus the leading term in the expansion is

$$e^{-\lambda h(x)} = e^{-\lambda h(x_0)} \left[\left(\frac{2\pi}{h''(x_0)\lambda} \right)^{1/2} \delta(x - x_0) + O\left(\frac{1}{\lambda^{3/2}} \right) \right], \lambda \to \infty,$$

$$(3.5.7)$$

the distributional Laplace formula.

Similarly, from the expansion

$$e^{-\lambda x^{2k}} \sim \sum_{n=0}^{\infty} \frac{\Gamma(\frac{2n+1}{2k})\delta^{(2n)}(x)}{k(2n)!\,\lambda^{\frac{2n+1}{2k}}}, \quad \text{as } \lambda \to \infty, \qquad (3.5.8)$$

we obtain Laplace's formula for a minimum x_0, where $h^{(j)}(x_0) = 0$, $0 \le j \le 2k - 1$, $h^{(2k)}(x_0) > 0$, as

$$e^{-\lambda h(x)} \sim e^{-\lambda h(x_0)} \sum_{n=0}^{\infty} \frac{\Gamma(\frac{2n+1}{2k})\delta^{(2n)}(\psi(x))}{k(2n)!\,\lambda^{\frac{2n+1}{2k}}}$$

$$\sim e^{-\lambda h(x_0)} \left[\frac{\Gamma(\frac{1}{2k})}{k} \left(\frac{h^{(2k)}(x_0)}{(2k)!\,\lambda} \right)^{1/2k} \delta(x - x_0) + O\left(\frac{1}{\lambda^{3/k}} \right) \right],$$

$$\text{as } \lambda \to \infty, \qquad (3.5.9)$$

in the space $\mathcal{D}(U)$, where U is a small neighborhood of x_0, and $\psi(x) = (h(x) - h(x_0))^{1/2k}$.

It is also a simple matter to obtain the asymptotic expansion of $e^{-\lambda h(x)}$ when the minimum of $h(x)$ is located at one of the endpoints of the interval of integration. In fact, suppose that near the endpoint $x = a$ we have $h(x) = h(a) + \psi(x)^{\alpha}$, where $\psi(x)$ is smooth, $\psi(x) > 0$, $x \ge a$ and $\alpha > 0$. Then from the expansion

$$H(x)e^{-\lambda x^{\alpha}} \sim \sum_{n=0}^{\infty} \frac{(-1)^n \Gamma(\frac{n+1}{\alpha})\delta^{(n)}(x)}{\alpha\,n!\,\lambda^{\frac{n+1}{\alpha}}}, \quad \text{as } \lambda \to \infty, \qquad (3.5.10)$$

we readily obtain

$$H(x - a)e^{-\lambda h(x)} \sim e^{-\lambda h(a)} \sum_{n=0}^{\infty} \frac{(-1)^n \Gamma(\frac{n+1}{\alpha})\delta^{(n)}(\psi(x))}{\alpha\,n!\,\lambda^{\frac{n+1}{\alpha}}} \qquad (3.5.11)$$

$$\sim e^{-\lambda h(a)} \left[\frac{\Gamma(\frac{1}{\alpha})\delta(x-a)}{\alpha \psi'(a)\lambda^{1/\alpha}} + O\left(\frac{1}{\lambda^{2/\alpha}}\right) \right], \quad \text{as} \quad \lambda \to \infty,$$

valid in $\mathcal{D}(U)$ for U a small neighborhood of a. Thus

$$
\int_a^\infty e^{-\lambda h(x)} \phi(x)dx \sim e^{-\lambda h(a)} \left[\left(\frac{\Gamma(1/\alpha)\phi(a)}{\alpha\psi'(a)} \right) \lambda^{-1/\alpha} \right.
$$
$$
+ \frac{\Gamma(2/\alpha)}{\alpha} \left(\frac{\psi''(a)\phi(a)}{(\psi'(a))^3} - \frac{\phi'(a)}{(\psi'(a))^2} \right) \lambda^{-2/\alpha}
$$
$$
\left. + O(\lambda^{-3/\alpha}) \right] \tag{3.5.12}
$$

if supp ϕ contains no other critical point of h.

Example 58. Laplace's formula provides another derivation of Stirling's approximation of $n!$. Indeed,

$$n! = \int_0^\infty e^{-t}t^n dt,$$

which on making the change $t = nx$ becomes

$$n! = n^{n+1} \int_0^\infty e^{-n(x-\ln x)}dx.$$

The minimum of the function $h(x) = x - \ln x$ is located at $x_0 = 1$. Since $h''(1) = 1$, we obtain

$$n! \sim n^{n+1}e^{-n} \left(\frac{2\pi}{n} \right)^{1/2} \sim \sqrt{2\pi} n^{n+1/2} e^{-n}.$$

Example 59. Let $\phi(x)$ be a smooth function which does not vanish near $x = b$. Then Laplace's formula, with $h(x) = \ln 1/x$, yields as $n \to \infty$

$$\int_0^b \phi(x)x^n dx \sim \frac{\phi(b)b^{n+1}}{n} \sim \frac{\phi(b)b^{n+1}}{n+1}. \tag{3.5.13}$$

It follows, in particular, that

$$\lim_{n\to\infty} \left| \int_0^b \phi(x)x^n dx \right|^{\frac{1}{n}} = b. \tag{3.5.14}$$

If the smooth function $\phi(x)$ is replaced by an integrable function $f(x)$ then (3.5.14) does not necessarily hold, but we have [10]

$$\overline{\lim_{n\to\infty}} \left| \int_0^b f(x)x^n dx \right|^{1/n} = b \tag{3.5.15}$$

As it is clear, the asymptotic evaluation of integrals of the type

$$I(\lambda) = \int_a^b f^\lambda(x)\phi(x)dx, \tag{3.5.16}$$

as $\lambda \to \infty$, can be obtained by using Laplace's formula by setting $h(x) = -\ln f(x)$. In particular, if $f(x)$ has a single maximum at the interior point $x = x_0$ and if $f''(x_0) < 0$ then

$$\int_a^b f^\lambda(x)\phi(x)dx \sim f(x_0)^{\lambda+1/2} \left(\frac{-2\pi}{\lambda f''(x_0)}\right)^{1/2} \phi(x_0), \quad \text{as } \lambda \to \infty. \tag{3.5.17}$$

Example 60. Use of (3.5.17) with $\phi(x) = 1$ shows that if f is positive and of class C^2 and has a single maximum at an interior point $x = x_0$ then its norm in the Lebesgue space L^p admits the approximation

$$\|f\|_p = \left(\int_a^b |f(x)|^p dx\right)^{1/p} \sim f(x_0) \left(\frac{-2\pi f(x_0)}{pf''(x_0)}\right)^{1/2p}, \tag{3.5.18}$$

if $p \gg 1$. Therefore,

$$\lim_{p\to\infty} \|f\|_p = \|f\|_\infty. \tag{3.5.19}$$

Although formula (3.5.18) does not hold if f is a continuous function, the density of the set of functions of class C^2 in the space of continous functions in $[a, b]$ shows that (3.5.19) remains valid if f is only required to be continuous.

Example 61. If n is an integer we have the formula

$$\int_0^\pi \sin^{2n} x dx = \frac{1 \cdot 3 \cdots (2n-1)}{2 \cdot 4 \cdots 2n}\pi. \tag{3.5.20}$$

If we now use (3.5.17) with $f(x) = \sin^2 x$ and $x_0 = \frac{\pi}{2}$, we obtain

$$\int_0^\pi \sin^{2n} x dx \sim \sqrt{\frac{\pi}{n}},$$

and thus

$$\frac{1 \cdot 3 \cdots (2n-1)}{2 \cdot 4 \cdots 2n} \sim \frac{1}{\sqrt{\pi n}}. \tag{3.5.21}$$

3.6 The Method of Steepest Descent

In this section we discuss the asymptotic approximation of integrals of the type

$$\int_C e^{\lambda f(z)} g(z) \, dz, \qquad \text{as } \lambda \to \infty, \tag{3.6.1}$$

where $f(z)$ and $g(z)$ are analytic functions defined in a complex region Ω and C is a contour in Ω.

Initially we take λ as a large positive real number, but later the case of complex λ with $|\lambda| >> 1$ will be considered.

If $f(z)$ happens to be real along C then the integral could be approximated by using Laplace's formula. The fact that $f(z)$ and $g(z)$ are analytic allows us to use the Cauchy theorem to deform C into other more appropriate contours. As it turns out, the best contours are the so-called steepest paths, where the imaginary part of $f(z)$ is constant: on such paths Laplace's formula is applicable.

If Ω is a region in the complex plane, we denote by $\mathcal{H}(\Omega)$ the space of holomorphic functions defined in Ω, equipped with the topology of uniform convergence on compacts of Ω. We would like to consider the integral (3.6.1) as the evaluation of certain functionals on the element $g \in \mathcal{H}(\Omega)$.

Let us consider the dual space $\mathcal{H}'(\Omega)$. Some of its simplest elements are the Dirac delta function and its derivatives $\delta^{(k)}(z - z_0)$, for $z_0 \in \Omega$. They are defined in the usual way:

$$< \delta^{(k)}(z - z_0), \, g(z) > \, = \, (-1)^k g^{(k)}(z_0). \tag{3.6.2}$$

Actually, series of the type

$$\sum_{n=0}^{\infty} a_n \delta^{(n)}(z - z_0)$$

belong to $\mathcal{H}'(\Omega)$ provided $a_n = O\left(\dfrac{r^n}{n!}\right)$ for some r smaller than the distance from z_0 to the boundary of Ω.

Observe, however, that there is not a notion of support in $\mathcal{H}(\Omega)$. The elements of $\mathcal{H}(\Omega)$ are analytic functions and thus their behavior in any small disc determines them completely. The same is true in $\mathcal{H}'(\Omega)$. Thus, it could be thought that the functional $\delta(z - z_0)$ is supported at $\{z_0\}$, but the formula

$$\delta(z - z_0) = \sum_{n=0}^{\infty} \frac{(z_1 - z_0)^n \delta^{(n)}(z - z_1)}{n!}, \tag{3.6.3}$$

valid if z_1 is near z_0, shows that $\delta(z-z_0)$ can also be thought to be supported at $\{z_1\}$. Furthermore, if C is a curve in Ω that encircles z_0 once in the counterclockwise direction, then we have the representation

$$\delta(z - z_0) = \frac{1}{2\pi i} \int_C \frac{\delta(z - \xi)d\xi}{\xi - z_0}. \qquad (3.6.4)$$

A particularly interesting class of functionals is formed by the $I(h; C)$, defined as

$$< I(h; C), g > = \int_C h(z)\, g(z)\, dz. \qquad (3.6.5)$$

Here h is analytic in Ω and C is a contour in Ω.

Observe that the functional $I(h; C)$ depends on C only through its homotopy class. That is, if the curve C_1 can be continuously deformed into the curve C_2, without leaving Ω and with fixed endpoints for open contours, then $I(h; C_1) = I(h; C_2)$.

Our aim is to study the asymptotic behavior of the functional

$$I(e^{\lambda f(z)}; C), \qquad \text{as } \lambda \to \infty, \qquad (3.6.6)$$

where $C = C_{ab}$ is an open contour from $z = a$ to $z = b$ in Ω. We shall do this by using the method of steepest descent.

Let us start with an example.

Example 62. Let us consider the expansion of $I(e^{\lambda z}; C)$, as $\lambda \to \infty$, where $C = C_{ab}$ is a curve from $z = a$ to $z = b$ in the *convex* region Ω. We first assume that $\Re e\, a < \Re e\, b$. Then we can deform C into a new contour $C' = C_1 \cup C_2$ consisting of two parts C_1 and C_2 such that if $\xi \in C_1$, then $\Re e\, \xi \leq m < \Re e\, b$, while C_2 is a horizontal segment from the point $b - \delta$ to b.

Then

$$I(e^{\lambda z}; C) = I(e^{\lambda z}; C_1) + I(e^{\lambda z}; C_2), \qquad (3.6.7)$$

but

$$I(e^{\lambda z}; C_1) = O(e^{\lambda m}), \qquad \text{as } \lambda \to \infty, \qquad (3.6.8)$$

while

$$< I(e^{\lambda z}; C_2), g(z) > = \int_{\Re e\, b - \delta}^{\Re e\, b} e^{\lambda(t + i \Im m\, b)} g(t + i \Im m\, b)\, dt$$

$$\sim e^{\lambda b} \sum_{n=0}^{\infty} \frac{(-1)^n g^{(n)}(b)}{\lambda^{n+1}}, \qquad \text{as } \lambda \to \infty, \qquad (3.6.9)$$

since according to Laplace's formula

$$H(t - \alpha) \, H(\beta - t) \, e^{\lambda t} \sim e^{\lambda \beta} \sum_{n=0}^{\infty} \frac{\delta^{(n)}(t - \beta)}{\lambda^{n+1}}, \qquad \text{as } \lambda \to \infty. \qquad (3.6.10)$$

Therefore, if $\Re e \, a < \Re e \, b$

$$I(e^{\lambda z}; \, C) \sim e^{\lambda b} \sum_{n=0}^{\infty} \frac{\delta^{(n)}(z - b)}{\lambda^{n+1}}, \qquad \text{as } \lambda \to \infty. \qquad (3.6.11)$$

The case when $\Re e \, a > \Re e \, b$ can be handled by observing that $I(e^{\lambda z}; C) = -I(e^{\lambda z}; -C)$, so that

$$I(e^{\lambda z}; \, C) \sim -e^{\lambda a} \sum_{n=0}^{\infty} \frac{\delta^{(n)}(z - a)}{\lambda^{n+1}}, \qquad \text{as } \lambda \to \infty. \qquad (3.6.12)$$

Finally, if $\Re e \, a = \Re e \, b$ then we can deform C into a contour $C' = C_{ad} \cup C_{db}$ with $\Re e \, d < \Re e \, b$. Using (3.6.11) and (3.6.12) we thus obtain

$$I(e^{\lambda z}; \, C) \sim e^{\lambda b} \sum_{n=0}^{\infty} \frac{\delta^{(n)}(z - b)}{\lambda^{n+1}} - e^{\lambda a} \sum_{n=0}^{\infty} \frac{\delta^{(n)}(z - a)}{\lambda^{n+1}}. \qquad (3.6.13)$$

In Example 62 the domain Ω was assumed to be convex. Actually, what is needed for (3.6.11) to remain valid, when $\Re e \, a < \Re e \, b$, is that the contour C can be deformed into one of the form $C_1 \cup C_2$ with $\Re e \, \xi \leq m < \Re e \, b$ if $\xi \in C_1$ and with C_2 a horizontal segment with right endpoint $z = b$. As we shall see, the analysis for the functional $I(e^{\lambda \, f(z)}; \, C)$ is quite similar if the contour C can be deformed into one where $\Re e \, f(z)$ has maxima only at one of the two endpoints. But before we study this situation we would like to show that formulas (3.6.11), (3.6.12) and (3.6.13) might fail for certain curves if Ω is not convex.

Example 63. Let us consider the region $\Omega = \mathbb{C} \backslash \{x\}$, where $x > 0$. Let C be a curve from -1 to 0 consisting of a contour from -1 to $x + 1$ in the upper half plane and a contour from $x + 1$ to 0 in the lower half plane. Then if $g \in \mathcal{H}(\Omega)$, using the calculus of residues we obtain

$$\int_C e^{\lambda z} g(z) \, dz = 2\pi i \, \Re es_{z=x}(e^{\lambda z} g(z)) + \int_{-1}^{0} e^{\lambda t} g(t) \, dt,$$

so that if the singular part of $g(z)$ at $z = x$ has the form

$$\sum_{n=1}^{\infty} \frac{a_n}{(z - x)^n},$$

then

$$\int_C e^{\lambda z} g(z)\, dz = 2\pi i e^{\lambda x} \sum_{n=1}^{\infty} \frac{a_n \lambda^{n-1}}{(n-1)!} + \int_{-1}^{0} e^{\lambda t} g(t)\, dt$$

$$= 2\pi i e^{\lambda x} \sum_{n=1}^{\infty} \frac{a_n \lambda^{n-1}}{(n-1)!} + O(1).$$

It follows that (3.6.11) does not hold in this case.

The expansion of $I(e^{\lambda f(z)}; C)$ when $f'(b) \neq 0$ and when the contour $C = C_{ab}$ can be deformed into another contour C' with $\Re e\, f(\xi) < \Re e\, f(b)$ for $\xi \in C'\backslash\{b\}$, may be obtained from (3.6.11) by a change of variables. Indeed, holomorphic changes of variables in the dual space $\mathcal{H}'(\Omega)$ are defined in the same way as with distributions: if the conformal map $\omega = F(z)$ is a bijection from Ω to the region Λ and $T \in \mathcal{H}'(\Lambda)$ then the functional $T(F(z))$ of $\mathcal{H}'(\Omega)$ is defined as

$$< T(F(z)),\, g(z) > \; = \; < T(\omega),\, g(F^{-1}(\omega))\frac{dz}{d\omega} > . \tag{3.6.14}$$

Observe in particular the familiar formulas

$$\delta(F(z)) = \frac{\delta(z - z_0)}{F'(z_0)}, \tag{3.6.15}$$

$$\delta'(F(z)) = \frac{\delta'(z - z_0)}{F'(z_0)} + \frac{F''(z_0)\, \delta(z - z_0)}{(F'(z_0))^2}, \tag{3.6.16}$$

which hold if $F(z_0) = 0$, $F'(z_0) \neq 0$.

Returning to $I(e^{\lambda f(z)}; C)$ we proceed as follows. Let Δ be a small disc centered at $f(b)$ such that the map $\omega = f(z)$ defines a bijection between a certain neighborhood Ω_1 of b and Δ. Next we deform C into a new contour $C_1 \cup C_2$ such that $\Re e\, (\xi) \leq m < \Re e\, f(b)$ if $\xi \in C_2$, while $f(C_2)$ is a contour contained in Δ. Then

$$I(e^{\lambda f(z)}; C_1) = O(e^{\lambda m}), \tag{3.6.17}$$

while

$$< I(e^{\lambda f(z)}; C_2),\, g(z) > \; = \; < I(e^{\lambda \omega};\, f(C_2)),\, g(f^{-1}(\omega))\frac{dz}{d\omega} >$$

$$\sim e^{\lambda f(b)} \sum_{n=0}^{\infty} < \delta^{(n)}(\omega - f(b)),\, g(f^{-1}(\omega))\frac{dz}{d\omega} > \lambda^{-n-1}$$

$$\sim e^{\lambda f(b)} \sum_{n=0}^{\infty} < \delta^{(n)}(f(z) - f(b)), \, g(z) > \lambda^{-n-1}.$$

Therefore,

$$I(e^{\lambda f(z)}; \, C) \sim e^{\lambda f(b)} \sum_{n=0}^{\infty} \frac{\delta^{(n)}(f(z) - f(b))}{\lambda^{n+1}}, \quad \text{as } \lambda \to \infty. \qquad (3.6.18)$$

The asymptotic formula (3.6.18) is valid when the main contribution to the integral comes from the endpoints of C. We now consider the case when the asymptotic development is obtained from interior critical points. The points $z_0 \in \Omega$ where $f'(z_0) = 0$ are called *saddle points* or *cols*. The graph of $| \, e^{\lambda f(z)} \, | = e^{\lambda \Re f(z)}$ as a function of $x = \Re z$ and $y = \Im z$ is called the *relief surface*. Clearly the cols are saddle points of the relief surface.

The curves along which $u = \Re f$ is constant are called *level curves*. The level curves are contour lines of the relief surface: on them the height $| \, e^{\lambda f(z)} \, |$ of the relief surface remains constant while the phase of $e^{\lambda f(z)}$ changes as rapidly as possible. On the other hand, the curves along which $v = \Im f$ remains constant are called *steepest paths*. The steepest paths are the gradient lines of the relief surface; on them $| \, e^{\lambda f(z)} \, |$ changes as rapidly as possible.

The saddle point z_0 is said to be of order m if $f'(z_0) = \cdots = f^{(m)}(z_0) = 0$ while $f^{(m+1)}(z_0) \neq 0$. At a saddle point of order m, $m + 1$ level curves meet at equal angles and these angles are bisected by $m+1$ steepest curves. These $m + 1$ level curves divide the relief surface into $m + 1$ valleys and $m + 1$ hills near the saddle.

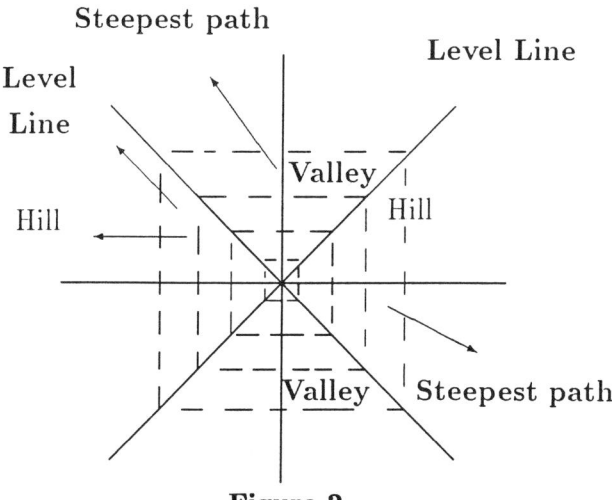

Figure 2

Figure 2 gives the geometry near the col $z = 0$ of the function $f(z) = z^2$. Since near any saddle point of the first order $z = z_0$ we have $f(z) = f(z_0) + \rho(z)^2$ for some conformal map ρ, it follows that locally the geometry near a col of the first order is not very different from this.

When studying the expansion of $I(e^{\lambda f(z)}; C)$ it is usually convenient to deform C into another contour consisting of several arcs of steepest paths. Let us first give an example.

Example 64. Let us consider the expansion of $I(e^{\lambda z^2}; C)$, where C is a curve from the point $z = a$ to the point $z = b$ of the convex region Ω. We assume that the saddle point $z = 0$ belongs to Ω.

The level lines through the col $z = 0$ are $x = y$ and $x = -y$, which divide the plane into two hills $(\frac{-\pi}{4} < \arg z < \frac{\pi}{4}$ and $\frac{3\pi}{4} < \arg z < \frac{5\pi}{4})$ and two valleys $(\frac{\pi}{4} < \arg z < \frac{3\pi}{4}$ and $\frac{5\pi}{4} < \arg z < \frac{7\pi}{4})$. The steepest lines through $z = 0$ are the lines $y = 0$ and $x = 0$. On $y = 0$ the function $|e^{\lambda z^2}| = e^{\lambda(x^2 - y^2)}$ has a minimum at $z = 0$, while on $x = 0$ the function $|e^{\lambda z^2}|$ has a maximum at $z = 0$.

It is not hard to see that the expansion of $I(e^{\lambda z^2}; C)$ can be obtained from the endpoint formula (3.6.18) in case a and b are on the same hill or different hills, when one belongs to a hill and the other to a valley or when they belong to the same valley. However, (3.6.18) cannot be applied when $z = a$ and $z = b$ are on different valleys. In such a case $\Re a^2$ and $\Re b^2$ are both negative, but on any curve C from a to b there are points ξ with $\Re \xi^2 \geq 0$. To obtain the development of $I(e^{\lambda z^2}; C)$ when a and b are on different valleys we deform C into the polygonal line $C_1 \cup C_2 \cup C_3$, where C_1 is a horizontal line from a to $i \Im a$, C_2 is a *steepest line*, from $i \Im a$ to $i \Im b$ and C_3 is a horizontal segment from $i \Im b$ to b. Since

$$I(e^{\lambda z^2}; C_j) = O(e^{-\lambda m}), \quad j = 1, 3, \tag{3.6.19}$$

where $-m = \max\{-(\Im a)^2, -(\Im b)^2\} < 0$, it is enough to obtain the expansion of $I(e^{\lambda z^2}; C_2)$. But

$$< I(e^{\lambda z^2}; C_2), g(z) > = \int_{C_2} e^{\lambda z^2} g(z) dz$$

$$= i \int_{\Im a}^{\Im b} e^{-\lambda t^2} g(it) \, dt$$

$$\sim \pm i \int_{-\infty}^{\infty} e^{-\lambda t^2} g(it) dt$$

$$\sim \pm \sum_{n=0}^{\infty} \frac{\Gamma(\frac{2n+1}{2}) i^{2n+1} g^{(2n)}(0)}{(2n)! \lambda^{\frac{2n+1}{2}}},$$

thus

$$I(e^{\lambda\,z^2};\,C) \sim \sigma i \sum_{n=0}^{\infty} \frac{(-1)^n \Gamma\left(\frac{2n+1}{2}\right) \delta^{(2n)}(z)}{(2n)!\lambda^{\frac{2n+1}{2}}}, \tag{3.6.20}$$

where $\sigma = \operatorname{sgn}(\Im m\, b - \Im m\, a) = \pm 1$.

The analysis near any first order col is very similar to the previous example. Actually, a simple change of variables in (3.6.20) gives the expansion in that case.

Let $f'(z_0) = 0$, $f''(z_0) \neq 0$, where $f = u + iv$ is analytic near $z = z_0$. Then, in a small neighborhood of $z = z_0$ we can write $f(z) = f(z_0) + \rho(z)^2$, where ρ is a bijection to a neighborhood of $z = 0$. Then if C is a curve from points $z = a$ and $z = b$ located on different valleys for $|\,e^{\lambda f(z)}\,|$ near $z = z_0$, we have

$$I(e^{\lambda f(z)}; C) = e^{\lambda f(z_0)} I(e^{\lambda \rho(z)^2}; C)$$

$$\sim \sigma i e^{\lambda f(z_0)} \sum_{n=0}^{\infty} \frac{(-1)^n \Gamma\left(\frac{2n+1}{2}\right) \delta^{(2n)}(\rho(z))}{(2n)!\lambda^{\frac{2n+1}{2}}},$$

and choosing the branch $\rho(z) = \sqrt{f(z) - f(z_0)}$ in such a way that $\sigma = \operatorname{sgn}(\Im m(f(b) - f(a))) = 1$, we obtain

$$I(e^{\lambda f(z)};\,C) \sim i e^{\lambda f(z_0)} \sum_{n=0}^{\infty} \frac{(-1)^n \Gamma\left(\frac{2n+1}{2}\right) \delta^{(2n)}(\sqrt{f(z) - f(z_0)})}{(2n)!\lambda^{\frac{2n+1}{2}}}. \tag{3.6.21}$$

The first order approximation

$$I(e^{\lambda f(z)}; C) \sim e^{\lambda f(z_0)} \sqrt{\frac{-2\pi}{\lambda f''(z_0)}} \delta(z - z_0), \tag{3.6.22}$$

is the distributional version of the so-called *saddle point approximation*

$$\int_C e^{\lambda f(z)} g(z)\,dz \sim e^{\lambda f(z_0)} \sqrt{\frac{-2\pi}{\lambda f''(z_0)}} g(z_0). \tag{3.6.23}$$

The asymptotic approximation of $I(e^{\lambda f(z)}; C)$ near a col of higher order can be obtained by using the same ideas. Let us start with the case $f(z) = z^k$, where C is a contour from the saddle point $z = 0$ to the point $z = b$, which is located in a valley. Observe that on each valley there is exactly one root of the equation $\omega^k = -1$: the ray with parametric equation $z = t\omega$, $t > 0$, is the steepest path leading to the saddle point on that valley. We denote ω as $\omega(b)$ when we want to indicate that ω and b belong to the same valley. We deform C into a contour consisting of a part C_1 from 0 to ω along the steepest path and a part C_2 from ω to b. Then, as $\lambda \to \infty$,

$$\int_C e^{\lambda z^k} g(z)dz = \int_{C_1} e^{\lambda z^k} g(z)dz + \int_{C_2} e^{\lambda z^k} g(z)dz$$

$$\sim \int_{C_1} e^{\lambda z^k} g(z)dz$$

$$\sim \omega \int_0^1 e^{-\lambda t^k} g(t\omega)dt$$

$$\sim \sum_{n=0}^{\infty} \frac{\Gamma\left(\frac{n+1}{k}\right) \omega^{n+1} g^{(n)}(0)}{kn!\lambda^{\frac{n+1}{k}}},$$

since

$$H(1-t)H(t)e^{-\lambda t^k} \sim H(t)e^{-\lambda t^k} \sim \sum_{n=0}^{\infty} \frac{(-1)^n \Gamma\left(\frac{n+1}{k}\right) \delta^{(n)}(t)}{kn!\lambda^{\frac{n+1}{k}}}.$$

Therefore

$$I(e^{\lambda z^k}; C) \sim \sum_{n=0}^{\infty} \frac{(-1)^n \Gamma\left(\frac{n+1}{k}\right) \omega(b)^{n+1} \delta^{(n)}(z)}{kn!\lambda^{\frac{n+1}{k}}}, \qquad (3.6.24)$$

for a path from $z = 0$ to $z = b$.

The case of a contour from points $z = a$ to $z = b$, located on different valleys follows from (3.6.24) as

$$I(e^{\lambda z^k}; C) \sim \sum_{n=0}^{\infty} \frac{(-1)^n \Gamma\frac{(n+1)}{k}(\omega(b)^{n+1} - \omega(a)^{n+1})\delta^{(n)}(z)}{kn!\, \lambda^{\frac{n+1}{k}}}. \qquad (3.6.25)$$

In the general case of a col of order k for the function $f(z)$ at $z = z_0$, we write $f(z) = f(z_0) + \rho(z)^k$ to obtain

$$I(e^{\lambda f(z)}; C) \sim e^{\lambda f(z_0)} \sum_{n=0}^{\infty} \frac{(-1)^n \Gamma\left(\frac{n+1}{k}\right) (\omega(\rho(b))^{n+1} \delta^{(n)}(\rho(z)))}{kn!\lambda^{\frac{n+1}{k}}} \qquad (3.6.26)$$

for a path from $z = z_0$ to the point $z = b$ on a valley, and

$$I(e^{\lambda z^k}; C) \sim e^{\lambda f(z_0)} \sum_{n=0}^{\infty} \frac{(-1)^n \Gamma\left(\frac{n+1}{k}\right) (\omega(\rho(b))^{n+1} - \omega(\rho(a))^{n+1})\delta^{(n)}(\rho(z))}{kn!\lambda^{\frac{n+1}{k}}}$$

$$(3.6.27)$$

for a path from the points $z = a$ to $z = b$, located on different valleys.

Example 65. Let us consider the integral

$$J(\lambda) = \frac{1}{2\pi i} \int_C \frac{e^{\lambda(z - \sqrt{z})}}{z} dz, \tag{3.6.28}$$

where the branch cut is placed along the negative real axis and where C is a path going from infinity in the fourth quadrant to infinity in the third quadrant.

Here $f(z) = z - \sqrt{z}$ and we have $f'(z) = 0$ only at $z = \frac{1}{4}$, which is a first order col. The level lines through $z = \frac{1}{4}$ are obtained from the equation $\Re e\, f(z) = \frac{1}{4}$: they are the two parabolas $y = \frac{1}{8} - 2x^2$ and $y = 2x^2 - \frac{1}{8}$. The steepest paths through the col are given by $\Im m\, f(z) = 0$. The positive real axis is a steepest path, where $f(z)$ has a minimum at $z = \frac{1}{4}$. The other steepest path is the parabola $P : x = \frac{1}{4} - y^2$, which goes from one valley to the other through the col. Therefore, it is convenient to deform the path of integration to P; fortunately, the contour C can be deformed to P without changing the value of the integral $J(\lambda)$, since the integrand is so small that there is no contribution at infinity.

Thus

$$J(\lambda) = \frac{1}{2\pi i} \int_C \frac{e^{\lambda(z - \sqrt{z})}}{z} dz$$

$$= \frac{1}{2\pi i} \int_P \frac{e^{\lambda(z - \sqrt{z})}}{z} dz$$

$$= \frac{2}{\pi} e^{\frac{-\lambda}{4}} \int_{-\infty}^{\infty} \frac{e^{-\lambda t^2}}{1 + 2it} dt$$

$$\sim \frac{2}{\pi} e^{\frac{-\lambda}{4}} \sum_{n=0}^{\infty} \frac{\Gamma(n + \frac{1}{2})(-4)^n}{\lambda^{n + \frac{1}{2}}}, \qquad \text{as } \lambda \to \infty, \tag{3.6.29}$$

where on P we used the parametrization $z = (\frac{1}{2} + it)^2$, $t \in \mathbb{R}$.

Up to now we have dealt with a real large parameter λ. The case when λ is a complex parameter with large absolute value can be handled by setting $\lambda = \xi s$, where $|\xi| = 1$ and $s = |\lambda| > 0$. We are thus lead to the consideration of the functional $I(e^{s\xi f(z)}; C)$. The function $\xi f(z)$ has the same saddle points as $f(z)$, but the multiplicative constant ξ produces a rotation on the pattern of hills and valleys. Since the expansion depends on the location of the endpoint of C with respect to hills and valleys, the development will have a change in form when the rotation makes the endpoint cross the level lines: this gives the Stokes lines for the development of $I(e^{\lambda f(z)}; C)$. Let us consider an example.

Example 66. Let us find the development of $I(e^{\lambda z^2}; [0, 1])$, as $|\lambda| \to \infty$, where $[0, 1]$ is the segment from $z = 0$ to $z = 1$ on the real axis. Writing $\lambda = \xi s$, with $|\xi| = 1$ and $s > 0$, we consider the function ξz^2. The level lines around the only col, $z = 0$, are the lines $\arg z = -\frac{1}{2}\arg \xi + \frac{\pi}{4} + \frac{n\pi}{2}$. It follows that as long as $|\arg \xi| < \pi$, that is, as long as $\Re \lambda > 0$, then the endpoint $z = 1$ belongs to a hill and thus

$$\int_0^1 e^{\lambda z^2} g(z)dz \sim e^{\lambda} g(1), \qquad \Re \lambda > 0. \tag{3.6.30}$$

On the other hand, if $\frac{\pi}{2} < \arg \xi < \frac{3\pi}{2}$, that is, if $\Re \lambda < 0$, then 1 belongs to a valley and the saddlepoint approximation

$$\int_0^1 e^{\lambda z^2} g(z)dz \sim \sqrt{\frac{-\pi}{4\lambda}}\, g(0), \qquad \Re \lambda < 0, \tag{3.6.31}$$

applies.

The rays $z = \pm t i$, $t > 0$, are the Stokes' lines for $I(e^{\lambda z^2}; [0, 1])$. Through them the approximation changes from $e^{\lambda}\delta(z - 1)$ to $\sqrt{\frac{-\pi}{4\lambda}}\,\delta(z)$.

We finish this section by discussing the asymptotic behavior of some important special functions, the Airy functions.

Example 67. The Airy functions can be defined as the solutions of the second order differential equation

$$A''(z) = z\, A(z). \tag{3.6.32}$$

It follows that they are entire functions of the complex variable. The two linearly independent solutions of (3.6.32) are usually taken as

$$A i(z) = \sum_{k=0}^{\infty} \frac{z^{3k}}{3^{2k+\frac{2}{3}} k!\, \Gamma(k + \frac{2}{3})} - \sum_{k=0}^{\infty} \frac{z^{3k+1}}{3^{2k+\frac{4}{3}} k!\, \Gamma(k + \frac{4}{3})}, \tag{3.6.33a}$$

$$Bi(z) = 3^{\frac{1}{2}}\left[\sum_{k=0}^{\infty} \frac{z^{3k}}{3^{2k+\frac{2}{3}} k!\, \Gamma(k + \frac{2}{3})} - \sum_{k=0}^{\infty} \frac{z^{3k+1}}{3^{2k+\frac{4}{3}} k!\, \Gamma(k + \frac{4}{3})}\right]. \tag{3.6.33b}$$

They can also be related to the Bessel functions of imaginary argument of order $\frac{1}{3}$.

For our purposes, it is convenient to introduce the Airy functions $A_+(z)$ and $A_-(z)$ defined by

$$A_{\pm}(z) = \frac{1}{2\pi i} \int_{L_{\pm}} e^{tz - \frac{t^3}{3}}\, dt, \tag{3.6.34}$$

where $L\pm$ are the contours indicated in the Figure 3.

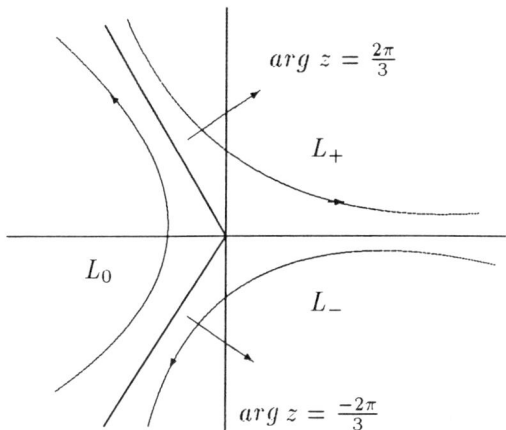

$arg \ z = \frac{2\pi}{3}$

L_+

L_0

L_-

$arg \ z = \frac{-2\pi}{3}$

Figure 3

We observe that [27], [102]

$$Ai(z) = \frac{1}{2\pi i} \int_{L_0} e^{tz - \frac{t^3}{3}} dt, \qquad (3.6.35)$$

and that these Airy functions are related by

$$Ai(z) = -A_+(z) - A_-(z), \qquad (3.6.36a)$$

$$Bi(z) = iA_+(z) - iA_-(z). \qquad (3.6.36b)$$

Let us now obtain the expansion of $Ai(z)$ as $\mid z \mid \to \infty$. First, we change the variables in (3.6.35) as $s = z^{\frac{1}{2}} t$, where the principal branch of the square root is taken, to obtain

$$Ai(z) = \frac{z^{\frac{1}{2}}}{2\pi i} \int_{L_0} e^{z^{\frac{3}{2}}(s - \frac{s^3}{3})} ds. \qquad (3.6.37)$$

We now apply the method of steepest descent. The saddle points are the roots of $f'(s) = 1 - s^2 = 0$, that is $s = \pm 1$; they are of the first order. The steepest lines through $s = -1$ are the real axis, which goes from hill to hill, and the branch of the hyperbola $y^2 - 3x^2 = 3$, which goes from valley to valley. Through $s = 1$ the situation is similar, but the real axis goes from valley to valley while the branch of the hyperbola goes from hill to hill. For the function $Ai(z)$, we deform L_0 to the branch of the hyperbola

$y^2 - 3x^2 = 3$, which we parametrize as $s - \frac{s^3}{3} = \frac{-2}{3} + \rho^2$, that is, as $\rho = (s+1)\sqrt{\frac{2}{3} - \frac{s}{3}}$. Thus

$$Ai(z) = \frac{z^{\frac{1}{2}}}{2\pi i} \int_{-\infty}^{\infty} e^{z^{\frac{3}{2}}(\frac{-2}{3}+\rho^2)} \frac{ds}{d\rho} d\rho$$

$$\sim \frac{e^{\frac{-2}{3}z^{\frac{3}{2}}}}{2\sqrt{\pi}z^{\frac{1}{4}}}. \tag{3.6.38}$$

This formula holds if z is real, but remains valid if $| \arg z^{\frac{3}{2}} | < \frac{3\pi}{2}$, that is, as long as $-\pi < \arg z < \pi$. The imaginary axis, which is the branch cut of $z^{\frac{1}{4}}$ and $z^{\frac{3}{2}}$, is the Stokes line.

Using the simple formulas

$$A_+(z) = e^{\frac{-2\pi i}{3}} Ai(ze^{\frac{-2\pi i}{3}}),$$

$$A_-(z) = e^{\frac{2\pi i}{3}} Ai\left(ze^{\frac{2\pi i}{3}}\right),$$

we derive the expansions

$$A_+(z) \sim \frac{e^{\frac{2}{3}z^{\frac{3}{2}}}}{2\sqrt{\pi}iz^{\frac{1}{4}}}, \qquad \frac{-\pi}{3} < \arg z < \frac{5\pi}{3}, \tag{3.6.39}$$

$$A_-(z) \sim \frac{-e^{\frac{2}{3}z^{\frac{3}{2}}}}{2\sqrt{\pi}i\,z^{\frac{1}{4}}}, \qquad \frac{\pi}{3} < \arg z < \frac{7\pi}{3}. \tag{3.6.40}$$

In particular, using (3.6.36a) we obtain the expansion of $Ai(z)$ along the Stokes line as

$$Ai(-x) \sim \frac{1}{\sqrt{\pi x}} \cos\left(\frac{2}{3}x^{\frac{3}{2}} - \frac{\pi}{4}\right), \qquad \text{as } x \to +\infty. \tag{3.6.41}$$

3.7 Expansion of Oscillatory Kernels

We shall now consider the moment asymptotic expansion for distributions of the space $\mathcal{O}'_\gamma$ and $\mathcal{O}'_C = \bigcap_{\gamma \in \mathbb{R}} \mathcal{O}'_\gamma$. The full moment expansion is valid in the space $\mathcal{O}'_C$, while a finite order version is valid in the space $\mathcal{O}'_\gamma$. The space $\mathcal{O}'_C$ contains not only kernels of rapid decay but also rapidly oscillating kernels, and thus many important particular cases are obtained. Actually, a simple change of variables permits us to extend the range of applicability of our results to a very large variety of situations.

Recall that a test function $\phi(x)$ belongs to $\mathcal{O}_\gamma(\mathbb{R})$ if it is smooth and if $\phi^{(j)}(x) = O(|x|^\gamma)$ as $|x| \to \infty$ for every $j \in \mathbb{N}$. The family of seminorms

$$\|\phi\|_{j,\gamma} = \sup\{\rho_\gamma(|x|)|\phi^{(j)}(x)| : x \in \mathbb{R}\}, \qquad (3.7.1)$$

where

$$\rho_\gamma(r) = \begin{cases} 1, & 0 \leq r \leq 1, \\ r^{-\gamma}, & r > 1, \end{cases} \qquad (3.7.2)$$

generates a topology for $\mathcal{O}_\gamma(\mathbb{R})$.

Lemma 3. *Let $X_q = \{\phi \in \mathcal{O}_\gamma : \phi^{(j)}(0) = 0, \ j < q\}$. Then if $q \leq \gamma$, for every continuous seminorm $\|\ \|_1$ there exists another continuous seminorm $\|\ \|_2$ such that if $\phi \in X_q$,*

$$\|\phi(\tfrac{x}{\lambda})\|_1 \leq \|\phi(x)\|_2 \lambda^{-q}, \qquad (3.7.3)$$

for $\lambda > \lambda_0$. In particular, $\|\phi(x/\lambda)\|_1 = O(\lambda^{-q})$ as $\lambda \to \infty$.

Proof. Let $\phi \in X_q$. Then there exists a constant K such that

$$|\phi(x)| \leq K|x|^q \ , \quad |x| \leq 1, \qquad (3.7.4a)$$

$$|\phi(x)| \leq K|x|^\gamma \ , \quad |x| \geq 1. \qquad (3.7.4b)$$

Since $q \leq \gamma$ it follows that if $\lambda \geq 1$ then

$$\|\phi(\tfrac{x}{\lambda})\|_{0,\gamma} \leq K/\lambda^q. \qquad (3.7.5)$$

Observing that

$$\|\phi(\tfrac{x}{\lambda})\|_{j,\gamma} = \lambda^{-j} \|\phi^{(j)}\left(\tfrac{x}{\lambda}\right)\|_{0,\gamma},$$

and that $\phi^{(j)} \in X_{q-j}$ if $j \leq q$, while $\phi^{(j)} \in X_0$ if $j > q$ it follows that

$$\|\phi(\tfrac{x}{\lambda})\|_{j,\gamma} = O\left(\frac{1}{\lambda^q}\right), \quad \text{if } j \leq q,$$

$$\|\phi(\tfrac{x}{\lambda})\|_{j,\gamma} = O\left(\frac{1}{\lambda^j}\right), \quad \text{if } j > q. \qquad \blacksquare$$

Hence,

Theorem 27. *Let $f \in \mathcal{O}_\gamma(\mathbb{R})$ and let $N = [\![\gamma]\!] - 1$. Then*

$$f(\lambda x) = \sum_{n=0}^{N} \frac{(-1)^n \mu_n \delta^{(n)}(x)}{n! \lambda^{n+1}} + O\left(\frac{1}{\lambda^{N+2}}\right), \quad as \ \lambda \to \infty, \qquad (3.7.6)$$

in the space $\mathcal{O}_\gamma(\mathbb{R})$.

Next, since $\mathcal{O}_C' = \cap \, \mathcal{O}_\gamma'$, we immediately obtain.

Theorem 28. *Let $f \in \mathcal{O}_C'(\mathbb{R})$. Then*

$$f(\lambda x) \sim \sum_{n=0}^{\infty} \frac{(-1)^n \mu_n \delta^{(n)}(x)}{n! \lambda^{n+1}}, \quad as \ \lambda \to \infty, \qquad (3.7.7)$$

in the space $\mathcal{O}_C(\mathbb{R})$.

Introducing, as before, the spaces $\mathcal{O}_{\gamma,p}\{\phi_n\}$, and $\mathcal{O}_{C,p}\{\phi_n\}$ we obtain the same results as in the spaces $\mathcal{E}'$ and $\mathcal{P}'$; that is,

Theorem 29. *If $f \in \mathcal{O}'_{\gamma,p}\{(\ln x)^r x^{\alpha_j}, \ 0 \le r \le k_j\}$, then*

$$f(\lambda x) \sim \sum_{j=0}^{N} \sum_{t=0}^{k_j} (-1)^{k_j - t} \left(\sum_{r=0}^{t} \binom{k_j - r}{t - r} \mu(t - r_j, \alpha_j) \, \delta_{k_j - r, j}(x)\right)$$

$$(\ln \lambda)^{k_j - t} \lambda^{\alpha_j - 1}, \quad as \ \lambda \to \infty, \qquad (3.7.8)$$

where $\delta_{r,j}(x) = (\ln x)^{-r} x^{-\alpha_j} \delta(x)$ and where $\mu(r, \alpha_j) = \ <f(x), (\ln x)^r x^{\alpha_j}>$ if $\alpha_{N+1} < \gamma$. If $f \in \mathcal{O}'_{C,p}\{(\ln x)^r x^{\alpha_j}\}$ then (3.7.8) holds for every N.

The moment asymptotic expansion holds in other cases as well. Let ϕ be smooth. Let $f \in \mathcal{D}'$ and write $f = f_0 + f_1$ where $f_0 \in \mathcal{E}'$ and 0 is not in the support of f_1. Then $f_1(x^\beta)$ is well defined for any $\beta > 1$, and if it happens that $f_1(x^\beta)$ is in $\mathcal{O}_C'$ then the change $x = u^\beta$ yields

$$< f(\lambda x), \phi(x) > \ = \ < f_0(\lambda x), \phi(x) > + \ < f_1(\lambda^{1/\beta} u^\beta), \psi(u) >,$$

where $\psi(u) = \beta u^{\beta - 1} \phi(u^\beta)$. It follows that the moment asymptotic expansion holds if $\beta(u)\psi(u) \in \mathcal{O}_C$ for any cut-off smooth function with $\beta(x) = 0$, $|x| \le \frac{1}{2}$, $\beta(x) = 1$, $|x| \ge 1$ (this is always the case if $\phi \in \mathcal{S}$). In particular, if $x\phi(x^2) \in \mathcal{O}_C$ then $\phi \in \mathcal{K}$, that is, if $\phi^{(n)}(x) = O(|x|^{\gamma - m})$ as $|x| \to \infty$ for some γ. Observe that $f \in \mathcal{K}'$ precisely when $f_1(x^2) \in \mathcal{O}_C'$. A typical element of $\mathcal{K}'$ is the oscillatory kernel e^{ix}. Of course we also have a corresponding expansion in the space $\mathcal{K}_p'\{(\ln x)^r x^{\alpha_j}\}$.

Example 68. If $\rho > 1$ the functions $H(\pm x)e^{\pm i|x|^\rho}$ as well as their combinations $e^{\pm i|x|^\rho}$ and $e^{\pm i\operatorname{sgn}(x)|x|^\rho}$ belong to $\mathcal{O}'_C$. Actually, as explained above, a simple change of variables shows that the moment asymptotic expansion holds as long as $\rho > 0$. Using the values

$$\int_0^\infty x^\alpha e^{ix^\rho}\,dx = \frac{1}{\rho}\Gamma\left(\frac{\alpha+1}{\rho}\right)e^{\frac{\pi i(\alpha+1)}{2\rho}}, \quad \alpha \neq -\rho, -2\rho, -3\rho, \cdots,$$

we readily obtain the expansions

$$H(x)e^{\pm i\lambda x^\rho} \sim \frac{1}{\rho}\sum_{n=0}^\infty \frac{(-1)^n\Gamma\left(\frac{n+1}{\rho}\right)e^{\pm\frac{\pi i(n+1)}{2\rho}}\delta^{(n)}(x)}{n!\,\lambda^{\frac{n+1}{\rho}}}, \tag{3.7.9}$$

and

$$H(-x)e^{\pm i\lambda|x|^\rho} \sim \frac{1}{\rho}\sum_{n=0}^\infty \frac{\Gamma\left(\frac{n+1}{\rho}\right)e^{\pm\frac{\pi i(n+1)}{2\rho}}\delta^{(n)}(x)}{n!\,\lambda^{\frac{n+1}{\rho}}}, \tag{3.7.10}$$

and consequently

$$e^{\pm i\lambda|x|^\rho} \sim \frac{2}{\rho}\sum_{n=0}^\infty \frac{\Gamma\left(\frac{2n+1}{\rho}\right)e^{\pm\frac{\pi i(n+1)}{2\rho}}\delta^{(2n)}(x)}{n!\,\lambda^{\frac{2n+1}{\rho}}}, \tag{3.7.11}$$

$$e^{\pm i\lambda\operatorname{sgn}(x)|x|^\rho} \sim \frac{1}{\rho}\sum_{n=0}^\infty \frac{\Gamma\left(\frac{n+1}{\rho}\right)\left[(-1)^n e^{\pm\frac{\pi i(n+1)}{2\rho}} + e^{\mp\frac{\pi i(n+1)}{2\rho}}\right]\delta^{(2n)}(x)}{n!\,\lambda^{\frac{2n+1}{\rho}}}. \tag{3.7.12}$$

Example 69. Formulas (3.7.9) through (3.7.12) permit us to obtain the asymptotic development of oscillatory integrals of the type

$$\Phi(\lambda) = \int_{-\infty}^\infty e^{i\lambda h(t)}\phi(t)\,dt, \quad \text{as} \quad \lambda \to \infty, \tag{3.7.13}$$

by simple substitution. The main contributions to the integral $\Phi(\lambda)$ come from the critical points of $h(t)$ if $\phi(t)$ is smooth. By using the standard arguments involving partitions of unity we can reduce matters to the case when $h(t)$ has a single critical point at $t = a$. If we suppose that for t near a we have $h(t) \sim h(a) \pm c|t-a|^\rho$ then we can write $h(t)$ as either $h(t) = h(a) \pm |\psi(t)|^\rho$ or $h(t) = h(a) \pm \operatorname{sgn}(t)|\psi(t)|^\rho$, where $\psi(t)$ is smooth, strictly increasing and $\psi(a) = 0$. Therefore, the asymptotic evaluation of the integral (3.7.13) requires the development of $H(\pm(t-a))e^{\pm i|\psi(t)|^\rho}$ (If the critical point is an endpoint) or of $e^{\pm i|\psi(t)|^\rho}$ or $e^{\pm i \operatorname{sgn}(t)|\psi(t)|^\rho}$ (If the critical point is interior). But this follows directly from (3.7.9) through (3.7.12) by setting $x = \psi(t)$. For instance, from (3.7.9) we obtain

$$H(t-a)e^{i\lambda h(t)} = H(t-a)e^{i\lambda[h(a)+|\psi(t)|^\rho]}$$

$$\sim \frac{e^{i\lambda h(a)}}{\rho} \sum_{n=0}^{\infty} \frac{(-1)^n \Gamma\left(\frac{n+1}{\rho}\right) e^{\frac{\pi i (n+1)}{2\rho}} \delta^{(n)}(\psi(t))}{n! \, \lambda^{\frac{n+1}{\rho}}} \tag{3.7.14}$$

$$= \frac{e^{i\lambda h(a)}}{\rho} \left[\frac{\Gamma\left(\frac{1}{\rho}\right) e^{\frac{\pi i}{2\rho}} \delta(x-a)}{n! \lambda^{\frac{1}{\rho}} \psi'(a)} + O\left(\frac{1}{\lambda^{\frac{2}{\rho}}}\right) \right].$$

A similar analysis can be applied in other cases.

A very important class of oscillatory generalized functions is that formed by the periodic distributions of zero mean. Any of these distributions belongs to $\mathcal{K}'$. Indeed, if f is a periodic generalized function with zero mean then for each n there exists another periodic generalized function with zero mean such that $g^{(n)} = f$ and if n is large enough, g is continuous. Therefore, the formula

$$< f, \phi > = (-1)^n < g, \phi^{(n)} > , \tag{3.7.15}$$

defines the values of f at ϕ if $\phi \in \mathcal{K}$ since if n is large enough then $\phi^{(n)}(x) = O(|x|^{-2})$ as $|x| \to \infty$, and thus $< g, \phi^{(n)} > = \int_{-\infty}^{\infty} g(x)\phi^{(n)}(x)dx$ is a convergent integral.

All the moments of such periodic distributions with zero mean vanish. Indeed, if $\phi(x) = x^k$, by taking $n > k$ in (3.7.15) it follows that

$$< f(x), x^k > = 0 , \quad k = 0, 1, 2, 3, \cdots . \tag{3.7.16}$$

When the mean of the periodic distribution f is a non-zero constant c then f does not belong to $\mathcal{K}'$, but we can write $f = c + f_0$, where $f_0 \in \mathcal{K}$. Recall that c is the constant in the Fourier series of f.

Summarizing, we have the following result

Theorem 30. *Let f be a periodic function of period p, with mean $c = \frac{1}{p}$ $< f(x), H(x-\alpha)H(\alpha+p-x) >$. Then*

$$f(\lambda x) = c + o(\lambda^{-\infty}) , \quad as \quad \lambda \to \infty, \tag{3.7.17}$$

in the space $\mathcal{K}(\mathbb{R}) \cap L^1(\mathbb{R})$. If $c = 0$ then (3.7.17) holds in $\mathcal{K}(\mathbb{R})$.

Example 70. Using Theorem 30 we obtain that

$$\hat{\phi}\,(\lambda) = \int_{-\infty}^{\infty} e^{i\lambda x}\phi(x)dx = o(\lambda^{-\infty})\,, \quad \text{as } \lambda \to \infty\,, \tag{3.7.18}$$

whenever $\phi \in \mathcal{K}$. That (3.7.18) holds if $\phi \in \mathcal{S}$ is clear since $\hat{\phi} \in \mathcal{S}$. However, (3.7.18) might cease to hold in spaces larger than $\mathcal{K}$. For instance, if $\phi(x) = e^{ix^2}$ then $\phi \in \mathcal{O}_M$, but $\hat{\phi}(\lambda) = (1+i)\sqrt{\frac{\pi}{2}}e^{-i\frac{\lambda^2}{4}}$ does not satisfy (3.7.18).

Example 71. The measure $f(x) = \sum_{n=-\infty}^{\infty}(-1)^n\delta(x-n)$ is periodic of period 2. It has zero mean. Therefore $f(\lambda x) = O(\lambda^{-\infty})$ as $\lambda \to \infty$ in the space $\mathcal{K}(\mathbb{R})$. Set $\varepsilon = \frac{1}{\lambda}$. Then if $\phi \in \mathcal{K}$ we have

$$\sum_{n=-\infty}^{\infty} (-1)^n\phi(n\varepsilon) = O(\varepsilon^{\infty})\,, \quad \text{as } \varepsilon \to 0. \tag{3.7.19}$$

Observe that the series $\sum_{n=-\infty}^{\infty}(-1)^n\phi(n\varepsilon)$ might be divergent if $\phi \in \mathcal{K}$, and in that case its interpretation is via (3.7.15). As we shall see in Chapter 5, however, the series is always Abel and Cesàro summable if $\phi \in \mathcal{K}$.

Example 72. The measure $\sum_{n=-\infty}^{\infty} \delta(x-n)$ is periodic of period 1, but its mean is 1. Therefore,

$$\sum_{n=-\infty}^{\infty} \delta(\lambda x - n) = 1 + O(\lambda^{-\infty})\,, \quad \text{as } \lambda \to \infty\,, \tag{3.7.20}$$

and thus setting $\varepsilon = \frac{1}{\lambda}$,

$$\sum_{n=-\infty}^{\infty} \phi(n\varepsilon) = \frac{1}{\varepsilon}\int_{-\infty}^{\infty} \phi(x)dx + O(\varepsilon^{\infty})\,, \quad \text{as } \varepsilon \to 0^+\,, \tag{3.7.21}$$

whenever $\phi \in \mathcal{K}(\mathbb{R}) \cap L^1(\mathbb{R})$. In particular, if $\phi(x) = \frac{1}{1+x^2}$ then (3.7.21) yields the relation

$$\sum_{n=1}^{\infty} \frac{1}{1+\varepsilon^2 n^2} = \frac{\pi}{2\varepsilon} - \frac{1}{2} + O(\varepsilon^{\infty})\,, \quad \text{as } \varepsilon \to 0^+\,. \tag{3.7.22}$$

The moment aymptotic expansion also holds in the space $\mathcal{O}'_M$. This can be obtained by using Fourier transform arguments.

Theorem 31. *If $f \in \mathcal{O}'_M(\mathbb{R})$ then the moment asymptotic expansion*

$$f(\lambda x) \sim \sum_{n=0}^{\infty} \frac{(-1)^n \mu_n \delta^{(n)}(x)}{n! \, \lambda^{n+1}}, \quad \text{as } \lambda \to \infty,$$

holds.

We finish this section by giving the asymptotic expansion of finite Fourier transforms.

Example 73. The asymptotic evaluation of finite Fourier transforms of the type

$$\Phi(\lambda) = F.p. \int_a^b e^{i\lambda x} \phi(x) dx \tag{3.7.23}$$

was given by Erdelyi [28] using integration by parts.

Here we assume that ϕ is smooth in (a, b) and has developments of the type

$$\phi(a + x) \sim \sum_{j=0}^{\infty} (A_j + B_j \ln x) x^{\alpha_j}, \quad \text{as } x \to 0, \tag{3.7.24a}$$

$$\phi(b - x) \sim \sum_{j=0}^{\infty} (C_j + D_j \ln x) x^{\alpha_j}, \quad \text{as } x \to 0, \tag{3.7.24b}$$

where $\Re e \alpha_n \nearrow \infty$. A simple argument involving neutralizers shows that we have $\Phi(\lambda) \sim \Phi_a(\lambda) + \Phi_b(\lambda)$, where $\Phi_a(\lambda) = e^{i\lambda a} F.p. \int_0^\infty e^{i\lambda x} \phi_1(x) dx$, $\Phi_b(\lambda) = e^{i\lambda b} F.p. \int_0^\infty e^{i\lambda x} \phi_2(x) dx$ and where $\phi_1(x)$ and $\phi_2(x)$ have the same developments, respectively, as those of $\phi(x + a)$ and $\phi(b - x)$, as $x \to 0$. But this follows as

$$\Phi_a(\lambda) \sim e^{i\lambda a} \sum_{j=0}^{\infty} [A_j \mu(\alpha_j) + B_j(\mu'(\alpha_j) - \mu(\alpha_j) \ln \lambda)] \lambda^{-\alpha_j - 1}, \tag{3.7.25a}$$

$$\Phi_b(\lambda) \sim e^{i\lambda b} \sum_{j=0}^{\infty} [C_j \nu(\alpha_j) + D_j(\nu'(\alpha_j) - \nu(\alpha_j) \ln \lambda)] \lambda^{-\alpha_j - 1}, \tag{3.7.25b}$$

as $\lambda \to \infty$, where

$$\mu(\alpha) = F.p \int_0^\infty x^\alpha e^{ix} dx = \Gamma(\alpha + 1) e^{\frac{\pi i(\alpha+1)}{2}},$$

$$\nu(\alpha) = F.p. \int_0^\infty x^\alpha e^{-ix} dx = \Gamma(\alpha + 1) e^{\frac{-\pi i(\alpha+1)}{2}},$$

and where the finite part values are used if $\alpha = -1, -2, -3 \cdots$.

3.8 The Expansion of $f(\lambda x)$ as $\lambda \to \infty$ In Other Cases

When $f(x)$ does not decay rapidly at infinity then the moment asymptotic expansion does not hold. For instance, if $f(x) = (x - x_0)_+^\alpha$, $\alpha \notin \mathbb{Z}$, equation (3.2.5) yields

$$f(\lambda x) = (\lambda x - x_0)_+^\alpha \sim \sum_{k=0}^{\infty} (-1)^n \binom{\alpha}{n} x_0^n \frac{x_+^{\alpha-n}}{\lambda^{n-\alpha}} \;, \qquad \text{as } \lambda \to \infty. \qquad (3.8.1)$$

Similarly,

$$\sum_{n=1}^{\infty} \delta(\lambda x - n) \sim H(x) + \sum_{n=0}^{\infty} \frac{(-1)^n \zeta(-n) \delta^{(n)}(x)}{n! \, \lambda^{n+1}} \;, \qquad \text{as } \lambda \to \infty, \qquad (3.8.2)$$

and even more clearly, $\ln |\lambda x| = \ln \lambda + \ln |x|$, where $\lambda > 1$.

By adding some restrictions on the behavior of $f(x)$ at infinity it is possible to obtain the expansion of $f(\lambda x)$ as $\lambda \to \infty$ in terms of homogeneous and associated generalized functions. Whereas the moment asymptotic development of a distribution of $\mathcal{P}'$ or $\mathcal{O}'_C$ contains only integral powers of λ, the related expansions of distributions of $\mathcal{S}'$ will contain not only arbitrary powers of λ but logarithmic terms as well.

In order to simplify the notation we shall deal with generalized functions whose support is bounded on the left. This amounts to considering the situation at $x = +\infty$ only; the analysis at $x = -\infty$ is completely analogous. We shall use the following notation: if $f \in \mathcal{S}'(\mathbb{R})$ we shall write $f(x) = O(g(x))$ as $x \to \infty$ if there is a constant A such that $f(x)$ is an ordinary function for $x > A$ and the order relation holds. Similar remarks apply to the notation $f(x) = o(g(x))$ as $x \to \infty$.

Lemma 4. *Let $f \in \mathcal{S}'(\mathbb{R})$ with support bounded on the left and such that $f(x) = O(x^\beta)$ as $x \to \infty$ where $\beta > -1$. Then*

$$f(\lambda x) = O(\lambda^\beta), \qquad \text{as } \lambda \to \infty \quad \text{in } \mathcal{S}'(\mathbb{R}). \qquad (3.8.3)$$

Proof. It is possible to find a constant M such that we can write $f = f_0 + f_1$, where supp f_0 is compact and where f_1 is an ordinary function with support in $[0, \infty)$ and with $|f_1(x)| \le M x^\beta$, $x \ge 0$. Since f_0 has compact support, we have

$$f_0(\lambda x) \sim \frac{\mu_0 \delta(x)}{\lambda} - \frac{\mu_1 \delta'(x)}{\lambda^2} + \cdots,$$

hence $f_0(\lambda x) = O(\lambda^{-1})$ and since $\beta > -1$, $f_0(\lambda x) = O(\lambda^\beta)$ as $\lambda \to \infty$.
On the other hand, if $\phi \in \mathcal{S}$ we have

$$| < f_1(\lambda x), \phi(x) > | \leq \int_0^\infty |f_1(\lambda x)||\phi(x)|dx \leq \left[M \int_0^\infty x^\beta \mid \phi(x) \mid dx \right] \lambda^\beta,$$

and hence $f_1(\lambda x) = O(\lambda^\beta)$ as $\lambda \to \infty$. ∎

A similar analysis yields the following lemma.

Lemma 5. *Let $f \in \mathcal{S}'(\mathbb{R})$ with support bounded on the left and such that $f(x) = o(x^\beta)$, as $x \to \infty$, where $\beta > -1$. Then*

$$f(\lambda x) = o(\lambda^\beta), \quad \text{as } \lambda \to \infty. \tag{3.8.4}$$

From Lemmas 4 and 5 we immediately obtain:

Lemma 6. *Let $f \in \mathcal{S}'(\mathbb{R})$ with support bounded on the left and such that $f(x) = b_1 x^{\beta_1} + \cdots + b_n x^{\beta_n} + o(x^{\beta_n})$ as $x \to \infty$, where $\beta_1 > \beta_2 > \cdots > \beta_n > -1$. Then*

$$f(\lambda x) = b_1 \lambda^{\beta_1} x_+^{\beta_1} + \cdots + b_n \lambda^{\beta_n} x_+^{\beta_n} + o(\lambda^{\beta_n}). \tag{3.8.5}$$

The situation for $\beta \leq -1$ is similar, except that the moment asymptotic expansion has to be taken into account.

Lemma 7. *Let $f \in \mathcal{S}'(\mathbb{R})$ with support bounded on the left and such that $f(x) = O(x^\beta)$ as $x \to \infty$, where $-(k+1) > \beta > -(k+2)$ for some $k \in \mathbb{N}$. Then*

$$f(\lambda x) = \sum_{j=0}^{k} \frac{(-1)^j \mu_j \delta^{(j)}(x)}{j! \, \lambda^{j+1}} + O(\lambda^\beta), \quad \text{as } \lambda \to \infty. \tag{3.8.6}$$

Proof. We can find a constant M and a decomposition $f = f_0 + f_1$, where f_0 has compact support, supp $f_1 \subseteq [0, \infty)$ and $|f_1(x)| \leq M x^\beta$, $x \geq 0$. Since supp f_0 is compact, $f_0(\lambda x)$ admits the moment expansion

$$f_0(\lambda x) = \sum_{j=0}^{k} \frac{(-1)^j \mu_j(f_0) \delta^{(j)}(x)}{j! \, \lambda^{j+1}} + O(\lambda^{-(k+2)}), \quad \text{as } \lambda \to \infty.$$

for any $k \in \mathbb{N}$. It thus suffices to show that (3.8.6) holds for f_1.

Let $\phi \in \mathcal{S}(\mathbb{R})$, then

$$< f_1(\lambda x) - \sum_{j=0}^{k} \frac{(-1)^j \mu_j(f_1) \delta^{(j)}(x)}{j! \lambda^{j+1}}, \phi(x) >$$

$$= < f_1(\lambda x), \phi(x) > - \sum_{j=0}^{k} \frac{\phi^{(j)}(0)}{j! \lambda^{j+1}} \int_0^\infty f_1(x) x^j dx$$

$$= \int_0^\infty f_1(\lambda x) \left[\phi(x) - \sum_{j=0}^{k} \frac{\phi^{(j)}(0)}{j!} x^j \right] dx.$$

Since $\phi \in \mathcal{S}(\mathbb{R})$ we can find a constant K such that

$$|\phi(x) - \sum_{j=0}^{k} \frac{\phi^{(j)}(0)}{j!} x^j| \leq K x^{k+1}, \qquad 0 \leq x \leq 1, \tag{3.8.7a}$$

$$|\phi(x) - \sum_{j=0}^{k} \frac{\phi^{(j)}(0)}{j!} x^j| \leq K x^k, \qquad x \geq 1. \tag{3.8.7b}$$

It follows that

$$| < f_1(\lambda x) - \sum_{j=0}^{k} \frac{(-1)^j \mu_j(f_1) \delta^{(j)}(x)}{j! \lambda^{j+1}}, \phi(x) > |$$

$$\leq \int_0^1 M K \lambda^\beta x^\beta x^{k+1} dx + \int_1^\infty M K \lambda^\beta x^\beta x^k dx$$

$$\leq M K \left[\frac{1}{k+2+\beta} - \frac{1}{k+1+\beta} \right] \lambda^\beta. \qquad \blacksquare$$

A similar argument shows that this lemma remains valid if we replace the big O by a little o.

In order to obtain the general expansion, we have to define the notion of generalized moments of certain distributions $f \in \mathcal{S}'(\mathbb{R})$. If $f \in \mathcal{S}'(\mathbb{R})$ has support bounded on the left and

$$f(x) = b_1 x^{\beta_1} + \cdots + b_n x^{\beta_n} + O(x^\beta), \quad \text{as } x \to \infty, \tag{3.8.8}$$

where $-(k+1) > \beta > -(k+2)$ and $\beta_1 > \beta_2 > \cdots > \beta_n > \beta$ then the generalized moments $\mu_j(f)$, $0 \le j \le k$, are defined as

$$\mu_j(f) = \; < f(x) - \sum_{i=1}^{n} b_i g_i(x), x^j >, \qquad (3.8.9)$$

where $g_j(x) = x_+^{\beta_j}$ if $\beta_j \neq -1, -2, \cdots$ and $g_j(x) = \mathcal{P}f\,(x^{\beta_j} H(x))$ if $\beta_j = -1, -2, \cdots$.

In the case where $f(x)$ is a locally integrable function with support bounded on the left, the above definition reduces to

$$\mu_j(f) = \; \text{F.p.} \int_{-\infty}^{\infty} f(x) x^j \, dx, \qquad (3.8.10)$$

the finite part being taken at $x = \infty$ where the integral becomes divergent.

Using this definition and the results obtained so far, we immediately obtain

Theorem 32. *Let $f \in \mathcal{S}'(\mathbb{R})$ with support bounded on the left. Suppose*

$$f(x) = b_1 x^{\beta_1} + \cdots + b_n x^{\beta_n} + O(x^{\beta}), \qquad as \; x \to \infty, \qquad (3.8.11)$$

where $\beta_1 > \beta_2 > \cdots > \beta_n > \beta$, and $-(k+1) > \beta > -(k+2)$. Then

$$f(x) = \sum_{j=1}^{n} b_j g_j(\lambda x) + \sum_{j=0}^{k} \frac{(-1)^j \mu_j \delta^{(j)}(\lambda x)}{j!} + O(\lambda^{\beta}), \qquad as \; \lambda \to \infty,$$

$$(3.8.12)$$

in the space $\mathcal{S}'(\mathbb{R})$, where $g_j(x) = x_+^{\beta_j}$ if $\beta_j \neq -1, -2, \cdots$ and $g_j(x) = \mathcal{P}f\,(x^{\beta_j} H(x))$ if $\beta_j = -1, -2 \cdots$.

We shall now consider several examples.

Example 74. Let us consider the asymptotic development of the Stieltjes integral

$$\Phi(s) = \int_{0}^{\infty} \frac{\phi(x) dx}{x + s}, \qquad (3.8.13)$$

as $s \to 0^+$ and $\phi \in \mathcal{S}(\mathbb{R})$. We have

$$\Phi(s) = \frac{1}{s} \int_{0}^{\infty} \frac{\phi(x) dx}{x/s + 1} = \left(\frac{1}{s}\right) < f(x/s), \phi(x) >, \qquad (3.8.14)$$

where

$$f(x) = \frac{H(x)}{x + 1}. \qquad (3.8.15)$$

Since $f(x) \sim \frac{1}{x} - \frac{1}{x^2} + \frac{1}{x^3} - \cdots$ as $x \to \infty$ and since the generalized moments vanish, i.e., $\mu_n = F.p. \int_0^\infty \frac{x^n}{1+x} dx = 0$, $n = 0, 1, 2, \cdots$ it follows that $f(\lambda x)$ has the expansion

$$f(\lambda x) \sim \mathcal{P}f\left(\frac{H(\lambda x)}{(\lambda x)}\right) - \mathcal{P}f\left(\frac{H(\lambda x)}{(\lambda x)^2}\right) + \mathcal{P}f\left(\frac{H(\lambda x)}{(\lambda x)^3}\right) - \cdots,$$

or

$$f(\lambda x) \sim \sum_{n=0}^\infty \left\{ \frac{-\ln \lambda}{n!} \frac{\delta^{(n)}(x)}{\lambda^{n+1}} + \frac{(-1)^n}{\lambda^{n+1}} \mathcal{P}f\left(\frac{H(x)}{x^{n+1}}\right) \right\}, \quad \text{as } \lambda \to \infty.$$
(3.8.16)

Hence

$$\Phi(s) \sim \sum_{n=0}^\infty \left\{ \frac{(-1)^n \phi^{(n)}(0)}{n!} \ln s + (-1)^n F.p. \int_0^\infty \frac{\phi(x)}{x^{n+1}} dx \right\} s^n,$$

$$\text{as } s \to 0^+.$$
(3.8.17)

Example 75. For the generalized Stieltjes integral

$$\Phi_\alpha(s) = \int_0^\infty \frac{\phi(x)}{(x+s)^\alpha} dx, \qquad \alpha \notin \mathbb{Z}$$
(3.8.18)

we have $\Phi_\alpha(1/\lambda) = \lambda^\alpha < f(\lambda x), \phi(x) >$, where $f(x) = H(x)(x+1)^{-\alpha}$. But

$$f(x) \sim \sum_{n=0}^\infty \binom{\alpha+n}{n} \frac{1}{x^{\alpha+n}}, \qquad \text{as} \qquad x \to \infty,$$

while

$$\mu_n = F.p. \int_0^\infty \frac{x^n}{(1+x)^\alpha} dx = \sum_{j=0}^n \binom{n}{j}(-1)^{n-j}\left(\frac{1}{j-\alpha+1}\right) = \frac{n+1}{\binom{\alpha-1}{n+1}}.$$

Therefore

$$f(\lambda x) \sim \sum_{n=0}^\infty \binom{\alpha+n}{n} \frac{x_+^{-(\alpha+n)}}{\lambda^{\alpha+n}} + \sum_{n=0}^\infty \frac{(-1)^n \Gamma(\alpha) \delta^{(n)}(x)}{\Gamma(\alpha-n-1)\lambda^{n+1}}, \qquad \text{as } \lambda \to \infty,$$
(3.8.19)

and consequently,

$$\Phi_\alpha(s) \sim \sum_{n=0}^{\infty} \left\{ \binom{\alpha+n}{n} \text{F.p.} \int_0^\infty \frac{\phi(x)}{x^{\alpha+n}} dx + \frac{\Gamma(\alpha)}{\Gamma(\alpha-n-1)} \phi^{(n)}(0)s^{1-\alpha} \right\} s^n,$$

$$\text{as } s \to 0^+. \tag{3.8.20}$$

Example 76. Using our results and making a change of variables it is sometimes possible to obtain the asymptotic behavior of $f(\varepsilon y)$ as $\varepsilon \to 0^+$. As an example let us consider the behavior of the Laplace transform [26]

$$\Phi(\varepsilon) = \int_0^\infty e^{-\varepsilon y} \phi(y) dy, \tag{3.8.21}$$

as $\varepsilon \to 0^+$. Let $\lambda = \dfrac{1}{\varepsilon}$ and let us change the variables, $y = \dfrac{1}{x}$, to obtain

$$\Phi\left(\frac{1}{\lambda}\right) = \int_0^\infty e^{\frac{-1}{\lambda x}} \phi\left(\frac{1}{x}\right) \frac{dx}{x^2}. \tag{3.8.22}$$

Let us now suppose that $x^{-2}\phi(\frac{1}{x}) = x^{-\alpha}\rho(x)$ where $\rho \in S$ and $\alpha \notin \mathbb{Z}$; in terms of ϕ. This means $y^2\phi(y) \sim y^\alpha \rho(0) + y^{\alpha-1}\rho'(0) + \frac{y^{\alpha-2}\rho''(0)}{2!} + \cdots$ as $y \to \infty$. Actually, the behavior of $\rho(x)$ at $x = \infty$ is not very important, but we assume $\rho \in S$ to fix the ideas. With this assumption, we need the expansion of $f(\lambda x)$ as $\lambda \to \infty$, where

$$f(x) = e^{\frac{-1}{x}} x_+^{-\alpha}. \tag{3.8.23}$$

Since

$$f(x) \sim \sum_{k=0}^{\infty} \frac{(-1)^k}{k! x^{k+\alpha}}, \quad \text{as } x \to \infty, \tag{3.8.24}$$

and since the moments are

$$\mu_n = \text{F.p.} \int_0^\infty e^{\frac{-1}{x}} x^{n-\alpha} dx = \Gamma(\alpha-n-1), \tag{3.8.25}$$

it follows that

$$e^{\frac{-1}{\lambda x}} x_+^{-\alpha} \sim \sum_{n=0}^{\infty} \frac{(-1)^n x_+^{-(n+\alpha)}}{n!\lambda^n}$$

$$+ \sum_{n=0}^{\infty} \frac{(-1)^n \Gamma(\alpha-n-1)\delta^{(n)}(x)}{n!\lambda^{n+1-\alpha}}, \quad \text{as } \lambda \to \infty. \tag{3.8.26}$$

Hence if $\phi(y) \sim a_0 y^{\alpha-2} + a_1 y^{\alpha-3} + a_2 y^{\alpha-4} + \cdots$ as $y \to \infty$, $\alpha \notin \mathbb{Z}$, then

$$\int_0^\infty e^{-\varepsilon y} \phi(y) dy \sim \sum_{n=0}^\infty \left[\frac{(-1)^n}{n!} \, F.p. \int_0^\infty \phi(y) y^n dy \right] \varepsilon^n$$

$$+ \sum_{n=0}^\infty \Gamma(\alpha - n - 1) a_n \varepsilon^{n+1-\alpha}, \quad \text{as} \quad \varepsilon \to 0^+.$$

$$(3.8.27)$$

In the case where $\alpha = k \in \mathbb{Z}$ the above analysis can be applied except that the moments become

$$\mu_n = F.p. \int_0^\infty e^{\frac{-1}{x}} x^{n-k} dx = \begin{cases} (k-n-2)!, & n \leq k-2, \\ \frac{(-1)^{n-k+1}}{(n-k+1)!} \psi(n-k+2), & n > k-2. \end{cases}$$

$$(3.8.28)$$

Hence, as $\lambda \to \infty$,

$$e^{\frac{-1}{\lambda x}} \mathcal{P}f \left(\frac{H(x)}{x^k} \right) \sim \sum_{n=0}^\infty \frac{(-1)^n}{n!} \mathcal{P}f \left(\frac{H(x)}{x^{n+k}} \right) \frac{1}{\lambda^n}$$

$$+ \sum_{n=0}^\infty \frac{(-1)^k \delta^{(n+k-1)}(x)}{(n+k-1)!} \left(\frac{\ln \lambda}{\lambda^n} \right)$$

$$+ \sum_{n=0}^{k-2} \frac{(-1)^n (k-n-2)!}{n!} \lambda^{k-n-1} \delta^{(n)}(x)$$

$$+ \sum_{n=k-1}^\infty \frac{(-1)^{k-1} \psi(n-k+2) \delta^{(n)}(x)}{n!(n-k+1)! \lambda^{n-k-1}}.$$

$$(3.8.29)$$

It follows that if

$$\phi(y) \sim a_0 y^{k-2} + a_1 y^{k-3} + a_2 y^{k-4} + \cdots, \quad \text{as} \quad y \to \infty,$$

then

$$\Phi(\varepsilon) \sim \sum_{n=0}^\infty \frac{(-1)^n}{n!} \left(F.p. \int_0^\infty \phi(y) y^n dy \right) \varepsilon^n$$

$$+ \sum_{n=0}^{k-2} \frac{(-1)^n (k-n-2)! a_n}{\varepsilon^{k-n-2}} + \sum_{n=k-1}^\infty \frac{(-1)^{k-n-1} \psi(n-k+2) a_n \varepsilon^{n-k+1}}{(n-k+1)!}$$

$$+ \ln \varepsilon \sum_{n=k+1}^\infty \frac{(-1)^{k-n-1} \psi(n-k+2) a_n \varepsilon^{n-k+1}}{(n-k+1)!}.$$

$$(3.8.30)$$

Other related results can be obtained by using these methods. For instance if $f \in \mathcal{S}'(\mathbb{R})$ has support bounded on the left and $f(x) = O(x^\beta \ln x)$

as $x \to \infty$, $\beta > -1$ then $f(\lambda x) = O(\lambda^\beta \ln \lambda)$, or if $f(x) = O(x^\beta \ln \ln x)$ as $x \to \infty$, $\beta > -1$, then $f(\lambda x) = O(\lambda^\beta \ln \ln \lambda)$ as $\lambda \to \infty$.

Example 77. Let p_n be the n-th prime and set $F(x) = \sum\limits_{n \leq x} p_n$. Then it follows from the prime number theorem [79] that

$$F(x) = \frac{1}{2} x^2 \ln x + O(x^2 \ln \ln x), \quad \text{as } x \to \infty, \tag{3.8.31}$$

and thus

$$F(\lambda x) = \frac{1}{2}(\lambda^2 \ln \lambda) x_+^2 + O(\lambda^2 \ln \ln \lambda), \quad \text{as } \lambda \to \infty. \tag{3.8.32}$$

Differentiation of this relation yields

$$\sum_{n=1}^{\infty} p_n \delta(\lambda x - n) = (\lambda \ln \lambda) x_+ + O(\lambda \ln \ln \lambda), \quad \text{as } \lambda \to \infty, \tag{3.8.33}$$

or

$$\sum_{n=1}^{\infty} p_n \phi(n\varepsilon) = -\frac{\ln \varepsilon}{\varepsilon^2} \int_0^\infty x\phi(x)dx + O\left(\frac{\ln \ln \frac{1}{\varepsilon}}{\varepsilon^2}\right), \quad \text{as } \varepsilon \to 0. \tag{3.8.34}$$

In particular [7], [95],

$$\sum_{n=1}^{\infty} p_n e^{-n\varepsilon} \sim -\frac{\ln \varepsilon}{\varepsilon^2}, \quad \text{as } \varepsilon \to 0. \tag{3.8.35}$$

Let now α_n be a sequence with $\Re e\, \alpha_n \nearrow \infty$ and let $f \in \mathcal{S}'\{x^{\alpha_n}\}$ be such that $f(x) \sim b_0 x^{\beta_0} + b_1 x^{\beta_1} + b_2 x^{\beta_2} + \cdots$ as $x \to \infty$, where $\beta_n \searrow -\infty$. Defining the generalized moments $\mu(\alpha_n)$ of f as

$$\mu(\alpha_n) = \, < f(x) - \sum_{j=0}^{m} b_j\, \mathcal{P}f\,(x^{\beta_j}), x^{\alpha_n} >, \tag{3.8.36}$$

where m is large enough to make $\beta_m + \alpha_n < -1$, we obtain the expansion

$$f(\lambda x) \sim \sum_{n=0}^{\infty} b_n \mathcal{P}f\,((\lambda x)^{\beta_n}) + \sum_{n=0}^{\infty} \frac{\mu(\alpha_n)\delta_n(x)}{\lambda^{\alpha_n+1}}, \quad \text{as } \lambda \to \infty, \tag{3.8.37}$$

in the space $\mathcal{S}'\{x^{\alpha_n}\}$, where $\delta_n(x) = x^{-\alpha_n} \delta(x)$.

Observe that in the space $S'\{x^{\alpha_n}\}$ we have

$$\mathcal{P}f((\lambda x)^\beta) = \lambda^\beta \mathcal{P}f(x^\beta), \quad \text{if } \beta \neq -\alpha_n - 1, \ \forall n, \tag{3.8.38a}$$

$$\mathcal{P}f((\lambda x)^{-1-\alpha_n}) = \lambda^{-1-\alpha_n} \mathcal{P}f(x^{-1-\alpha_n}) + \lambda^{-1-\alpha_n} \ln \lambda \delta_n(x). \tag{3.8.38b}$$

Example 78. Let us obtain the asymptotic expansion of the principal value integral

$$\Psi(t) = \mathcal{P}.v. \int_0^\infty \frac{\psi(x)dx}{x-t}, \tag{3.8.39}$$

as $t \to 0^+$, where $\psi \in S\{x^{\alpha_n}\}$. Setting $t = \frac{1}{\lambda}$, (3.8.39) can be written as $\Psi\left(\frac{1}{\lambda}\right) = \lambda < f(\lambda x), \Psi(x) >$, where $f(x) = \mathcal{P}.v. \left(\frac{H(x)}{x-1}\right)$.

Since

$$f(x) \sim \frac{1}{x} + \frac{1}{x^2} + \frac{1}{x^3} + \cdots, \quad \text{as } x \to \infty, \tag{3.8.40}$$

and

$$F.p. \ (\mathcal{P}.v. \int_0^\infty \frac{x^\alpha}{x-1}dx) = \pi \cot \pi \alpha, \ \alpha \notin \mathbb{Z}, \tag{3.8.41a}$$

$$F.p. \ (\mathcal{P}.v. \int_0^\infty \frac{x^\alpha}{x-1}dx) = 0, \qquad \alpha \in \mathbb{Z}, \tag{3.8.41b}$$

we immediately obtain the expansion of $f(\lambda x)$ as $\lambda \to \infty$. We shall consider two cases according as to whether (3.8.41a) or (3.8.41b) has to be used.

Case I. $\psi(x) \sim a_1 x^{\alpha_1} + a_2 x^{\alpha_2} + \cdots$ as $x \to 0^+$, where $\alpha_i \notin \mathbb{Z}$. Then

$$f(\lambda x) \sim \sum_{n=1}^\infty \frac{1}{\lambda^n} \mathcal{P}f\left(\frac{1}{x^n}\right) + \sum_{n=1}^\infty \frac{(\pi \cot \pi \alpha_n)\delta_n(x)}{\lambda^{\alpha_n+1}}, \quad \text{as } \lambda \to \infty, \tag{3.8.42}$$

and thus

$$\mathcal{P}.v. \int_0^\infty \frac{\psi(x)dx}{x-t} \sim \sum_{n=0}^\infty \left(F.p. \int_0^\infty \frac{\psi(x)}{x^{n+1}}dx\right) t^n$$

$$+ \sum_{n=1}^\infty (\pi \cot \pi \alpha_n)a_n t^{\alpha_n}, \quad \text{as } t \to 0^+. \tag{3.8.43}$$

Formula (3.8.43) permits us to observe the special role played by the sequence $\alpha_n = n - \frac{1}{2}$, $n = 1, 2, 3, \cdots$. In this case, if

$$\psi(x) = \sqrt{x}\psi_1(x), \qquad \psi_1(x) \sim a_0 + a_1 x + a_2 x^2 + \cdots \text{ as } x \to 0, \tag{3.8.44}$$

then

$$\mathcal{P}.v. \int_0^\infty \frac{\sqrt{x}\psi_1(x)dx}{x - t} \sim \sum_{n=0}^\infty F.p. \left(\int_0^\infty \frac{\psi_1(x)}{x^{n+\frac{1}{2}}}dx \right) t^n, \quad \text{as } t \to 0^+.$$

(3.8.45)

This shows the fact that the operator $\int_a^b \frac{(x - a)^\alpha(b - x)^\beta \phi(x)dx}{x - t}$ sends the space $\mathcal{E}[a, b]$ to itself if and only if $\alpha = n + \frac{1}{2}$, $\beta = m + \frac{1}{2}$ for some $n, m \in \mathbb{N}$ [36], [38].

Case II. All the positive integers $\{0, 1, 2, \cdots\}$ are among the α'_n. By changing the notation, if needed, we could assume

$$\psi(x) \sim \sum_{n=0}^\infty b_n x^n + \sum_{n=1}^\infty a_n x^{\alpha_n}, \quad x \to 0,$$

(3.8.46)

where $\alpha_n \notin \mathbb{Z}$. Since in this case we have

$$\mathcal{P}f\left(\frac{1}{(\lambda x)^n}\right) = \frac{1}{\lambda^n}\mathcal{P}f\left(\frac{1}{x^n}\right) + \frac{\ln\lambda}{\lambda^n}(x^{-n-1}\delta(x)),$$

(3.8.47)

we obtain

$$f(\lambda x) \sim \sum_{n=1}^\infty \frac{1}{\lambda^n}\mathcal{P}f\left(\frac{1}{x^n}\right) + \ln\lambda \sum_{n=1}^\infty \frac{1}{\lambda^n}(x^{-n-1}\delta(x))$$

$$+ \sum_{n=1}^\infty \frac{(\pi\cot\pi\alpha_n)\delta_{\alpha_n}(x)}{x^{\alpha_n+1}},$$

(3.8.48)

$$\mathcal{P}.v. \int_0^\infty \frac{\psi(x)}{x - t}dx \sim \sum_{n=0}^\infty \left\{ \int_0^\infty \frac{\psi(x)}{x^{n+1}}dx \right\} t^n$$

$$- \ln t \left\{ \sum_{n=0}^\infty b_n t^n \right\} + \sum_{n=1}^\infty (\pi\cot\pi\alpha_n)a_n t^{\alpha_n}, \quad \text{as } t \to 0^+.$$

(3.8.49)

3.9 Asymptotic Separation of Variables

We have devoted the greater part of this chapter to studying the asymptotic development of distributions of the type $f(\lambda x)$ as $\lambda \to \infty$. We have identified a large class of distributions for which the moment asymptotic expansion holds, but we have also obtained the development in many cases when the moment expansion is not valid. All of these expansions are given in terms of homogeneous and associated homogenous functions. The purpose of this section is to show that this is the only possible situation [46]: if $f(\lambda x)$ admits the development

$$f(\lambda x) \sim \rho_1(\lambda)h_1(x) + \rho_2(\lambda)h_2(x) + \cdots, \quad \text{as } \lambda \to \infty, \qquad (3.9.1)$$

in terms of an asymptotic sequence $\{\rho_n(\lambda)\}$ then all the terms are homogeneous or associated homogeneous generalized functions.

Let us start with the generalized functions that admit the asymptotic separation of variables

$$f(\lambda x) = \rho(\lambda)h(x) + o(\rho(\lambda)), \quad \text{as } \lambda \to \infty, \qquad (3.9.2)$$

which corresponds to the first term in (3.9.1). This case has been studied by Vladimirov, Drozhinov and Zavyalov [112]. In their notation, (3.9.2) means that the generalized function f admits the *quasiasymptote* h with respect to the gauge function $\rho(\lambda)$. The function $\rho(\lambda)$ is assumed to be positive for $\lambda >> 1$ while $h \neq 0$.

We now show that (3.9.2) imposes very strong restrictions on both ρ and h. Let us start with the function ρ. Since $h \neq 0$, we can find $\phi \in \mathcal{D}$ with $< h, \phi > = 1$ and thus

$$< f(\lambda x), \phi(x) > = \rho(\lambda) + o(\rho(\lambda)), \quad \text{as } \lambda \to \infty. \qquad (3.9.3)$$

But also, if $a > 0$ then

$$\rho(\lambda a) = < f(\lambda a x), \phi(x) > + o(\rho(\lambda)), \qquad (3.9.4)$$

and so

$$\lim_{\lambda \to \infty} \frac{\rho(\lambda a)}{\rho(\lambda)} = \lim_{\lambda \to \infty} \frac{< f(\lambda a x), \phi(x) >}{\rho(\lambda)}$$

$$= \lim_{\lambda \to \infty} \frac{< f(\lambda x), \phi(\frac{x}{a}) >}{a\rho(\lambda)}$$

$$= < h(ax), \phi(x) > .$$

It follows that the function

$$C\,(a) = \lim_{\lambda \to \infty} \frac{\rho(\lambda a)}{\rho(\lambda)} = < h(ax), \phi(x) > \qquad (3.9.5)$$

exists and is smooth for $a > 0$. But it is easy to see that

$$C\,(ab) = C\,(a)\,C\,(b), \qquad (3.9.6)$$

and consequently,

$$C(a) = a^\beta \tag{3.9.7}$$

for some $\beta \in \mathbb{R}$. Hence the function $\rho(\lambda)$ is a regularly varying function of order β, according to the next definition.

Definition. *A function $\rho(\lambda)$ that satisfies*

$$\lim_{\lambda \to \infty} \frac{\rho(\lambda a)}{\rho(\lambda)} = a^\beta \tag{3.9.8}$$

for $a > 0$, is called a regularly varying function of order β. When $\beta = 0$, that is, when $\rho(\lambda a) = \rho(\lambda) + o(\rho(\lambda))$, the term slowly varying function is used.

The regularly varying functions were introduced by Karamata [72]. They play a very important role in the study of Tauberian theorems [112] and in many other areas of mathematics [98]. Examples of regularly varying functions of order β include the functions λ^β, $\lambda^\beta \mid \ln \lambda \mid^\alpha$ and $\lambda^\beta(2+\sin \ln \lambda)$. Regularly varying functions of order β can be said to be asymptotically homogeneous functions of order β.

The analysis so far shows that if $f(\lambda x)$ has a quasiasymptote $h(x)$ with respect to $\rho(\lambda)$ then $\rho(\lambda)$ is a regularly varying function of some order β. But then $h(x)$ is homogeneous of degree β, since

$$h(tx) \sim \frac{f(\lambda t x)}{\rho(\lambda)} \sim \frac{\rho(\lambda t)}{\rho(\lambda)} \frac{h(\lambda t x)}{\rho(\lambda t)} \sim t^\beta h(x).$$

In particular, if supp h is known to be contained in $[0, \infty)$, it follows that $h = c h_\beta$, where c is a constant and where $h_\beta = x_+^\beta$ if $\beta \neq -1, -2, -3, \cdots$ while $h_\beta(x) = \delta^{(k)}(x)$ if $\beta = -(1 + k)$, $k = 0, 1, 2, \cdots$. Summarizing, we have the following result.

Lemma 8. *Let $f \in \mathcal{D}'(\mathbb{R})$ admit the asymptotic separation of variables*

$$f(\lambda x) = \rho(\lambda)h(x) + o(\rho(\lambda)), \quad as \ \lambda \to \infty.$$

Then $\rho(\lambda)$ is regularly varying of some order $\beta \in \mathbb{R}$ and $h(x)$ is homogeneous of order β. In case the support of f is bounded on the left then $h(x) = c h_\beta(x)$ for some constant c.

Next, let us consider the case when $f(\lambda x)$ admits a two term expansion

$$f(\lambda x) = \rho_1(\lambda)h_1(x) + \rho_2(\lambda)h_2(x) + o(\rho_2(\lambda)), \tag{3.9.9}$$

where $\rho_2 = o(\rho_1)$. If $h_1 = h_2$ then (3.9.9) degenerates into a single term expansion and thus we suppose that $h_1 \neq h_2$.

We already know that ρ_1 is regularly varying of some order β_1 and that h_1 is homogeneous of degree β_1. Considering a test function ϕ_0 that satisfies $< h_1, \phi_0 > = 0$, $< h_2, \phi_0 > = 1$, the argument used before shows that ρ_2 is also regularly varying, of some order β_2. Since $\rho_2 = o(\rho_1)$, we should have $\beta_2 \leq \beta_1$. The cases $\beta_2 = \beta_1$ and $\beta_2 < \beta_1$ are essentially different. Indeed, it follows from (3.9.9) that if $< h_1(x), \phi(x) > = 0$ then

$$< h_2(\lambda x), \phi(x) > = \lambda^{\beta_2} < h_2(x), \phi(x) >,$$

and thus

$$h_2(\lambda x) = \lambda^{\beta_2} h_2(x) + a(\lambda) h_1(x) \qquad (3.9.10)$$

for some function $a(\lambda)$.

When $\beta_1 < \beta_2$, the relation (3.9.10) can only hold if $a(\lambda) = 0$ and it follows that h_2 is homogeneous of degree β_2. When $\beta_1 = \beta_2$, then (3.9.10) shows that

$$a(\lambda\mu) = \lambda^{\beta_1} a(\mu) + \mu^{\beta_1} a(\lambda) \qquad (3.9.11)$$

and since $a(1) = 0$, it follows that $a(\lambda) = c\lambda^{\beta_1} \ln \lambda$ for some c. Therefore, h_2 is associated homogeneous of order 1 and degree β_1.

When $\beta_2 < \beta_1$ the expansion (3.9.9) conveys some more information. Indeed, (3.9.9) yields

$$\rho_1(\lambda a) = \rho_1(\lambda) a^{\beta_1} + o(\rho_2(\lambda)), \qquad (3.9.12)$$

the order relation being uniform for a in compact subsets of $(0, \infty)$. Considering intervals of the form $[\lambda, 2\lambda]$, $[2\lambda, 4\lambda], \cdots, [2^{n-1}\lambda, 2^n \lambda]$, we can see that for each $\varepsilon > 0$ there exists λ_0 such that

$$\left| \frac{\rho_1(\lambda)}{\lambda^{\beta_1}} - \frac{\rho_1(\mu)}{\mu^{\beta_1}} \right| \leq \frac{\varepsilon \rho_2(\lambda_0)}{\lambda_0^{\beta_1}}, \quad \lambda, \mu \geq \lambda_0. \qquad (3.9.13)$$

Using the Cauchy criterion, there exists the limit $c = \lim_{\lambda \to \infty} \lambda^{-\beta_1} \rho_1(\lambda)$, and by letting $\mu \to \infty$ in (3.9.13) it follows that $\rho_1(\lambda) = c\lambda^{\beta_1} + o(\rho_2(\lambda))$. Therefore (3.9.9) can be replaced by the equivalent expansion

$$f(\lambda x) = c\lambda^{\beta_1} h_1(x) + \rho_2(\lambda) h_2(x) + o(\rho_2(\lambda)), \qquad (3.9.14)$$

where h_1 and h_2, respectively, are homogeneous, of degrees β_1 and β_2,

The ideas given for a two term expansion can be generalized to an N term expansion of the form

$$f(\lambda x) = \sum_{j=1}^{N} \rho_j(\lambda) h_j(x) + o(\rho_N(x)), \qquad (3.9.15)$$

where $\{\rho_n(\lambda)\}$ is an asymptotic sequence as $\lambda \to \infty$. It is easy to see that the same argument used before gives that the $\rho_j(\lambda)$ are regularly varying functions, of orders $\beta_1 \geq \beta_2 \geq \cdots \geq \beta_N$. Therefore, the expansion can be arranged in blocks of the form

$$f(\lambda x) = \sum_{i=1}^{n} \sum_{j=1}^{m_i} \rho_{ij}(\lambda) h_{ij}(x) + o(\rho_{n\,m_n}(\lambda)), \qquad (3.9.16)$$

where $\rho_{i(j+1)}(\lambda) = o(\rho_{ij}(\lambda))$ and where $\rho_{ij}(\lambda)$ is a regularly varying function of order β_i, $\beta_1 > \beta_2 > \cdots > \beta_n$.

Theorem 33. *Let $f \in \mathcal{D}'(\mathbb{R})$ have the asymptotic expansion (3.9.16). Then if $i < n$,*

$$\rho_{ij}(\lambda) = \lambda^{\beta_i} p_{ij}(\ln \lambda) + o(\rho_{nm_n}(\lambda)), \qquad (3.9.17)$$

where p_{ij} is a polynomial of degree $m_i - j$ of the form

$$p_{ij}(t) = \sum_{r=0}^{m_i - j} c_{ir}\, t^r, \qquad (3.9.18)$$

for some constants $\{c_{ir}\}$. The functions $h_{ij}(x)$ are associated homogeneous of order j and degree β_i. The expansion (3.9.16) can be rewritten as

$$f(\lambda x) = \sum_{i=1}^{n-1} G_i(\lambda x) + \sum_{j=1}^{m_n} \rho_{nj}(\lambda) h_{nj}(x) + o(\rho_{nm_n}(\lambda)), \qquad (3.9.19)$$

where the G_i are associated homogeneous of order m_i and degree β_i.

The case when f has support bounded on the left deserves special mention.

Theorem 34. *Let $f \in \mathcal{D}'(R)$ have support bounded on the left and admit the expansion (3.9.16) and consequently (3.9.19). Then if $\beta_i \neq -1, -2, -3, \cdots$*

$$G_i(x) = \sum_{j=1}^{m_i} d_{ij}\, x_+^{\beta_i} (\ln x)^{j-1} \qquad (3.9.20)$$

while

$$G_i(x) = d_{i1}\, \delta^{(k)}(x) + \sum_{j=2}^{m_i} d_{ij}\, \mathcal{P}f\left(\frac{H(x)(\ln x)^{j-2}}{x^{k+1}}\right), \qquad (3.9.21)$$

if $\beta_i = -(k+1)$, $k = 0, 1, 2, \cdots$. Similarly

$$h_{nj}(x) = \sum_{r=1}^{j} d_{njr} x_+^{\beta_n} (\ln x)^{r-1}, \tag{3.9.22}$$

if $\beta_n \neq -1, -2, -3, \cdots$ and

$$h_{nj} = d_{nj1} \delta^{(k)}(x) + \sum_{r=2}^{j} d_{njr} \, \mathcal{P}f\left(\frac{H(x)(\ln x)^{r-2}}{x^{k+1}}\right), \tag{3.9.23}$$

if $\beta_n = -(k+1)$, $k = 0, 1, 2, \cdots$.

These theorems show that except for the last block, the terms in the asymptotic separation of variables of $f(\lambda x)$ are all associated homogeneous functions. The results for the last block cannot be improved, however, as the next example shows.

Example 79. Let us consider the generalized function $f(x) = H(x - 1)\sqrt{\ln x}$. Then

$$f(\lambda x) \sim (\ln \lambda)^{\frac{1}{2}} H(x) + \frac{1}{2}(\ln \lambda)^{\frac{-1}{2}} H(x) \ln x - \frac{1}{8}(\ln \lambda)^{\frac{-3}{2}} H(x)(\ln x)^2 + \cdots,$$
$$\tag{3.9.24}$$

as $\lambda \to \infty$. Here the asymptotic sequence is $\{(\ln \lambda)^{\frac{1}{2}-n}\}$, which is infinite but which consists of regularly varying functions all of order 0. The sequence of functions $H(x), H(x) \ln x, H(x)(\ln x)^2, \cdots$ consists of associated homogeneous functions, but the sequence $(\ln \lambda)^{\frac{1}{2}}, (\ln \lambda)^{\frac{-1}{2}}, (\ln \lambda)^{\frac{-3}{2}}, \cdots$ does not.

We would like to observe that we have given our results under the hypothesis that $f \in \mathcal{D}'(\mathbb{R})$ and not $f \in \mathcal{S}'(\mathbb{R})$ as in Section 3.8. However, any generalized function that admits a quasiasymptote, $f(\lambda x) = \rho(\lambda) h(x)$, must belong to $\mathcal{S}(\mathbb{R})$ and the asymptotic relation holds there. In particular, since convergence in $\mathcal{S}'(\mathbb{R})$ implies uniform convergence of some primitives of some order, we obtain the following result [112].

Theorem 35. *Let $f \in \mathcal{S}'(\mathbb{R})$ with support bounded on the left. Then f admits the asymptotic separation of variables*

$$f(\lambda x) \sim \rho(\lambda) h(x), \tag{3.9.25}$$

if and only if there exist N such that

$$\lim_{\lambda \to \infty} \frac{F_N(\lambda)}{\lambda^N \rho(\lambda)} = c \neq 0, \tag{3.9.26}$$

where $F_N(\lambda)$ is the primitive of order N of f with support bounded on the left.

If f is a locally integrable function with support in the interval $[0, \infty)$ then F_N is given by

$$F_N(x) = \int_0^x \frac{(x-t)^{N-1}}{(N-1)!} f(t)dt, \quad x > 0. \tag{3.9.27}$$

Therefore condition (3.9.26) means that the average $x^{-N}F_N(x)$ of f has an ordinary asymptote of the form $c\rho(\lambda)$.

We would also like to indicate that using regularly varying functions as gauges allows us to give results similar to those of Theorem 32.

Theorem 36. *Let $\rho(x)$ be a locally integrable regularly varying function of order β, with support bounded on the left.*

(a) If $\beta > -1$, then in $\mathcal{S}(\mathbb{R})$,

$$\rho(\lambda x) = \rho(\lambda)x_+^\beta + o(\rho(\lambda)), \quad as \ \lambda \to \infty. \tag{3.9.28}$$

(b) If $-(k+2) < \beta < -(k+1)$ for some $k = 0, 1, 2, \cdots$ then in $\mathcal{S}(\mathbb{R})$

$$\rho(\lambda x) = \sum_{j=0}^k \frac{(-1)^j \mu_j \delta^{(j)}(x)}{j! \lambda^{j+1}} + \rho(\lambda)x_+^\beta + o(\rho(\lambda)), \tag{3.9.29}$$

where $\mu_j = \ <\rho(x), x^j>\ $ are the moments.

(c) If $\beta = -(k+1)$ for some $k = 0, 1, 2, \cdots$ then in $\mathcal{S}(\mathbb{R})$

$$\rho(\lambda x) = \sum_{j=0}^{k-1} \frac{(-1)^j \mu_j \delta^{(j)}(x)}{j! \lambda^{j+1}} + \sigma(\lambda)(-1)^k \delta^{(k)}(x)$$

$$+ \rho(\lambda)\mathcal{P}f\left(\frac{H(x)}{x^{k+1}}\right) + o(\rho(\lambda)), \tag{3.9.30}$$

where $\sigma(\lambda)$ is the regularly varying function of order β given by

$$\sigma(\lambda) = \frac{1}{k! \lambda^{k+1}} \int_{-\infty}^\lambda \rho(t)t^k\, dt. \tag{3.9.31}$$

Proof. The proof of (a) is similar but simpler, so we shall consider only (b) and (c). Since adding a compactly supported distribution to ρ does not

alter the formulas, we assume that supp $\rho \subseteq [0, \infty)$. For (b) we write that
if $\phi \in \mathcal{S}$

$$\int_0^\infty \rho(\lambda x)\phi(x)dx = \sum_{j=0}^k \frac{\mu_j \phi^{(j)}(0)}{j!\lambda^{j+1}} + \int_0^\infty \rho(\lambda x)\left[\phi(x) - \sum_{j=0}^k \frac{\phi^{(j)}(0)x^j}{j!}\right]dx,$$

and observe that there exists a constant M such that

$$\left|\phi(x) - \sum_{j=0}^k \frac{\phi^{(j)}(0)x^j}{j!}\right| \leq M \mid x \mid^{k+1}, \quad \mid x \mid \leq 1,$$

$$\left|\phi(x) - \sum_{j=0}^k \frac{\phi^{(j)}(0)x^j}{j!}\right| \leq M \mid x \mid^k, \quad \mid x \mid \geq 1,$$

to conclude that we can apply the Lebesgue limit theorem:

$$\lim_{\lambda \to \infty} \frac{1}{\rho(\lambda)} \int_0^\infty \rho(\lambda x)\left[\phi(x) - \sum_{j=0}^k \frac{\phi^{(j)}(0)x^j}{j!}\right]dx$$

$$= \int_0^\infty x^\beta \left[\phi(x) - \sum_{j=0}^k \frac{\phi^{(j)}(0)x^j}{j!}\right]dx.$$

But

$$< x_+^\beta, \phi(x) > = \int_0^\infty x^\beta \left[\phi(x) - \sum_{j=0}^k \frac{\phi^{(j)}(0)x^j}{j!}\right]dx$$

if $\phi \in \mathcal{S}$ and $-(k+1) > \beta > -(k+2)$, and thus (3.9.29) follows.

To prove (3.9.30) we observe that if $\phi^{(k)}(0) = 0$, the Lebesgue convergence argument just used gives

$$\int_0^\infty \rho(\lambda x)\phi(x)dx = \sum_{j=0}^{k-1} \frac{\mu_j \phi^{(j)}(0)}{j!\lambda^{j+1}}$$

$$+ \int_0^\infty \rho(\lambda x)\left[\phi(x) - \sum_{j=0}^k \frac{\phi^{(j)}(0)x^j}{j!}\right]dx$$

$$= \sum_{j=0}^{k-1} \frac{\mu_j \phi^{(j)}(0)}{j!\lambda^{j+1}} + \rho(\lambda) \int_0^\infty x^{-k-1}\left[\phi(x) - \sum_{j=0}^k \frac{\phi^{(j)}(0)x^j}{j!}\right]dx + o(\rho(\lambda)).$$

$$(3.9.32)$$

Thus, in the general case $\displaystyle\int_0^\infty \rho(\lambda x)\phi(x)dx$

$$= \int_0^\infty \rho(\lambda x)\left[\phi(x) - \frac{H(1-x)\phi^{(k)}(0)x^k}{k!}\right]dx$$
$$+ \int_0^\infty \frac{H(1-x)\phi^{(k)}(0)x^k \rho(\lambda x)}{k!}dx$$
$$= \int_0^\infty \rho(\lambda x)\left[\phi(x) - H(1-x)\frac{\phi^{(k)}(0)x^k}{k!}\right]dx + \phi^{(k)}(0)\sigma(\lambda),$$

and applying (3.9.32) to the first term,

$$\int_0^\infty \rho(\lambda x)\phi(x)dx = \sum_{j=0}^{k-1}\frac{\mu_j \phi^{(j)}(0)}{j!\lambda^{j+1}} + \phi^{(k)}(0)\sigma(\lambda)$$
$$+ \; F.p. \int_0^\infty \frac{\phi(x)dx}{x^{k+1}}\rho(\lambda) + o(\rho(\lambda))$$

which is (3.9.30). ∎

Using the same ideas, we can prove the following theorem.

Theorem 37. *Let $\rho(x)$ be a locally integrable regularly varying function of order β. Let $\alpha_j \nearrow \infty$, with $\alpha_n = \beta$. Then in $S\{x^{\alpha_j}\}$*

$$\rho(\lambda x) = \sum_{j=0}^{n-1}\frac{\mu(\alpha_j)(x^{-\alpha_j}\delta(x))}{\lambda^{\alpha_j+1}} + \sigma(\lambda)(x^{-\alpha_n}\delta(x))$$
$$+ \rho(\lambda)\mathcal{P}f(x^\beta H(x)) + o(\rho(\lambda)),$$

$$(3.9.33)$$

where $\mu(\alpha_j) = <\rho(x), x^{\alpha_j}>$ are the moments and where

$$\sigma(\lambda) = \lambda^\beta \; F.p. \int_0^\lambda t^{-\beta-1}\rho(t)dt. \qquad (3.9.34)$$

Let us consider an example of these results.

Example 80. Let $\rho(x)$ be a continuously differentiable increasing regularly varying function of order $\alpha > 0$. Since $[\![\rho(x)]\!] = \rho(x) + o(\rho(x))$, as $x \to \infty$, it follows that

$$[\![\rho(\lambda x)H(\lambda x)]\!] = \rho(\lambda)x_+^\alpha + o(\rho(\lambda)), \quad \text{as } \lambda \to \infty. \qquad (3.9.35)$$

Differentiation of this relation gives

$$[\![\rho(0)]\!]\delta(x) + \sum_{n=n_0}^{\infty} \delta\left(x - \frac{x_n}{\lambda}\right) = \alpha\rho(\lambda)x_+^{\alpha-1} + o(\rho(\lambda)), \tag{3.9.36}$$

where x_n are the solutions of

$$\rho(x_n) = n. \tag{3.9.37}$$

Evaluating (3.9.36) at a test function $\phi \in \mathcal{S}(\mathbb{R})$ and setting $\lambda = \dfrac{1}{\varepsilon}$, where $\varepsilon \to 0^+$, yields

$$\sum_{n=n_0}^{\infty} \phi(\varepsilon x_n) = \alpha\rho\left(\frac{1}{\varepsilon}\right)\int_0^{\infty} x^{\alpha-1}\phi(x)dx + o\left(\rho\left(\frac{1}{\varepsilon}\right)\right). \tag{3.9.38}$$

When $\rho(x) = x^\alpha$, $x > 0$, (3.9.38) gives

$$\sum_{n=1}^{\infty} \phi(\varepsilon n^{\frac{1}{\alpha}}) = \frac{\alpha}{\varepsilon^\alpha}\int_0^{\infty} x^{\alpha-1}\phi(x)dx + o\left(\frac{1}{\varepsilon^\alpha}\right), \quad \text{as } \varepsilon \to 0^+, \tag{3.9.39}$$

a formula that shall be improved in Chapter 5.

If $\rho(x) = x\ln x$, $x \geq 1$, then we obtain

$$\sum_{n=0}^{\infty} \phi(\varepsilon x_n) = \frac{-\ln\varepsilon}{\varepsilon}\int_0^{\infty} \phi(x)dx + o\left(\frac{-\ln\varepsilon}{\varepsilon}\right), \quad \text{as } \varepsilon \to 0^+, \tag{3.9.40}$$

where $x_n \ln x_n = n$, $n = 0, 1, 2, \cdots$. On the other hand, if $\rho(x) = \ln x$, $x \geq 1$, then (3.9.38) yields

$$\sum_{n=0}^{\infty} \phi(\varepsilon e^n) = o(\ln\varepsilon), \quad \text{as } \varepsilon \to 0^+. \tag{3.9.41}$$

The results of this section can be readily extended to the study of the local behavior of generalized functions. Actually, Lojasiewicz [82] defined the value of distribution $f \in \mathcal{D}'(\mathbb{R})$ at the point x_0 as the limit

$$f(x_0) = \lim_{\varepsilon \to 0} f(x_0 + \varepsilon x), \tag{3.9.42}$$

if the limit exists in $\mathcal{D}'(\mathbb{R})$; that is, if

$$\lim_{\varepsilon \to 0} < f(x_0 + \varepsilon x), \phi(x) > = f(x_0)\int_{-\infty}^{\infty} \phi(x)dx, \tag{3.9.43}$$

for each $\phi \in \mathcal{D}(\mathbb{R})$.

More generally, one could try to look for a representation of the form

$$f(x_0 + \varepsilon x) \sim \sigma(\varepsilon)g(x), \quad \text{as } \varepsilon \to 0, \qquad (3.9.44)$$

in the space $\mathcal{D}'(\mathbb{R})$. As before, σ is positive and g non-null. Then it is easy to show that $\sigma(\varepsilon)$ has to be *regularly varying at the origin,* in the sense that

$$\lim_{\varepsilon \to 0} \frac{\sigma(a\varepsilon)}{\sigma(\varepsilon)} = a^\beta, \qquad (3.9.45)$$

for some β. Then g is homogeneous of order β. Results for N term expansions, $f(x_0 + \varepsilon x) = \sigma_1(\varepsilon)g_1(x) + \cdots + \sigma_N(\varepsilon)g_N(x) + o(\sigma(\varepsilon))$ are also easy to obtain.

Example 81. The generalized function $f(x) = \sin\frac{1}{x}$ is oscillatory near $x = 0$. However, it is easy to see that $f(0)$ exists and equals 0.

On the other hand, if $g(x) = \sin(\ln x)$, then with $\varepsilon_n = e^{-n\pi}$ we have $g(\varepsilon_n x) = (-1)^n g(x)$ and thus $g(\varepsilon x)$ cannot have a quasiasymptote as $\varepsilon \to 0$.

CHAPTER 4

The Asymptotic Expansion of Multi-Dimensional Generalized Functions

4.1 Introduction

In this chapter we continue our study of the asymptotic development of distributions and the corresponding analysis of integrals with a large parameter. We presently consider the multi-dimensional situation.

We start with the Taylor expansions in Section 4.2. The multi-dimensional moment asymptotic expansion is given in Section 4.3. These results, in turn, find application in the development of Laplace and Fourier type integrals which are considered, respectively, in Sections 4.4 and 4.5. The moment expansion is the basic tool in the analysis, but the complicated geometries possible for the multi-dimensional regions and their corresponding boundaries make the analysis much more complex than that of Chapter 3.

An example where the moment asymptotic expansion does not hold is given in Section 4.6. Section 4.7 is devoted to the study of partial expansions of the type $f(\lambda x, y)$ as $\lambda \to \infty$. Many interesting results are obtained. In particular, if (r, w) are polar coordinates, the partial expansion of $f(\lambda r, w)$ as $\lambda \to \infty$ provides very sharp expansions of $f(\lambda x)$ as $\lambda \to \infty$. Section 4.8 provides an application in quantum mechanics. Using our methods, the asymptotic relation between the two different definitions of the quantum mechanical twisted product is obtained.

4.2 Taylor Expansion in Several Variables

In this section we study the Taylor expansion of generalized functions of several variables as well as some related interesting asymptotic developments [45].

Any smooth function near $x \in \mathbb{R}^n$ admits the Taylor expansion

$$\phi(x_0 + \varepsilon y) \sim \sum_{N=0}^{\infty} (D^N \phi(x_0) \parallel y^N) \frac{\varepsilon^N}{N!}, \quad \text{as } \varepsilon \to 0, \qquad (4.2.1)$$

$$\sim \sum_{|k|=0}^{\infty} \frac{D^k \phi(x_0) y^k}{k!} \varepsilon^{|k|}, \quad \text{as } \varepsilon \to 0.$$

It follows that if $f(x)$ is a distribution of any of the standard spaces $\mathcal{D}'(\mathbb{R}^n), \mathcal{E}'(\mathbb{R}^n), \mathcal{P}'(\mathbb{R}^n)$, etc., then we have the following asymptotic Taylor expansion

$$f(x + \varepsilon y) \sim \sum_{N=0}^{\infty} (D^N f(x) \parallel y^N) \frac{\varepsilon^N}{N!}, \quad \text{as } \varepsilon \to 0, \tag{4.2.2}$$

$$\sim \sum_{|k|=0}^{\infty} \frac{D^k f(x) y^k}{k!} \varepsilon^{|k|}, \quad \text{as } \varepsilon \to 0.$$

The interpretation of (4.2.2) is in the weak or distributional sense: it means that for any test function $\phi(x)$ we have

$$< f(x+\varepsilon y), \phi(x) > \sim \sum_{|k|=0}^{N} \frac{< D^k f(x), \phi(x) > y^k}{k!} \varepsilon^{|k|} + O(\varepsilon^{N+1}), \quad \text{as } \varepsilon \to 0.$$

For instance, if $f(x) = \delta(x)$, relation (4.2.2) becomes

$$\delta(x + \varepsilon y) \sim \sum_{N=0}^{\infty} \frac{1}{N!} (D^N \delta(x) \parallel y^N) \varepsilon^N, \quad \text{as } \varepsilon \to 0. \tag{4.2.3}$$

Evaluation of this relation at a test function ϕ gives (4.2.1) again.

Example 82. Let us consider the Taylor expansion of $\mathcal{P}f(|\,x\,|^{-k})$. Using the formulas for the derivatives of r^{-k} given in Chapter 2 we obtain the following asymptotic expansion:

$$\mathcal{P}f\left(\frac{1}{|\,x + \varepsilon y\,|^k}\right) \sim \mathcal{P}f\left(\frac{1}{|\,x\,|^k}\right)$$

$$+ \left\{ -k x \cdot y \mathcal{P}f\left(\frac{1}{|\,x\,|^{k+2}}\right) - \frac{c_{m,n}}{(2m)!k} D_y \nabla^{2m} \delta(x) \right\} \varepsilon$$

$$+ \left\{ \{k(k+2)(x \cdot y)^2 \mathcal{P}f\left(\frac{1}{|\,x\,|^{k+4}}\right) \right.$$

$$- k\,|\,y\,|^2\, \mathcal{P}f\left(\frac{1}{|\,x\,|^{k+2}}\right)$$

$$- \frac{k c_{m+1,n}}{(2m+2)!(k+2)}\,|\,y\,|^2\, \nabla^{2m+2} \delta(x)$$

$$\left. - \frac{c_{m,n}}{(2m)!}\left(\frac{1}{k} + \frac{1}{k+2}\right) D_y^2 \nabla^{2m} \delta(x) \right\} \varepsilon^2 + \cdots ,$$

$$\text{as } \varepsilon \to 0, \tag{4.2.4}$$

where $D_y = y_i D_i$ stands for the derivative in the y direction and where the constants $c_{m,n}$ are given by

$$c_{m,n} = \frac{2\Gamma(m+1/2)\pi^{\frac{n-1}{2}}}{\Gamma(m+n/2)}. \tag{4.2.5}$$

In particular, if $n = 3$ and $k = 1$ we obtain

$$\frac{1}{|\,x+\varepsilon y\,|} \sim \frac{1}{|\,x\,|} - x \cdot y \; \mathcal{P}f\left(\frac{1}{|\,x\,|^3}\right)\varepsilon$$

$$+ \left\{3(x \cdot y)^2\,\mathcal{P}f\left(\frac{1}{|\,x\,|^5}\right) - |\,y\,|^2\,\mathcal{P}f\left(\frac{1}{|\,x\,|^3}\right) - \frac{4\pi}{3}\,|\,y\,|^2\,\delta(x)\right\}\varepsilon^2 + \cdots,$$

$$\text{as } \varepsilon \to 0. \tag{4.2.6}$$

Thus, if $\phi \in \mathcal{D}(\mathbb{R}^3)$ we have

$$\int \frac{\phi(x)}{|\,x+\varepsilon y\,|}dx \sim \int \frac{\phi(x)}{|\,x\,|}dx - \left(F.p.\int \frac{x \cdot y\phi(x)}{|\,x\,|^3}dx\right)\varepsilon$$

$$+ \left\{F.p.\int \frac{[3(x \cdot y)^2 - |\,y\,|^2]\,|\,x\,|^2]\phi(x)}{|\,x\,|^5}dx - \frac{4\pi}{3}\,|\,y\,|^2\,\phi(0)\right\}\varepsilon^2 + \cdots,$$

$$\text{as } \varepsilon \to 0.$$

Some interesting expansions are obtained by using a change of variables in (4.2.2). Indeed, if $\Psi(x)$ is a smooth transformation of $\mathbb{R}^n$ to itself with a non-singular Jacobian, then replacing x by $\Psi(x)$ in (4.2.2) yields

$$f(\Psi(x)+\varepsilon y) \sim \sum_{N=0}^{\infty}(D^N f(\Psi(x))\|y^N)\frac{\varepsilon^N}{N!}, \quad \text{as } \varepsilon \to 0. \tag{4.2.7}$$

In particular, if we apply (4.2.7) to a distribution $g(x_1)$, $g \in \mathcal{D}'(\mathbb{R})$, that depends only on x_1 but not on $x_2, \cdots, x_n$, with $y = (1, 0, \cdots, 0)$ we obtain

$$g(\psi(x)+\varepsilon) \sim \sum_{k=0}^{\infty}\frac{g^{(k)}(\psi(x))}{k!}\varepsilon^k, \quad \text{as } \varepsilon \to 0, \tag{4.2.8}$$

where $\psi = \psi_1$ is the first component of Ψ. We emphasize that in order for (4.2.8) to apply, the gradient $D\psi$ cannot vanish at any point.

Example 83. If $D\psi$ does not vanish near the hypersurface defined by $\psi = 0$ then taking $g(x) = \delta(x)$ in (4.2.8) yields the Gel'fand–Shilov expansion [53]

$$\delta(\psi(x)+\varepsilon) \sim \sum_{k=0}^{\infty}\frac{\delta^{(k)}(\psi(x))}{k!}\varepsilon^k, \quad \text{as } \varepsilon \to 0. \tag{4.2.9}$$

When we replace the fixed vector y by a smooth function $\rho(x)$ of the vector x then we obtain the expansion

$$f(x + \varepsilon\rho(x)) \sim \sum_{N=0}^{\infty} (D^N f(x)\|\rho^N(x))\frac{\varepsilon^N}{N!}$$

$$\sim \sum_{|k|=0}^{\infty} \frac{D^k f(x)\rho^k(x)}{k!}\varepsilon^{|k|}, \tag{4.2.10}$$

where we assume that ε is so small that the matrix $(\delta_{ij} + \varepsilon\frac{\partial\rho_i}{\partial x_j})$ is non-singular.

The proof of (4.2.10) is as follows. If ϕ is a test function, then

$$< f(x + \varepsilon\rho(x)), \phi(x) > = < f(y), \psi(y) >, \tag{4.2.11}$$

where

$$y = x + \varepsilon\rho(x), \tag{4.2.12}$$

$$\psi(y) = \frac{\phi(x)}{\det (I + \varepsilon D\rho)}. \tag{4.2.13}$$

If we can show that $\psi(y)$ has the expansion

$$\psi(y) \sim \sum_{|k|=0}^{\infty} (-1)^{|k|} D^k(\phi(y)\rho^k(y))\varepsilon^{|k|}, \tag{4.2.14}$$

then (4.2.10) would follow. But (4.2.10) certainly holds whenever $f(x)$ is smooth, and thus (4.2.14) holds distributionally. Since $\psi(y)$ has a pointwise expansion then the distributional expansion (4.2.14) should also hold pointwise.

Example 84. If we take $f(x) = \delta(x - a)$, we obtain the expansion

$$\delta(x - a + \varepsilon\rho(x)) \sim \sum_{|k|=0}^{\infty} \frac{D^k\delta(x - a)\rho^k(x)}{k!}\varepsilon^{|k|}, \tag{4.2.15}$$

as $\varepsilon \to 0$.

Example 85. Let $\psi(x)$ be a function with $D\psi \neq 0$ near the surface $\psi(x) = 0$. Then using (4.2.10) with $f(x) = \delta(x_1)$ and changing x_1 to $\psi(x)$, we obtain

$$\delta(\psi(x) + \varepsilon\rho(x)) \sim \sum_{k=0}^{\infty} \frac{\delta^{(k)}(\psi(x))\rho^k(x)}{k!}\varepsilon^k \quad \text{as } \varepsilon \to 0. \tag{4.2.16}$$

This formula is the asymptotic version of a result of Caboz, Codaccioni and Constantinescu [16], who consider the case when $\psi(x) = \mathbb{Q}(x) - c$, where $\mathbb{Q}(x)$ is a positive definite quadratic form and $c > 0$.

Actually, if $\psi_1(x), \cdots, \psi_r(x)$ are such that the matrix $(D\psi_1, \cdots, D\psi_r)$ has rank r near the manifold $\psi_1(x) = 0, \cdots, \psi_r(x) = 0$ defined by them then (4.2.16) admits the generalization

$$\delta(\psi_1(x) + \varepsilon\rho_1(x), \cdots, \psi_r(x) + \varepsilon\rho_r(x))$$

$$\sim \sum_{k \in \mathbb{N}^r} \frac{D^k \delta(\psi_1(x), \cdots, \psi_r(x))\rho^k(x)}{k!}\varepsilon^{|k|}, \qquad (4.2.17)$$

for the delta functions concentrated on perturbations of an $(n-r)$-dimensional manifold.

The Taylor expansion (4.2.2) as well as its generalization (4.2.10) are not convergent, in general. However, if $f \in \mathcal{E}'(\mathbb{R}^n)$ and ϕ is real-analytic in a neighborhood of supp f then the Taylor series of ϕ converges uniformly in supp f for ε small enough and it follows that

$$< f(x + \varepsilon y), \phi(x) > = \sum_{N=0}^{\infty} < D^N f(x) \| y^N, \phi(x) > \frac{\varepsilon^N}{N!}, \qquad (4.2.18)$$

for $| \varepsilon | < r_0$.

If $\rho(x)$ is also real-analytic in a neighborhood of supp f then (4.2.14) is convergent if $| \varepsilon | < r_1$. Therefore,

$$< f(x + \varepsilon\rho(x)), \phi(x) > = \sum_{N=0}^{\infty} < D^N f(x) \| \rho^N, \phi(x) > \frac{\varepsilon^N}{N!}. \qquad (4.2.19)$$

4.3 The Multi-Dimensional Moment Asymptotic Expansion

In this section we study the multi-dimensional moment asymptotic expansion. As was the case with the asymptotic developments in one variable, the moment asymptotic expansion is the basic tool for the distributional expansions in several variables.

If f is a generalized function of any of the spaces $\mathcal{E}'(\mathbb{R}^n), \mathcal{P}'(\mathbb{R}^n), \mathcal{O}'_C(\mathbb{R}^n), \mathcal{O}'_M(\mathbb{R}^n)$ or $\mathcal{K}'(\mathbb{R}^n)$ then it has well-defined moments $\mu_k = \mu_k(f)$, given by

$$\mu_k = < f(x), x^k > = < f(x_1, \ldots, x_n), x_1^{k_1} \ldots x_n^{k_n} >, \qquad (4.3.1)$$

for $k \in \mathbb{N}^n$. In general, distributions of the spaces $\mathcal{D}'(\mathbb{R}^n)$ or $\mathcal{S}'(\mathbb{R}^n)$ do not possess moments of all orders.

We proved in the Chapter 3 that the moment asymptotic expansion holds for a large class of distributions, including distributions of fast decay and distributions of rapid oscillation. The same result is true in several variables.

Theorem 38. *Let $\mathcal{A}(\mathbb{R}^n)$ be any of the spaces $\mathcal{E}, \mathcal{P}, \mathcal{O}_C, \mathcal{O}_M$ or $\mathcal{K}$. Then if $f \in \mathcal{A}'(\mathbb{R}^n)$,*

$$f(\lambda x) \sim \sum_{|k|=0}^{\infty} \frac{(-1)^k \mu_k D^k \delta(x)}{\lambda^{|k|+n}}, \quad as \ \lambda \to \infty, \tag{4.3.2}$$

where μ_k are the moments of the generalized function f,

$$\mu_k = \ <f(x), x^k>, \tag{4.3.3}$$

in the sense that if $\phi \in \mathcal{A}(\mathbb{R}^n)$ then

$$<f(\lambda x), \phi(x)> = \sum_{|k|=0}^{N} \frac{\mu_k D^k \phi(0)}{k! \lambda^{|k|+n}} + O(\frac{1}{\lambda^{N+n+1}}), \quad as \ \lambda \to \infty. \tag{4.3.4}$$

Proof. Using the procedures of Chapter 3, we can show that if $\phi \in X_q$, where $X_q = \{\phi \in \mathcal{A}(\mathbb{R}^n) : D^k \phi(0) = 0 \ \text{for} \ |k| < q\}$, then

$$\|\phi(x/\lambda)\| = O\left(\frac{1}{\lambda^q}\right), \text{as} \ \lambda \to \infty, \tag{4.3.5}$$

for any continuous seminorm $\| \ \|$ of $\mathcal{A}(\mathbb{R}^n)$.

Now let $\phi \in \mathcal{A}(\mathbb{R}^n)$ and let

$$P_N(x) = \sum_{|k|=0}^{N} \frac{D^k \phi(0)}{k!} x^k, \tag{4.3.6}$$

be its Taylor polynomial of order N. Observe that $\phi_N = \phi - P_N$ belongs to X_{N+1}.

Then we have

$$<f(\lambda x), \phi(x)> = \sum_{|k|=0}^{N} \frac{\mu_k D^k \phi(0)}{k! \lambda^{|k|+n}} + R_N(\lambda), \tag{4.3.7}$$

where the remainder $R_N(\lambda)$ is given by $R_N(\lambda) = <f(\lambda x), \phi_N(x)>$. But using (4.3.5),

$$| R_N(\lambda) | = \frac{1}{\lambda^n} | <f(x), \phi_N(x/\lambda)> | = O(\frac{1}{\lambda^{n+N+1}}),$$

since $\|\phi\| = |<f(x), \phi(x)>|$ is a continuous seminorm of $\mathcal{A}(\mathbb{R}^n)$. ∎

Let us give some examples.

Example 86. Let $f(x) = \delta(S_1)$ be the delta function of the unit sphere of $\mathbb{R}^n$. Then $f(\lambda x) = \frac{1}{\lambda}\delta(S_{1/\lambda})$, where $S_r = \{x : | x | = r\}$ is the sphere of radius r. Also $\mu_j = 0$ unless $j = 2k$, in which case

$$\mu_{2k_1,\cdots,2k_n} = \frac{2\Gamma(k_1 + 1/2) \cdots \Gamma(k_n + 1/2)}{\Gamma(N + n/2)}.$$

Use of these values in (4.3.2) yields the asymptotic expansion

$$\delta(S_{1/\lambda}) \sim \sum_{k=0}^{\infty} \frac{c_{k,n} \nabla^{2k} \delta(x)}{(2k)! \lambda^{n-1+2k}}, \quad \text{as } \lambda \to \infty, \tag{4.3.8}$$

where $c_{k,n} = \dfrac{2\Gamma(k + 1/2)\pi^{\frac{n-1}{2}}}{\Gamma(k + n/2)}$.

Example 87. Let R be a bounded set of positive measure in $\mathbb{R}^n$ and let $H_R(x)$ be its characteristic function, given by:

$$H_R(x) = \begin{cases} 1, & x \in R, \\ 0, & x \notin R. \end{cases} \tag{4.3.9}$$

Then

$$H_R(\lambda x) \sim \sum_{|k|=0}^{\infty} \frac{(-1)^{|k|} \mu_k D^k \delta(x)}{k! \lambda^{|k|+n}}, \quad \text{as } \lambda \to \infty, \tag{4.3.10}$$

where

$$\mu_k = \int_R x^k dx, \tag{4.3.11}$$

so that, in particular, $\mu_0 = | R |$, the volume of R.

If the point q is the center of mass of R, defined as

$$q = \frac{1}{| R |} \int_R x dx, \tag{4.3.12}$$

then we have

$$\delta(\lambda \boldsymbol{x} - \boldsymbol{q}) \sim \sum_{|\boldsymbol{k}|=0}^{\infty} \frac{(-1)^{|\boldsymbol{k}|} \boldsymbol{q}^{\boldsymbol{k}} \boldsymbol{D}^{\boldsymbol{k}} \delta(\boldsymbol{x})}{\boldsymbol{k}! \lambda^{|\boldsymbol{k}|+n}}, \tag{4.3.13}$$

and thus

$$H_R(\lambda \boldsymbol{x}) = |R| \, \delta(\lambda \boldsymbol{x} - \boldsymbol{q}) + O(\lambda^{-n-3}). \tag{4.3.14}$$

This formula shows that in many cases the volume R can be replaced in the far field by a point charge at $\boldsymbol{q}$ of mass $|R|$.

Example 88. Let us consider the function $f(\boldsymbol{x}) = e^{-|\boldsymbol{x}|^2}$. Since f belongs to $\mathcal{P}'(\mathbb{R}^n)$, the moment asymptotic expansion holds for f. Its moments are given as $\mu_j = 0$ unless $j = 2k$, in which case

$$\mu_{2k_1, \cdots, 2k_n} = \Gamma(\frac{2k_1 + 1}{2}) \cdots \Gamma(\frac{2k_n + 1}{2}). \tag{4.3.15}$$

Thus, as $\lambda \to \infty$

$$e^{-\lambda |\boldsymbol{x}|^2} = e^{-|\lambda^{1/2} \boldsymbol{x}|^2} \sim \sum_{N=0}^{\infty} \frac{\pi^{n/2}}{2^{2N}} \sum_{|\boldsymbol{k}|=N} \frac{\boldsymbol{D}^{2\boldsymbol{k}} \delta(\boldsymbol{x})}{\boldsymbol{k}! \lambda^{N+n/2}}, \tag{4.3.16}$$

in the space $\mathcal{P}(\mathbb{R}^n)$. Therefore, if $\phi \in \mathcal{P}(\mathbb{R}^n)$ then

$$\int_{\mathbb{R}^n} e^{-\lambda |\boldsymbol{x}|^2} \phi(\boldsymbol{x}) d\boldsymbol{x} \sim \sum_{n=0}^{\infty} \frac{\pi^{n/2}}{2^{2N}} \sum_{|\boldsymbol{k}|=N} \frac{\boldsymbol{D}^{2\boldsymbol{k}} \phi(0)}{\boldsymbol{k}!} \cdot \frac{1}{\lambda^{N+n/2}}$$

$$\sim \pi^{\frac{n}{2}} \left[\frac{\phi(0)}{\lambda^{n/2}} + \frac{\nabla^2 \phi(0)}{4\lambda^{n/2+1}} + (\nabla^4 \phi(0) + \sum_{i \neq j} \frac{\partial^4 \phi(0)}{\partial^2 x_i \partial^2 x_j}) \frac{1}{32\lambda^{n/2+2}} + \cdots \right],$$

$$\text{as } \lambda \to \infty. \tag{4.3.17}$$

The same technique applies to regularizations of divergent integrals.

Example 89. Let us consider the generalized function $f_q(\boldsymbol{x}) = \mathcal{P}f(|\boldsymbol{x}|^q) e^{-|\boldsymbol{x}|^2}$, where $\mathcal{P}f(r^q)$ is the Hadamard finite part regularization defined in Chapter 2. Its moments are given as

$$\mu_j^q = <f_q(\boldsymbol{x}), \boldsymbol{x}^j> = \left(F.p. \int_0^{\infty} e^{-r^2} r^{n-1+q|j|} dr \right) \left(\int_{S_1} \boldsymbol{w}^j d\sigma(\boldsymbol{w}) \right),$$

where $S_1 = \{\boldsymbol{x} \in \mathbb{R}^n : |\boldsymbol{x}| = 1\}$ is the unit sphere in $\mathbb{R}^n$ and $d\sigma$ is the Lebesgue measure on the sphere.

It follows that $\mu_j = 0$ unless $j = 2k$, in which case

$$\mu_{2k}^q = \frac{\Gamma(\mid k \mid +\frac{n+q}{2})\Gamma(k_1 + 1/2) \cdots \Gamma(k_n + 1/2)}{\Gamma(\mid k \mid +n/2)},$$

$$q \neq -n, \ -n - 2, \ -n - 4, \ldots \qquad (4.3.18)$$

while if $q = -n - 2m, m : 0, 1, 2, \cdots$ then

$$\mu_{2k}^q = \frac{(-1)^{m-\mid k \mid}\psi(m- \mid k \mid +1)\Gamma(k_1 + 1/2) \cdots \Gamma(k_n + 1/2)}{(m- \mid k \mid)!\Gamma(\mid k \mid +n/2)}, \ 0 \leq \mid k \mid \leq m,$$

$$(4.3.19)$$

$$\mu_{2k}^q = \frac{\Gamma(\mid k \mid -m)\Gamma(k_1 + 1/2) \cdots \Gamma(k_n + 1/2)}{\Gamma(\mid k \mid +n/2)}, \ \mid k \mid \geq m + 1, \quad (4.3.20)$$

where $\psi(z) = \frac{\Gamma'(z)}{\Gamma(z)}$ is the digamma function.

Therefore, if $q \neq -n - 2m$

$$\mathcal{P}f(\mid \lambda x \mid^q)e^{-\lambda\mid x\mid^2} \sim \sum_{N=0}^{\infty} \frac{\pi^{n/2}\Gamma\left(N + \frac{n+q}{2}\right)}{2^{2N}\Gamma(N + n/2)} \sum_{\mid k \mid=N} \frac{D^{2k}\delta(x)}{k!\lambda^{N+n/2}}, \ \text{as } \lambda \to \infty.$$

$$(4.3.21)$$

Since

$$\mathcal{P}f(\mid \lambda x \mid^q) = \lambda^q \mathcal{P}f(\mid x \mid^q) \qquad (4.3.22)$$

if $\lambda > 0$ and $q \neq -n - 2m$, it follows that if $\phi \in \mathcal{P}(\mathbb{R}^n)$ then

$$F.p. \int_{\mathbb{R}^n} \mid x \mid^q e^{-\lambda\mid x\mid^2} \phi(x)dx \sim \sum_{N=0}^{\infty} \frac{\pi^{n/2}\Gamma(N + \frac{n+q}{2})}{2^{2N}\Gamma(N + n/2)} \sum_{\mid k \mid=N} \frac{D^{2k}\phi(0)}{k!\lambda^{N+q+n/2}},$$

$$(4.3.23)$$

as $\lambda \to \infty$.

When $q = -n - 2m, m : 0, 1, 2, \cdots$, then

$$\mathcal{P}f(\mid \lambda x \mid^{-n-2m})e^{-\lambda\mid x\mid^2}$$

$$\sim \sum_{N=0}^{m} \frac{(-1)^{m-N}\psi(m - N + 1)\pi^{n/2}}{2^{2n}(m - N)!\Gamma(N + n/2)} \sum_{\mid k \mid=N} \frac{D^{2k}\delta(x)}{k!\lambda^{N+n/2}}$$

$$+ \sum_{N=m+1}^{\infty} \frac{\Gamma(N - m)\pi^{n/2}}{2^{2N}\Gamma(N + n/2)} \sum_{\mid k \mid=N} \frac{D^{2k}\delta(x)}{k!\lambda^{N+n/2}}, \qquad (4.3.24)$$

but since

$$\mathcal{P}f\left(\frac{1}{\mid \lambda x \mid^{n+2m}}\right) = \frac{1}{\lambda^{n+2m}}\mathcal{P}f\left(\frac{1}{\mid x \mid^{n+2m}}\right) - \frac{c_{m,n}\ln \lambda}{(2m)!\lambda^{n+2m}}\nabla^{2m}\delta(x),$$

$$(4.3.25)$$

where

$$c_{m,n} = \frac{2\Gamma(m+1/2)\pi^{\frac{n-1}{2}}}{\Gamma(m+n/2)},$$
(4.3.26)

we obtain that if $\phi \in \mathcal{P}(\mathbb{R}^n)$ then

$$F.p. \int_{\mathbb{R}^n} \frac{e^{-\lambda|x|^2}\phi(x)}{|x|^{n+2m}}dx \sim \frac{c_{m,n}\ln\lambda}{(2m)!}\nabla^{2m}(e^{-\lambda|x|^2}\phi(x))|_{x=0}$$

$$+ \sum_{N=0}^{m} \frac{(-1)^{m-N}\psi(m-N+1)}{2^{2N}(m-N)!\Gamma(N+n/2)} \sum_{|k|=N} \frac{D^{2k}\phi(0)}{k!\lambda^{N-n/2-2m}}$$

$$+ \sum_{N=m+1}^{\infty} \frac{\Gamma(N-m)\pi^{n/2}}{2^{2N}\Gamma(N+n/2)} \sum_{|k|=N} \frac{D^{2k}\phi(0)}{k!\lambda^{N-n/2-2m}}, \qquad \text{as } \lambda \to \infty. \quad (4.3.27)$$

Let us now consider some oscillatory generalized functions.

Example 90. Let us find the asymptotic development of the generalized function $e^{i\lambda(x_1^2+\cdots+x_r^2-x_{r+1}^2-\cdots-x_n^2)}$ as $\lambda \to \infty$. Recall the expansions

$$e^{i\lambda x^2} \sim \sum_{n=0}^{\infty} \frac{\Gamma\left(\frac{2n+1}{2}\right)e^{\frac{\pi i(2n+1)}{4}}\delta^{(2n)}(x)}{(2n)!\lambda^{\frac{2n+1}{2}}}, \qquad \text{as } \lambda \to \infty, \quad (4.3.28)$$

$$e^{-i\lambda x^2} \sim \sum_{n=0}^{\infty} \frac{\Gamma\left(\frac{2n+1}{2}\right)e^{-\frac{\pi i(2n+1)}{4}}\phi^{(2n)}(x)}{(2n)!\lambda^{\frac{2n+1}{2}}}, \qquad \text{as } \lambda \to \infty. \quad (4.3.29)$$

Then

$$e^{i\lambda(x_1^2+\cdots+x_r^2-x_{r+1}^2-\cdots-x_n^2)} = e^{i\lambda x_1^2}\cdots e^{i\lambda x_r^2}e^{-i\lambda x_{r+1}^2}\cdots e^{-i\lambda x_n^2}$$

$$\sim \sum_{|k|=0}^{\infty} \frac{e^{\frac{\pi i(n-2r)}{2}}e^{\frac{\pi i}{2}(k_1+\cdots+k_r-k_{r+1}-\cdots-k_n)}\Gamma(k_1+1/2)\cdots\Gamma(k_n+1/2)D^{2k}\delta(x)}{(2k)!\lambda^{|k|+n/2}}.$$

(4.3.30)

The asymptotic expansion of periodic functions of several variables can also be obtained. Let f be a function or distribution defined in $\mathbb{R}^n$ and let $P = \{p : f(x+p) = f(x)\}$ be its set of periods. We say that f is n-periodic if the set P generates $\mathbb{R}^n$ or, equivalently, if there are n linearly independent periods $p_1, \cdots, p_n$. In this case we let V be the parallelepiped $V = \{\sum_{i=1}^{n} t_i p_i : 0 \le t_i \le 1\}$. Then, as in the one-dimensional situation, we can show that each n-periodic generalized function having $p_1, \cdots, p_n$ as periods has a "mean" value, which is given by

$$c = \frac{1}{|V|} \int_V f(x)dx, \qquad (4.3.31)$$

in the case where f is locally integrable.

We have then an expansion similar to that in one dimension.

Theorem 39. *Let $f(x)$ be an n-periodic generalized function defined in $\mathbb{R}^n$ with mean c. Then*

$$f(\lambda x) = c + o(\lambda^{-\infty}), \quad as\ \lambda \to \infty, \qquad (4.3.32)$$

in the space $\mathcal{S}(\mathbb{R}^n)$.

Observe that (4.3.32) will usually hold in spaces larger than $\mathcal{S}(\mathbb{R}^n)$. For instance, if $f(x,y) = e^{ix}$, then

$$< f(\lambda x, \lambda y), \phi(x, y) > = \int_{-\infty}^{\infty} \int_{-\infty}^{\infty} e^{i\lambda x} \phi(x, y)dx dy$$

$$= \int_{-\infty}^{\infty} e^{i\lambda x} \int_{-\infty}^{\infty} \phi(x, y)dy dx,$$

so that

$$< f(\lambda x, \lambda y), \phi(x, y) > = o(\lambda^{-\infty}),$$

as long as $\Phi(x) = \int_{-\infty}^{\infty} \phi(x, y)dy$ belongs to $\mathcal{K}(\mathbb{R})$. However, the space where (4.3.32) holds depends on f.

Example 91. Let us consider the distribution

$$f(x) = \sum_{k \in \mathbb{Z}^n} \delta(x - k_1 p_1 - \cdots - k_n p_n), \qquad (4.3.33)$$

where $p_1, \cdots, p_n$ is a base of $\mathbb{R}^n$. Then f is n-periodic, with mean $|V|^{-1}$.

Thus $f(\lambda x) = |V|^{-1} + o(\lambda^{-\infty})$ and setting $\lambda = \frac{1}{\varepsilon}$ we obtain

$$\sum_{k \in \mathbb{Z}} \phi(\varepsilon(k_1 p_1 + \cdots + k_n p_n)) = \frac{1}{|V| \varepsilon^n} \int_{\mathbb{R}^n} \phi(x)dx + o(\varepsilon^{\infty}), \qquad (4.3.34)$$

as $\varepsilon \to 0^+$.

4.4 Laplace's Formula

We shall now consider the asymptotic evaluation of multi-dimensional integrals of the type

$$I(\lambda) = \int_R e^{-\lambda h(x)} \phi(x) dx, \qquad (4.4.1)$$

where $h(x)$ is a *real* function and R is a region of $\mathbb{R}^n$.

We assume that both $h(x)$ and $\phi(x)$ are smooth in a region containing the closure $\overline{R}$ of the region R. As far as the region R is concerned we assume that its boundary ∂R consists of a finite union of smooth hypersurfaces. Observe that the smoothness of the region of integration was not a question in the one-dimensional situation since the only one-dimensional regions are the open intervals.

It was shown by Focke [48] that the main contribution to $I(\lambda)$ for $\lambda >> 1$ comes from the vicinity of the minima of $h(x)$ in $\overline{R}$. The isolated critical points of h where minima occur are classified into three types according to their location. The interior critical points are of type I. The critical points of the smooth parts of the boundary are of type II while those situated on the non-smooth parts of the boundary, namely edges and corners, are of type III. The asymptotic expansion takes a somewhat different form depending on the type of isolated critical point. Observe, however, that minima of h can also be located along more complicated subsets of R, such as lower dimensional manifolds; the asymptotic analysis in such cases could be rather complex.

Let us start with a type I minimum. Suppose that x_0 is an interior minimum of the function h. We wish to find an asymptotic approximation of the integral

$$I(\lambda) = \int_{\mathbb{R}^n} e^{-\lambda h(x)} \phi(x) dx, \qquad (4.4.2)$$

where $\phi \in \mathcal{D}(U)$ and U is a small neighborhood of x_0 that contains no other critical points of h.

Since x_0 is an interior minimum of h, it follows that $\frac{\partial h}{\partial x_i} |_{x_0} = 0$ for $1 \leq i \leq n$. If the Hessian matrix $A = \left[\frac{\partial^2 h}{\partial x_i \partial x_j} |_{x_0} \right]$ is positive definite we say that x_0 is non-degenerate and we suppose that this is the case.

Under these circumstances it is possible to find a local change of variables $y = \Psi(x) = (\psi_i(x))$ with $\Psi(x_0) = 0$, $\frac{\partial(\psi_1, \cdots, \psi_n)}{\partial(x_1, \cdots, x_n)} > 0$ and with

$$h(x) = h(x_0) + |\Psi(x)|^2, \qquad (4.4.3)$$

for x near x_0.

Then, if we use (4.3.16), we obtain

$$e^{-\lambda h(\boldsymbol{x})} = e^{-\lambda h(\boldsymbol{x}_0)} e^{-\lambda |\Psi(\boldsymbol{x})|^2}$$

$$\sim e^{-\lambda h(\boldsymbol{x}_0)} \sum_{N=0}^{\infty} \frac{\pi^{n/2}}{2^{2N}} \sum_{|\boldsymbol{k}|=N} \frac{D^{2\boldsymbol{k}} \delta(\Psi(\boldsymbol{x}))}{\boldsymbol{k}! \lambda^{n/2+N}}, \quad \text{as} \ \ \lambda \to \infty, \tag{4.4.4}$$

in the space $\mathcal{D}(U)$.

Since

$$\delta(\Psi(\boldsymbol{x})) = \frac{\delta(\boldsymbol{x} - \boldsymbol{x}_0)}{\det(D\Psi(\boldsymbol{x}_0))} = \frac{2^{n/2} \delta(\boldsymbol{x} - \boldsymbol{x}_0)}{\sqrt{\det D^2 h(\boldsymbol{x}_0)}}, \tag{4.4.5}$$

the leading term of expansion (4.4.4) takes the form

$$e^{-\lambda h(\boldsymbol{x})} = e^{-\lambda h(\boldsymbol{x}_0)} \left[\left(\frac{2\pi}{\lambda}\right)^{n/2} (\det A)^{-1/2} \delta(\boldsymbol{x} - \boldsymbol{x}_0) + O\left(\frac{1}{\lambda^{n/2+1}}\right) \right],$$
$$\tag{4.4.6}$$

as $\lambda \to \infty$ in the space $\mathcal{D}(U)$. This is the distributional Laplace asymptotic formula. Evaluation at ϕ gives the standard Laplace formula

$$I(\lambda) = \int_{\mathbb{R}^n} e^{-\lambda h(\boldsymbol{x})} \phi(\boldsymbol{x}) d\boldsymbol{x} \sim e^{-\lambda h(\boldsymbol{x}_0)} \left[\left(\frac{2\pi}{\lambda}\right)^{n/2} \frac{\phi(\boldsymbol{x}_0)}{\sqrt{\det A}} + O\left(\frac{1}{\lambda^{n/2+1}}\right) \right],$$
$$\tag{4.4.7}$$

as $\lambda \to \infty$.

The distributional framework also permits us to obtain the development of the integral $I(\lambda)$ about degenerate critical points. In order to show how this can be done, we shall need the moment asymptotic expansion of the function $e^{-x^{2k}}$, $k = 1, 2, 3, \cdots$, which belongs to $\mathcal{P}(\mathbb{R})$. It is given by

$$e^{-\lambda x^{2k}} \sim \sum_{n=0}^{\infty} \frac{\Gamma\left(\frac{2n+1}{2k}\right) \delta^{(2n)}(x)}{k(2n)! \lambda^{\frac{2n+1}{2k}}}, \quad \text{as} \ \ \lambda \to \infty. \tag{4.4.8}$$

Example 92. Let us consider the integral

$$I(\lambda) = \int_{-\infty}^{\infty} \int_{-\infty}^{\infty} e^{-\lambda(x^2 + y^4)} \phi(x, y) dx dy, \tag{4.4.9}$$

where $\phi \in \mathcal{P}(\mathbb{R}^2)$. Use of (4.4.8) yields

$$e^{-\lambda(x^2 + y^4)} = e^{-\lambda x^2} e^{-\lambda y^4} \sim \left[\sum_{n=0}^{\infty} \frac{\Gamma\left(\frac{2n+1}{2}\right) \delta^{(2n)}(x)}{(2n)! \lambda^{\frac{2n+1}{2}}} \right] \left[\sum_{m=0}^{\infty} \frac{\Gamma\left(\frac{2m+1}{4}\right) \delta^{(2m)}(y)}{2(2m)! \lambda^{\frac{2m+1}{4}}} \right]$$

$$\sim \frac{\Gamma(1/2)\Gamma(1/4)\delta(x)\delta(y)}{2\lambda^{3/4}} + \frac{\Gamma(1/2)\Gamma(3/4)\delta(x)\delta''(y)}{4\lambda^{5/4}}$$

$$+ \left[\frac{\Gamma(1/2)\Gamma(5/4)\delta(x)\delta^{(4)}(y)}{48} + \frac{\Gamma(3/2)\Gamma(1/4)\delta''(x)\delta(y)}{4\lambda^{5/4}} \right] \frac{1}{\lambda^{7/4}} + \cdots$$

and it follows that

$$I(\lambda) \sim \frac{\Gamma(1/2)\Gamma(1/4)\phi(0,0)}{2\lambda^{3/4}} + \frac{\Gamma(1/2)\Gamma(3/4)\frac{\partial^2 \phi}{\partial y^2}(0,0)}{4\lambda^{5/4}}$$

$$+ \left[\frac{\Gamma(1/2)\Gamma(5/4)\frac{\partial^4 \phi}{\partial y^4}(0,0)}{48} + \frac{\Gamma(3/2)\Gamma(1/4)\frac{\partial^2 \phi}{\partial y^2}(0,0)}{4} \right] \frac{1}{\lambda^{7/4}} + \cdots$$

as $\lambda \to \infty$.

A similar analysis can be applied to integrals of the type

$$I(\lambda) = \int_{\mathbb{R}^n} e^{-\lambda h(\boldsymbol{x})} \phi(\boldsymbol{x}) d\boldsymbol{x},$$

where $\phi \in \mathcal{D}(U)$, and U is a small neighborhood of the critical point $\boldsymbol{x}_0$, if we can find a local change of variables $\boldsymbol{y} = \boldsymbol{\Psi}(\boldsymbol{x})$ with $\boldsymbol{\Psi}(\boldsymbol{x}_0) = 0$, $\frac{\partial(\psi_1, \cdots, \psi_n)}{\partial(x_1, \cdots, x_n)} > 0$ such that

$$h(\boldsymbol{x}) = h(\boldsymbol{x}_0) + \psi_1^{2k_1}(\boldsymbol{x}) + \cdots + \psi_n^{2k_n}(\boldsymbol{x}).$$

If $k_1 \leq k_2 \leq \cdots \leq k_n$, we obtain

$$e^{-\lambda h(\boldsymbol{x})} \sim e^{-\lambda h(\boldsymbol{x}_0)} \left[\frac{\Gamma(1/2k_1) \cdots \Gamma(1/2k_n)\delta(\boldsymbol{x} - \boldsymbol{x}_0)}{k_1 \cdots k_n(\det \boldsymbol{D}\,\boldsymbol{\Psi}(\boldsymbol{x}_0))\lambda^{\frac{1}{2k_1} + \cdots + \frac{1}{2k_n}}} \right. \tag{4.4.10}$$

$$\left. + O\left(\frac{1}{\lambda^{\frac{1}{2k_1} + \cdots + \frac{1}{2k_{n-1}} + \frac{3}{2k_n}}} \right) \right], \quad \text{as} \lambda \to \infty.$$

Let us now consider a type II critical point. Thus, let $\boldsymbol{x}_0$ be an isolated local minimum of $h(\boldsymbol{x})$ in $\overline{R}$ located on a smooth part Σ of the boundary ∂R. We say that $\boldsymbol{x}_0$ is non-degenerate if the following two conditions hold:

(a) the exterior normal derivative of h is negative at $\boldsymbol{x}_0$.

(b) The point $\boldsymbol{x}_0$ is a non-degenerate type I minimum of the restriction of h to Σ.

This means that if $(v_1, \cdots, v_{n-1})$ is any local coordinate system on Σ, then at $x = x_0$ the partial derivatives $\dfrac{\partial h}{\partial v_\alpha}$ vanish and the Hessian matrix $\left(\dfrac{\partial^2 h}{\partial v_\alpha \partial v_\beta}\right)$ is positive definite.

When x_0 is a non-degenerate type II minimum, then we can find a local change of variables $y = \boldsymbol{\Psi}(x) = (\psi_i(x))$ with $\boldsymbol{\Psi}(x_0) = 0$ such that the inequality $\psi_1(x) > 0$ represents the region R in the vicinity of x_0 and such that

$$h(x) = h(x_0) + \psi_1(x) + \psi_2^2(x) + \cdots + \psi_n^2(x). \tag{4.4.11}$$

Observe that under these circumstances the surface Σ is given by $\psi_1(x) = 0$ near $x = x_0$.

Then if $H_R(x)$ is the characteristic function of the region R, we have

$$e^{-\lambda h(x)} H_R(x) = e^{-\lambda h(x_0)} e^{-\lambda \psi_1(x)} H(\psi_1(x)) e^{-\lambda \psi_2^2(x)} \cdots e^{-\lambda \psi_n^2(x)}$$

$$\sim e^{-\lambda h(x_0)} \left[\sum_{k=0}^{\infty} \frac{(-1)^k \delta^{(k)}(\psi_1)}{\lambda^{k+1}} \right] \prod_{j=2}^{n} \left[\sum_{k=0}^{\infty} \frac{\Gamma(\frac{2k+1}{2}) \delta^{(2k)}(\psi_j)}{(2k)! \lambda^{\frac{2k+1}{2}}} \right]. \tag{4.4.12}$$

The first order approximation is given by

$$e^{-\lambda h(x)} H_R(x) = e^{-\lambda h(x_0)} \left[\frac{\pi^{\frac{n-1}{2}} \delta(\boldsymbol{\Psi}(x))}{\lambda^{\frac{n+1}{2}}} + O\left(\frac{1}{\lambda^{\frac{n+3}{2}}}\right) \right],$$

or

$$e^{-\lambda h(x)} H_R(x) = e^{-\lambda h(x_0)} \left[\frac{2^{\frac{n}{2}} \pi^{\frac{n-1}{2}} \delta(x - x_0)}{\sqrt{\det A} \lambda^{\frac{n+1}{2}}} + O\left(\frac{1}{\lambda^{\frac{n+3}{2}}}\right) \right], \tag{4.4.13}$$

where A is the Hessian matrix $D^2(h - \psi_1 + \psi_1^2)\big|_{x_0}$.

Evaluation of (4.4.13) at a test function ϕ yields Laplace's formula for a non-degenerate type II minimum as

$$\int_R e^{-\lambda(x)} \phi(x) dx \sim e^{-\lambda(x_0)} \frac{2^{\frac{n}{2}} \pi^{\frac{n-1}{2}} \phi(x_0)}{\sqrt{\det A} \lambda^{\frac{n+1}{2}}}. \tag{4.4.14}$$

Degenerate type II points can be handled in certain cases. If we can find a local change of variables $\boldsymbol{\Psi}(x) = (\psi_i(x))$ with $\boldsymbol{\Psi}(x_0) = 0$ such that the inequality $\psi_1(x) > 0$ represents R near $x = x_0$ and such that

$$h(x) = h(x_0) + \psi_1(x)^k + \psi_2(x)^{2k_2} + \cdots + \psi_n(x)^{2k_n}, \tag{4.4.15}$$

then

$$e^{-\lambda h(x)} = e^{-\lambda h(x_0)} e^{-\lambda \psi_1(x)^k} H(\psi_1(x)) e^{-\lambda \psi_2(x)^{2k_2}} \cdots e^{-\lambda \psi_n(x)^{2k_n}}$$

$$\sim e^{-\lambda h(x_0)} \frac{\Gamma(\frac{1}{k})\Gamma(\frac{1}{2k_2}) \cdots \Gamma(\frac{1}{2k_n})\delta(\psi_1) \cdots \delta(\psi_n)}{k \cdot k_2 \cdots k_n \quad \lambda^{\frac{1}{k} + \frac{1}{2k_2} + \cdots + \frac{1}{2k_n}}}. \qquad (4.4.16)$$

In particular, suppose x_0 is a non-degenerate local minimum of h in a region R_1 that contains $\overline{R}$. If x_0 happens to be located in the smooth part of the boundary ∂R, then it would be a degenerate type II point and the previous formula, with $k = 2$, $k_2 = \cdots = k_n = 1$, would be applicable. Hence

$$e^{-\lambda h(x)} H_R(x) \sim \frac{1}{2} e^{-\lambda h(x_0)} \left(\frac{2\pi}{\lambda}\right)^{\frac{n}{2}} (\det A)^{\frac{-1}{2}} \delta(x - x_0), \qquad (4.4.17)$$

where $A = D^2 h\big|_{x_0}$. Therefore, when a non-degenerate type I critical point becomes a type II minimum then Laplace's formula (4.4.6) has to be modified by multiplying by the factor $\frac{1}{2}$.

The analysis of type III critical points can be rather complex, since there are many possibilities for the geometry of the non-smooth part of the boundary. We illustrate the ideas by considering the case of a corner in the boundary of a two-dimensional region.

Let $x_0 = (x_1^o, x_2^o)$ be a corner of the boundary ∂R. In such a case we can find a neighborhood V of x_0 and two smooth functions $\rho_1(x), \rho_2(x)$ defined in V, with non-vanishing Jacobian $\frac{\partial(\rho_1, \rho_2)}{\partial(x_1, x_2)}$, such that the inequalities $\rho_1(x) > 0$, $\rho_2(x) > 0$ describe the set $V \cap R$. The boundary ∂R in the vicinity of x_0 will consist of two parts, the curve $C_1 : \rho_1(x) = 0$, $\rho_2(x) \geq 0$ and the curve $C_2 : \rho_2(x) = 0$, $\rho_1(x) \geq 0$. The curves C_1 and C_2 meet at x_0 and they make an angle α. We suppose that $0 < \alpha < \pi$.

We denote by $T_i(x)$ the unit tangent vector to $x \in C_i$ in the direction of movement toward x_0. At $x = x_0$ there are two unit tangent vectors T_1 and T_2. They make an angle α, so that

$$T_1 \cdot T_2 = \cos \alpha. \qquad (4.4.18)$$

Observe that the derivative of h with respect to arc length along C_1 and C_2 is given by the directional derivative $\frac{dh}{ds} = Dh \cdot T_i$, $i = 1, 2$. At the corner there are two derivative values $\left(\frac{dh}{ds}\right)_i = Dh \cdot T_i$, $i = 1, 2$.

Suppose now that the corner x_0 is an isolated minimum of $h(x)$ in $\overline{R}$ which is non-degenerate in the sense that $\left(\frac{dh}{ds}\right)_i < 0$, $i = 1, 2$. It follows that near x_0 the function $h(x)$ can be written as

$$h(x) = h(x_0) + c_1 \rho_1(x) + \rho_2(x) + O(\rho_1^2(x) + \rho_2^2(x)), \qquad (4.4.19)$$

where $c_1 > 0$, $c_2 > 0$. A further change of variables will allow us to replace $\rho_1(x), \rho_2(x)$ by two new functions $\psi_1(x), \psi_2(x)$ such that, as before, the inequalities $\psi_1(x) > 0$, $\psi_2(x) > 0$ locally describe R, and the curves C_i are given by $\psi_i(x) = 0$, $\psi_j(x) \geq 0$, $j \neq i$, but with

$$h(x) = h(x_0) + c_1\psi_1(x) + c_2\psi_2(x). \tag{4.4.20}$$

It follows that

$$e^{-\lambda h(x)} H_R(x) = e^{-\lambda(h(x_0)+c_1\psi_1(x)+c_2\psi_2(x))} H(\psi_1(x))H(\psi_2(x))$$

$$\sim e^{-\lambda h(x_0)} \left(\sum_{k=0}^{\infty} \frac{(-1)^k \delta^{(k)}(\psi_1)}{c_1^{k+1}\lambda^{k+1}} \right) \left(\sum_{j=0}^{\infty} \frac{(-1)^j \delta^{(j)}(\psi_2)}{c_2^{j+1}\lambda^{j+1}} \right),$$

or

$$e^{-\lambda h(x)} H_R(x) \sim e^{-\lambda h(x_0)} \sum_{m=0}^{\infty}(-1)^m \sum_{k+j=m} \frac{\delta^{(k)}(\psi_1)\delta^{(j)}(\psi_2)}{c_1^k c_2^j} \cdot \frac{1}{\lambda^{m+2}}.$$

$$\tag{4.4.21}$$

The first term of this asymptotic series is given by $\dfrac{e^{-\lambda h(x_0)}\delta(\psi_1)\delta(\psi_2)}{c_1 c_2 \lambda^2}$.

But an easy computation shows that $c_i = \dfrac{1}{|\,D\psi_i\,|\sin\alpha}\left(\dfrac{dh}{ds}\right)_i$, while

$\left|\dfrac{\partial(\psi_1,\psi_2)}{\partial(x_1,x_2)}\right| = |\,D\psi_1\,||\,D\psi_2\,|\sin\alpha$. Thus

$$e^{-\lambda h(x)} H_R(x) \sim \frac{e^{-\lambda h(x_0)}\sin\alpha \delta(x - x_0)}{\left(\frac{dh}{ds}\right)_1 \left(\frac{dh}{ds}\right)_2 \lambda^2}. \tag{4.4.22}$$

This is the distributional version of the Laplace formula

$$\int_R e^{-\lambda h(x)}\phi(x)dx \sim \frac{e^{-\lambda h(x_0)}\sin\alpha\phi(x_0)}{\left(\frac{dh}{ds}\right)_1 \left(\frac{dh}{ds}\right)_2 \lambda^2}, \quad \text{as } \lambda \to \infty, \tag{4.4.23}$$

valid if $\phi \in \mathcal{D}(U)$, where U is any neighborhood of x_0 such that the function $h(x)$ has a global minimum in $\bar{R} \cap \bar{U}$ at x_0.

We now give several examples of these ideas.

Example 93. Let us consider the expansion of the generalized function

$$F(\lambda, x) = F(\lambda, x; x_0) = e^{-\lambda|x-x_0|^2} H_R(x), \tag{4.4.24}$$

for $\lambda >> 1$. We wish to analyze how the expansion changes as the point x_0 moves from the region R to its complement. We assume that the boundary ∂R is smooth.

Case I. $x_0 \in R$.

Here x_0 is the global minimum of $h(x) = |x - x_0|^2$ in $\bar{R}$. It is of type I. Since the Hessian matrix $D^2 h \mid x_0$ is twice the identity matrix, we obtain from (4.4.6) the approximation

$$F(\lambda, x) \sim \left(\frac{\pi}{\lambda}\right)^{\frac{n}{2}} \delta(x - x_0), \quad \text{as } \lambda \to \infty. \tag{4.4.25}$$

Case II. $x_0 \in \partial R$.

In this situation x_0 is still the global minimum of $h(x)$ in $\bar{R}$. It is a degenerate type II point. Formula (4.4.17) is applicable and we obtain

$$F(\lambda, x) \sim \frac{1}{2}\left(\frac{\pi}{\lambda}\right)^{n} \delta(x - x_0), \quad \text{as } \lambda \to \infty. \tag{4.4.26}$$

Case III. $x_0 \notin \bar{R}$

Now the minima of $|x - x_0|^2$ in $\bar{R}$ are located on the boundary. Suppose first that the minimum is achieved at a unique point y_1. Then y_1 is a non-degenerate type II point. Thus

$$F(\lambda, x) \sim \frac{e^{-\lambda d^2} 2^{\frac{n}{2}} \delta(x - y_1)}{\lambda^{\frac{n+1}{2}} \sqrt{\det A_1}}, \tag{4.4.27}$$

where A_1 is the Hessian matrix of $h - \psi - \psi^2$ at $x = y_1$, the region R being given locally by $\psi_1(x) > 0$, and where d is the distance from x_0 to R. When the minimum is achieved at finitely many points $y_1, \cdots, y_k$ then (4.4.27) has to be replaced by the formula

$$F(\lambda, x) \sim \frac{e^{-\lambda d^2} 2^{\frac{n}{2}}}{\lambda^{\frac{n+1}{2}}} \sum_{j=1}^{k} \frac{\delta(x - y_j)}{\sqrt{\det A_j}}. \tag{4.4.28}$$

Example 94. Let us consider the expansion of the generalized function $F(\lambda, x)$ of Example 93 when R is the region $R = \{x_0 + rw : r > d, w \in E\}$, where E is a given open subset of the unit sphere $\{w \in \mathbb{R}^n : |w| = 1\}$. This corresponds to Case III above, but now all the points of the form $x_0 + dw$, $w \in \bar{E}$, are minima for $h(x) = |x - x_0|^2$.

If $\phi \in \mathcal{P}(\mathbb{R}^n)$ then

$$\int_R e^{-\lambda |x - x_0|^2} \phi(x) dx = \int_d^\infty e^{-\lambda r^2} r^{n-1} \int_E \phi(x_0 + rw) d\sigma(w)$$

$$\sim \frac{e^{-\lambda d^2} d^{n-2}}{2\lambda} \int_E \phi(\boldsymbol{x}_0 + d\boldsymbol{w})d\sigma(\boldsymbol{w}),$$

and so

$$e^{-\lambda|\boldsymbol{x}-\boldsymbol{x}_0|} H_R(\boldsymbol{x}) \sim \frac{e^{-\lambda d^2} d^{n-2}}{2\lambda} \delta(\Sigma), \quad \text{as } \lambda \to \infty, \tag{4.4.29}$$

where the surface Σ is the part of ∂R given by $\Sigma = \{\boldsymbol{x}_0 + d\boldsymbol{w} : \boldsymbol{w} \in E\}$. Actually, (4.4.29) remains valid for regions more general than this R: it suffices to require that the boundary of the region and the closed ball $\{\boldsymbol{x}_0 + \boldsymbol{z} :| \boldsymbol{z} |\leq d\}$ meet along the closed surface $\bar{\Sigma}$.

Observe that the leading term in (4.4.28) is of the order $\frac{e^{-\lambda d^2}}{\lambda^{\frac{n+1}{2}}}$ while in (4.4.29) it is of the order $\frac{e^{-\lambda d^2}}{\lambda}$.

Example 95. We now consider the expansion of $F(\lambda, \boldsymbol{x})$ when the relevant critical point is a corner in a two-dimensional region. We suppose that the region R is given by the inequalities $\psi_1(x,y) > 0$, $\psi_2(x,y) > 0$, while the boundary ∂R consists of the parts $C_1 : \psi_1 = 0$, $\psi_2 \geq 0$ and $C_2 : \psi_2 = 0$, $\psi_1 \geq 0$. We assume that the corner is located at the origin. Let T_1 and T_2 be the unit tangent vectors to the curves C_i at the corner. Let α be the angle at the corner, where $0 < \alpha < \pi$.

If $\boldsymbol{x}_0$ belongs to the sector determined by T_1 and T_2 then the minimum of $| \boldsymbol{x} - \boldsymbol{x}_0 |^2$ occurs at the corner $\boldsymbol{x} = \boldsymbol{0}$. Since $\left(\frac{dh}{ds}\right)_i = -2\boldsymbol{x}_0 \cdot T_i$, formula (4.4.22) yields the approximation

$$F(\lambda, \boldsymbol{x}) \sim \frac{e^{-\lambda d^2} \sin \alpha \delta(\boldsymbol{x})}{4(\boldsymbol{x}_0 \cdot T_1)(\boldsymbol{x}_0 \cdot T_2)\lambda^2}. \tag{4.4.30}$$

When $\boldsymbol{x}_0 = \boldsymbol{0}$ we have the coincidence of types I and III. Then (4.4.30) cannot be applied. To obtain the approximation we proceed as follows. Let $\theta = \alpha_i(r)$, $i = 1, 2$, be the description of the curves C_i in polar coordinates (r, θ) near the origin, so that locally the region is given by the inequalities $\alpha_1(r) < \theta < \alpha_2(r)$, $r > 0$. Observe that $\alpha = \alpha_2(0) - \alpha_1(0)$ since $\alpha_i(0)$ is the angle the curve C_i makes with the positive real axis.

Then if $\phi \in \mathcal{D}(U)$, where U is a small neighborhood of the origin, we have

$$< F(\lambda, \boldsymbol{x}), \theta(\boldsymbol{x}) >= \int_0^\infty e^{-\lambda r^2} r \int_{\alpha_1(r)}^{\alpha_2(r)} \phi(r \cos \theta, r \sin \theta)d\theta dr$$

$$\sim \frac{\alpha \phi(0)}{2\lambda},$$

therefore

$$F(\lambda, x) \sim \frac{\alpha \delta(x)}{2\lambda}, \quad \text{as } \lambda \to \infty. \tag{4.4.31}$$

Example 96. An interesting example is provided by asymptotic approximation of the sum

$$S(s, n) = \sum_{k=0}^{2n} (-1)^{k+n} \binom{2n}{k}^s, \tag{4.4.32}$$

as $n \to \infty$ [21]. Here both s and n are integers.

It is known that $S(1, n) = 0, S(2, n) = \frac{(2n)!}{(n!)^2}$ and $S(3, n) = \frac{(3n)!}{(n!)^3}$, but no formula for $S(s, n)$ is known if $s > 3$. In particular, our asymptotic formula (4.4.36) shows that the guess $\frac{(sn)!}{(n!)^s}$ is not correct.

To use our theory, we observe that $S(s, n)$ is equal to the coefficient of $z_1^0 \cdots z_r^0$ in the product $(-1)^n (1 + z_1)^{2n} \cdots (1 + z_r)^{2n} [1 - \frac{1}{z_1 \cdots z_r}]^{2n}$, where $r = s - 1$. Thus, using the Cauchy theorem,

$$S(s, n) =$$

$$\frac{(-1)^n}{(2\pi i)^r} \int_{|z_1|=1} \cdots \int_{|z_r|=1} (1+z_1)^{2n} \cdots (1+z_r)^{2n} [1 - \frac{1}{z_1 \cdots z_r}]^{2n} \frac{dz_1}{z_1} \cdots \frac{dz_r}{z_r},$$

$$\tag{4.4.33}$$

and setting $z_j = e^{2i\theta_j}$, $|\theta_j| < \frac{\pi}{2}$,

$$S(s, n) = \frac{2^{rn+2n}}{\pi^r} \int_{-\frac{\pi}{2}}^{\frac{\pi}{2}} \cdots \int_{-\frac{\pi}{2}}^{\frac{\pi}{2}} [\cos\theta_1 \cdots \cos\theta_r \sin(\theta_1 + \cdots + \theta_r)]^{2n} d\theta_1 \cdots d\theta_r.$$

$$\tag{4.4.34}$$

Let

$$g(\theta_1, \cdots, \theta_r) = \cos\theta_1 \cdots \cos\theta_r \sin(\theta_1 + \cdots + \theta_r). \tag{4.4.35}$$

We need to find the minima of $-\log|g(\theta_1, \cdots, \theta_r)|$ or, what is the same, the maxima of $|g(\theta_1, \cdots, \theta_r)|$ on $|\theta_j| < \frac{\pi}{2}$. By symmetry, it is enough to work in the region where $\theta_1 + \cdots + \theta_r > 0$ and thus $g(\theta_1, \cdots, \theta_r) > 0$. But $\frac{\partial g}{\partial \theta_j} = (-\tan\theta_j + \cot(\theta_1 + \cdots + \theta_r))g$ and it follows that the local maxima are at $(\theta_1, \cdots, \theta_r) = \frac{k\pi}{2s}(1, \cdots, 1)$, $k = 1, 3, 5, \cdots$, the global maximum is thus at $(\frac{\pi}{2s}, \cdots, \frac{\pi}{2s})$. The Hessian matrix $\left(\frac{\partial^2 h}{\partial \theta_i \partial \theta_j}\right) = \left(\frac{\partial}{\partial \theta_i}\left(\frac{1}{g}\frac{\partial g}{\partial \theta_j}\right)\right)$ is given by $(\delta_{ij} \sec^2\theta_j + \csc^2(\theta_1 + \cdots + \theta_r))$ which at $(\frac{\pi}{2s}, \cdots, \frac{\pi}{2s})$ yields $A = Dh\big|_{x_0} = (\delta_{ij} + 1)\cos^{-2}\left(\frac{\pi}{2s}\right)$.

Thus

$$S(s,n) \sim \frac{2^{2rn+2n}}{\pi^r} \cdot 2 \left(\frac{2\pi}{2n}\right)^{\frac{r}{2}} (\det A)^{\frac{-1}{2}} e^{2nh(x_0)},$$

or

$$S(s,n) \sim \left(2\cos\left(\frac{\pi}{2s}\right)\right)^{2ns+s-1} 2^{2-s} (\pi n)^{\frac{1-s}{2}} \sqrt{s}, \quad \text{as} \quad n \to \infty \qquad (4.4.36)$$

if the value $\det A = (r+1)\cos^{-2r}\left(\frac{\pi}{2s}\right)$ is used.

4.5 Fourier Type Integrals

We now turn our attention to the study of the asymptotic development of oscillatory integrals of the type

$$\int_R e^{i\lambda h(x)} \phi(x) dx, \qquad (4.5.1)$$

where R is a region in $\mathbb{R}^n$, and where h and ϕ are smooth in a neighborhood of the closure $\overline{R}$ of the region.

The analysis of the asymptotic behavior of the integral (4.5.1) is similar to that of Section 4.4, but somewhat more complicated. In the present case the main contributions to the integral for large λ come from the vicinities of the *critical points* of h. The critical points are classified into three types, depending on whether they belong to the interior of R or to the smooth or non-smooth parts of the boundary ∂R.

The interior or type I critical points are the stationary points of h, that is, points where the gradient Dh vanishes. In fact, if $\phi \in \mathcal{D}(\mathbb{R}^n)$ and if Dh does not vanish in the support of ϕ then

$$\int_{\mathbb{R}^n} e^{i\lambda h(x)} \phi(x) dx = O(\lambda^{-\infty}), \quad \text{as} \quad \lambda \to \infty, \qquad (4.5.2)$$

since a change of variables reduces the integral to a Fourier transform of the type $\int_{\mathbb{R}^n} e^{i\lambda x_1} \psi(x) dx$ with $\psi \in \mathcal{D}(\mathbb{R}^n)$.

The type II critical points are those points in the smooth parts of the boundary where the restriction of h to the boundary is stationary, i.e., those points where the gradient Dh is normal to the boundary.

Finally, the type III critical points are those belonging to the non-smooth parts of the boundary. The classification and characterization in this case are quite complicated, as complicated as the geometry of the non-smooth parts can be. As we shall see, a corner in a two-dimensional region is always a critical point. However, points located along an edge of the boundary of a three-dimensional region might or might not be critical points. Recall also that the endpoints in the one-dimensional case are always critical points.

Let us start with an isolated type I critical point $x_0 \in R$. We assume that x_0 is non-degenerate in the sense that the Hessian matrix $A = D^2 h|_{x_0}$ is non-singular. Observe that the matrix A defines a quadratic form $Q(x) = x \cdot Ax$. This quadratic form can be diagonalized by a change of variables $y = Tx$ so that $Q(T^{-1}y) = y_1^2 + \cdots + y_r^2 - y_{r+1}^2 - \cdots - y_n^2$. The number of positive and negative squares, r and $n - r$, are invariants of the matrix A; their difference, $2r - n$, is known as the signature of A and is denoted by $sig\ (A)$. Using Morse theory [87] a similar result can be obtained for h itself. Namely, there is a change of variables $y = \Psi(x) = (\psi_i(x))$, with $\Psi(x_0) = 0$ and with a non-vanishing Jacobian such that:

$$h(x) = h(x_0) + \psi_1^2(x) + \cdots + \psi_r^2(x) - \psi_{r+1}^2(x) - \cdots - \psi_n^2(x). \quad (4.5.3)$$

The asymptotic expansion of $e^{i\lambda h(x)}$ in the space $\mathcal{D}'(U)$, where U is a neighborhood of $x = x_0$ that contains no other critical points of h, follows by substitution as

$$e^{i\lambda h(x)} = e^{i\lambda h(x_0)} e^{i\lambda \psi_1^2(x)} \cdots e^{i\lambda \psi_r^2(x)} e^{-i\lambda \psi_{r+1}^2(x)} \cdots e^{-i\lambda \psi_n^2(x)}$$

$$\sim e^{i\lambda h(x_0)} \prod_{j=1}^{r} \sum_{k=0}^{\infty} \frac{\Gamma\left(\frac{2k+1}{2}\right) e^{\frac{\pi i}{4}(2k+1)} \delta^{(2k)}(\psi_j)}{(2k)! \lambda^{\frac{2k+1}{2}}}$$

$$\prod_{j=r+1}^{n} \sum_{k=0}^{\infty} \frac{\Gamma\left(\frac{2k+1}{2}\right) e^{\frac{-\pi i}{4}(2k+1)} \delta^{(2k)}(\psi_j)}{(2k)! \lambda^{\frac{2k+1}{2}}}. \quad (4.5.4)$$

The first order approximation is

$$\frac{e^{i\lambda h(x_0)} \pi^{n/2} e^{\frac{\pi i}{4}(2r-n)} \delta(\psi_1) \cdots \delta(\psi_n)}{\lambda^{n/2}},$$

and thus

$$e^{i\lambda h(x)} \sim e^{i\lambda h(x_0)} \left(\frac{2\pi}{\lambda}\right)^{n/2} \frac{e^{\frac{\pi i}{4} sig\ A} \delta(x - x_0)}{\sqrt{|\det A|}}, \quad (4.5.5)$$

since the Jacobian $|\det D\Psi(x_0)|$ is easily seen to be equal to $2^{-n/2}\sqrt{|\det A|}$. We emphasize that (4.5.4) and (4.5.5) hold in $\mathcal{D}'(U)$ if U contains no other critical points of h. Evaluation at a test function $\phi \in \mathcal{D}(U)$ gives

$$\int_{\mathbb{R}^n} e^{i\lambda h(x)} \phi(x) dx \sim e^{i\lambda h(x_0)} \left[\left(\frac{2\pi}{\lambda}\right)^{n/2} \frac{e^{\frac{\pi i}{4} sig A} \phi(x_0)}{\sqrt{|\det A|}} + O\left(\frac{1}{\lambda^{\frac{n+2}{2}}}\right)\right]. \quad (4.5.6)$$

We now pass to the type II critical points. Let $x_0 \in \Sigma$, where Σ is a smooth part of ∂R. We say that x_0 is non-degenerate if the following two conditions hold. First, the gradient $Dh\,|_{x_0}$ is a non-zero multiple of n, the exterior unit normal at x_0. Second, if $v_1, \cdots, v_{n-1}$ is any local coordinate system in Σ near $x = x_0$ then h as a function of $v_1, \cdots, v_{n-1}$ has a non-degenerate type I critical point at x_0. Under these circumstances we can find a change of variables $y = \Psi(x) = (\psi_i(x))$ with $\Psi(x_0) = 0$ such that the inequality $\psi_1(x) > 0$ locally represents R, while

$$h(x) = h(x_0) + c\psi_1(x) + \psi_2^2(x) + \cdots + \psi_r^2(x) - \psi_{r+1}^2(x) - \cdots - \psi_n^2(x), \quad (4.5.7)$$

where c is a non-zero constant that can be found from the relation $\frac{dh}{dn}\big|_{x_0} = c\frac{d\psi_1}{dn}\big|_{x_0}$ or , equivalently, $Dh(x_0) = cD\psi_1(x_0)$.

Therefore,

$$e^{i\lambda h(x)} H_R(x) \sim e^{i\lambda h(x_0)} e^{i\lambda c\psi_1(x)} H(\psi_1(x)) \prod_{j=2}^{r} e^{i\lambda \psi_j^2(x)} \prod_{j=r+1}^{n} e^{-i\lambda \psi_j^2(x)}$$

$$\sim e^{i\lambda h(x_0)} \sum_{k=0}^{\infty} (-1)^k \left(\frac{ci}{\lambda}\right)^{k+1} \delta^{(k)}(\psi_1)$$

$$\times \prod_{j=2}^{r} \sum_{k=0}^{\infty} \frac{\Gamma\left(\frac{2k+1}{2}\right) e^{\frac{\pi i}{4}(2k+1)} \delta^{(2k)}(\psi_j)}{(2k)! \lambda^{\frac{2k+1}{2}}} \cdot \prod_{j=r+1}^{n} \sum_{k=0}^{\infty} \frac{\Gamma\left(\frac{2k+1}{2}\right) e^{\frac{-\pi i}{4}(2k+1)} \delta^{(2k)}(\psi_j)}{(2k)! \lambda^{\frac{2k+1}{2}}}.$$

$$(4.5.8)$$

The first order approximation is

$$e^{ih(x)} H_R(x) \sim \frac{e^{i\lambda h(x_0)} 2^{\frac{n}{2}} \pi^{\frac{n-1}{2}} e^{\frac{\pi i}{4}(2r-n-1)} ci \delta(x - x_0)}{\lambda^{\frac{n+1}{2}} \sqrt{|\det A|}}, \quad (4.5.9)$$

where A is the Hessian matrix $D^2(h - \psi_1 - \psi_1^2)$ at $x = x_0$. Hence, if $\phi \in \mathcal{D}(U)$, where U contains no other critical points of h then

$$\int_R e^{i\lambda h(x)} \phi(x) dx \sim \frac{e^{i\lambda h(x_0)} 2^{\frac{n}{2}} \pi^{\frac{n-1}{2}} e^{\frac{\pi i}{4}(2r-n-1)} ci \phi(x_0)}{\lambda^{\frac{n+1}{2}} \sqrt{|\det A|}}. \quad (4.5.10)$$

We shall not study degenerate critical points in detail. However we would like to briefly consider the situation when a non-degenerate type I critical point belongs to the boundary of a region. As we showed in the previous section, for Laplace type integrals the contribution from such a point is precisely one half of the contribution when the point is interior. In the present case this is no longer true, as can be seen from the following examples.

Example 97. Let us consider the asymptotic development as $\lambda \to \infty$ of the generalized function

$$F_n(\lambda, x, y) = e^{i\lambda xy} H(y - x^n), \qquad (4.5.11)$$

where n is an even positive integer.

In this case the only critical point is $(0,0)$ which is a degenerate type II critical point for the region $y > x^n$, but which would be a non-degenerate type I if in the interior of any region. There are two level lines through $(0,0)$, the coordinate axes $x = 0$ and $y = 0$. The boundary of the region is the line $y = x^n$, which is tangent to the level line $y = 0$ at $(0,0)$.

To obtain an asymptotic approximation we first show that if $\phi \in \mathcal{D}(\mathbb{R}^2)$

$$< F_n(\lambda, x, y), \phi(x, y) > = \phi(0,0) < F_n(\lambda, x, y), \phi_0(x, y) > + O\left(\frac{1}{\lambda^{1 + \frac{1}{n+1}}}\right),$$
$$(4.5.12)$$

where ϕ_0 is identically equal to 1 in a neighborhood of $(0,0)$. Indeed, differentiation of (4.5.11) yields

$$\frac{\partial F_n}{\partial x} = i\lambda y F_n - nx^{n-1} e^{i\lambda xy} \delta(y - x^n), \qquad (4.5.13a)$$

$$\frac{\partial F}{\partial y} = i\lambda x F_n + e^{i\lambda xy} \delta(y - x^n). \qquad (4.5.13b)$$

The one-dimensional theory shows that the line integrals are $O\left(\frac{1}{\lambda^{\frac{1}{n+1}}}\right)$. Since F_n is clearly $O(1)$ it follows that both xF_n and yF_n are $O\left(\frac{1}{\lambda}\right)$, but writing $\phi = \phi(0,0)\phi_0 + x\phi_1 + y\phi_2$ yields

$$< F_n, \phi > = \phi(0,0) < F_n, \phi_0 > + < xF_n, \phi_1 > + < yF_n, \phi_2 > .$$
$$(4.5.14)$$

Thus if we show that $< F_n, \phi_0 > = O\left(\frac{1}{\lambda}\right)$ it would follow that F_n is also $O\left(\frac{1}{\lambda}\right)$, and using (4.5.13a, b) again that xF_n and yF_n are $O\left(\frac{1}{\lambda^{1 + \frac{1}{n+1}}}\right)$. Therefore we would obtain (4.5.12).

We have

$$< F_n, \phi > \sim \int_{-\infty}^{\infty} \int_{x^n}^{\infty} e^{i\lambda xy} \, dy \, dx$$

$$\sim \int_{-\infty}^{\infty} e^{i\lambda x^{n+1}} \left[\pi\delta(\lambda x) + \frac{i}{\pi x}\right] dx$$

$$\sim \frac{\pi}{\lambda}\left(1 - \frac{1}{n+1}\right),$$

and so (4.5.12) holds and we obtain the approximation

$$F_n(\lambda, x, y) = \frac{\pi}{\lambda}\left(1 - \frac{1}{n+1}\right)\delta(x)\delta(y) + O\left(\frac{1}{\lambda^{1+\frac{1}{n+1}}}\right). \qquad (4.5.15)$$

Observe that

$$e^{i\lambda xy} \sim \frac{2\pi}{\lambda}\delta(x)\delta(y) \qquad (4.5.16)$$

and that the leading term in (4.5.15) is obtained by multiplying this by the factor $\frac{1}{2}\left(1 - \frac{1}{n+1}\right)$.

Example 98. Let us now consider the generalized function

$$F_0(\lambda, x, y) = e^{i\lambda xy}H(y). \qquad (4.5.17)$$

Here the boundary is the x-axis: it consists entirely of type II critical points. The point $(0, 0)$ is a degenerate type II point. Using the topological tensor product ideas that we will explain in Section 4.7 we can obtain the expansion of $F_0(\lambda, x, y)$ as

$$e^{i\lambda xy}H(y) \sim 2\pi \sum_{k=0}^{\infty} \frac{i^k \delta^{+(k)}(x)\delta^{(k)}(y)}{k!\lambda^{k+1}}, \qquad (4.5.18)$$

where $\delta^+(x) = \frac{1}{2}\delta(x) - \frac{1}{2\pi i}Pf\left(\frac{1}{x}\right)$ is a Heisenberg delta. Presently we prove (4.5.18) only when applied to test functions of the type $\phi(x)\psi(y)$ with $\phi, \psi \in \mathcal{S}$:

$$< e^{i\lambda xy}H(y), \phi(x)\psi(y) > = \int_0^\infty \hat{\phi}(\lambda y)\psi(y)dy,$$

but $\hat{\phi}(y)$ belongs to $\mathcal{S}$ and thus we can use the moment expansion,

$$\int_0^\infty \hat{\phi}(\lambda y)\psi(y)dy \sim \frac{\mu_0\psi(0)}{\lambda} + \frac{\mu_1\psi'(0)}{\lambda^2} + \frac{\mu_2\psi''(0)}{2!\lambda^3} + \cdots,$$

where the moments are given by

$$\mu_k = \int_0^\infty \hat{\phi}(y)y^k dy$$

$$= 2\pi i^k < \delta^{(+)(k)}(x), \phi(x) >,$$

and thus (4.5.18) follows in this case.

We would like to mention that there are regions R and smooth functions $h(x)$ with no critical points of any type. In such cases $e^{i\lambda h(x)} H_R(x) = O(\lambda^{-\infty})$ as $\lambda \to \infty$. A typical example is the generalized function $e^{i\lambda x_1} H(x_2)$; actually, except for a change of variables, this is the only example. Notice, however that if R is bounded then there should be at least one critical point in the boundary ∂R.

Next, let us consider a particular case of a type III critical point, a corner in a two-dimensional region. Thus, let $x_0 = (x_1^0, x_2^0)$ be a corner of the boundary ∂R. Let $\rho_1(x), \rho_2(x)$ be a smooth function such that the inequalities $\rho_1(x) > 0$, $\rho_2(x) > 0$ locally describe R and such that near x_0 the boundary consists of the two curves $C_1 : \rho_1(x) = 0$, $\rho_2(x) > 0$ and $C_2 : \rho_2(x) = 0$, $\rho_1(x) > 0$, which make an angle α at $x = x_0$, $0 < \alpha < \pi$. Let T_i be the unit tangent vectors to the curves C_i in the direction of movement toward x_0.

We say that x_0 is non-degenerate if the two line derivative values $\left(\dfrac{dh}{ds}\right)_i = Dh \cdot T_i$, $i = 1, 2$, do not vanish at x_0. In such a case the functions $\rho_1(x)$ and $\rho_2(x)$ can be replaced by functions $\psi_1(x)$ and $\psi_2(x)$ that represent R in a similar fashion and such that

$$h(x) = h(x_0) + c_1 \psi_1(x) + c_2 \psi_2(x), \qquad (4.5.19)$$

where $c_j = \dfrac{1}{|D\psi_j| \sin \alpha} \left(\dfrac{dh}{ds}\right)_j$ are non-zero constants. Therefore,

$$e^{i\lambda h(x)} H_R(x) = e^{i\lambda(h(x_0) + c_1 \psi_1(x) + c_2 \psi_2(x))} H(\psi_1(x)) H(\psi_2(x))$$

$$\sim e^{i\lambda h(x_0)} \left(\sum_{k=0}^{\infty} \frac{i^k \delta^{(k)}(\psi_1)}{c_1^{k+1} \lambda^{k+1}}\right) \left(\sum_{k=0}^{\infty} \frac{i^k \delta^{(k)}(\psi_2)}{c_2^{k+1} \lambda^{k+1}}\right). \qquad (4.5.20)$$

The leading term takes the form

$$e^{i\lambda h(x)} H_R(x) \sim \frac{e^{i\lambda h(x_0)} \sin \alpha \delta(x - x_0)}{\left(\frac{dh}{ds}\right)_1 \left(\frac{dh}{ds}\right)_2 \lambda^2}, \qquad (4.5.21)$$

and thus

$$\int_R e^{i\lambda h(x)} \phi(x) dx \sim \frac{e^{i\lambda h(x_0)} \sin \alpha \phi(x_0)}{\left(\frac{dh}{ds}\right)_1 \left(\frac{dh}{ds}\right)_2 \lambda^2} \qquad (4.5.22)$$

if $\phi \in \mathcal{D}(U)$, where U contains no other critical points of h.

We now give several examples of these methods.

Example 99. In Chapter 3 we obtained the asymptotic expansion as $\lambda \to \infty$ of finite Fourier transforms of the type $\int_a^b e^{i\lambda x} \phi(x) dx$ where ϕ is smooth in $[a, b]$. Let us now consider the corresponding multi-dimensional situation. Let R be a region in $\mathbb{R}^n$ and let $\phi \in \mathcal{S}(\mathbb{R}^n)$; we would like to study the asymptotic develoment of $\hat{f}(z)$ as $|z| \to \infty$, where $f = \phi H_R$. Write $z = \lambda u$, with $|u| = 1$. Then

$$\hat{f}(\lambda u) = \int_R e^{i\lambda u \cdot x} \phi(x) dx. \tag{4.5.23}$$

Here $h(x) = u \cdot x$. Clearly there are no interior critical points. Type II critical points are those boundary points where u is normal to ∂R. The condition for non-degeneracy is that the Gaussian curvature be non-zero at the critical point. Recall that the Gaussian curvature of a surface given by $x_n = \phi(x_1, \cdots, x_{n-1})$ at a point $x_0 = (x_1^0, ..., x_n^0)$, where $D\phi = 0$ is the determinant of the Hessian matrix $D^2\phi$; the general case can be reduced to this one by a rotation of the coordinate axis.

If only a finite number $y_1, \cdots, y_k$ of non-degenerate type II critical points are present, (4.5.10) gives

$$\hat{f}(\lambda u) \sim \sum_{j=1}^{k} \frac{e^{i\lambda u \cdot y_j} (2\pi)^{\frac{n-1}{2}} e^{\frac{i\pi}{4}(2r_j - n - 1)} (u \cdot n_j) i\phi(y_j)}{|K_j| \lambda^{\frac{n+1}{2}}}, \tag{4.5.24}$$

where n_j is the exterior unit normal at y_j, so that $u \cdot n_j = \pm 1$, and where K_j is the Gaussian curvature at y_j.

The possibility that no critical points exist is not ruled out. For instance if $R = \{x \in \mathbb{R}^n : x_1 > 0\}$ then there are no critical points unless $n = \pm(1, 0, \cdots, 0)$ and in those cases $\hat{f}(\lambda u) = o(\lambda^{-\infty})$ as $\lambda \to \infty$. When $u = \pm(1, 0, \cdots, 0)$ then the whole boundary consists of critical points and the result is

$$\hat{f}(\pm\lambda, 0, \cdots, 0) \sim \frac{\pm i}{\lambda} \int_{-\infty}^{\infty} \int_{-\infty}^{\infty} \phi(0, x_2, \cdots, x_n) dx_2 \cdots dx_n.$$

Example 100. Let us consider the asymptotic approximation as $|z| \to \infty$ of the Fourier transform $\mathcal{F}\{\phi\delta(\Sigma); z\}$ of a single layer. We suppose that the Σ is a smooth surface with non-vanishing Gaussian curvature everywhere and that $\phi \in \mathcal{D}(\Sigma)$. Locally Σ bounds a region R and the function ϕ can be extended to R as a smooth function. The formulas for the distributional derivatives of Chapter 2 yield

$$\frac{\partial}{\partial x_j}(\phi H_R) = \frac{\partial \phi}{\partial x_j} H_R + \phi n_j \delta(\Sigma). \tag{4.5.25}$$

Let $|\boldsymbol{u}| = 1$ and let $\boldsymbol{y}$ be a point on Σ where $\boldsymbol{u}$ is normal to Σ. Suppose supp ϕ is small enough so that $\boldsymbol{y}$ is the only place on $\Sigma \cap$ supp ϕ where this happens. Then if we apply the Fourier transform to (4.5.25) we get

$$\mathcal{F}\{\phi n_j \delta(\Sigma); \lambda \boldsymbol{u}\} = \lambda u_j \mathcal{F}\{\phi H_R; \lambda \boldsymbol{u}\} - \mathcal{F}\left\{\frac{\partial \phi}{\partial x_j} H_R; \lambda \boldsymbol{u}\right\}$$

$$= \frac{e^{i\lambda \boldsymbol{u} \cdot \boldsymbol{y}} (2\pi)^{\frac{n-1}{2}} e^{\frac{\pi i}{4}(2r-n-1)} (\boldsymbol{u} \cdot \boldsymbol{n}) i \phi(\boldsymbol{y}) u_j}{|K| \lambda^{\frac{n-1}{2}}} + O\left(\frac{1}{\lambda^{\frac{n+1}{2}}}\right).$$

Multiplying by u_j summing on j and recalling that $\boldsymbol{u} \cdot \boldsymbol{n} = \pm 1$ at $\boldsymbol{y}$ we obtain

$$\mathcal{F}\{\phi \delta(\Sigma); \lambda \boldsymbol{u}\} \sim \left(\frac{2\pi}{\lambda}\right)^{\frac{n-1}{2}} \frac{e^{i\lambda \boldsymbol{u} \cdot \boldsymbol{y}} e^{\frac{i\pi}{4}(2r-n-1)} i \phi(\boldsymbol{y})}{|K|}. \tag{4.5.26}$$

The estimate $\Phi(\boldsymbol{z}) = \mathcal{F}\{\phi \delta(\Sigma); \boldsymbol{z}\} = O(|\boldsymbol{z}|^{\frac{-(n-1)}{2}})$ valid for surfaces with non-vanishing Gaussian curvature, is the basic step for the derivation of some restriction theorems for Fourier transforms [106], [107]. Without going into the details, the main idea is the following. Since the derivatives of Φ are also Fourier transforms of layers on Σ, the estimate $D^k \Phi(\boldsymbol{z}) = O(|\boldsymbol{z}|^{-(\frac{n-1}{2})})$ holds for each $\boldsymbol{k}$ and thus Φ belongs to the space $\mathcal{O}_{\frac{n-1}{2}}$. It follows from the identity

$$< \hat{f}(\boldsymbol{x}), \phi \delta(\Sigma) > = < f(\boldsymbol{z}), \Phi(\boldsymbol{z}) >, \tag{4.5.27}$$

that if $f \in \mathcal{O}'_{\frac{n-1}{2}}$ then its Fourier transform admits a restriction to Σ.

This result is interesting since in general $\hat{f}(\boldsymbol{x})$ is a distribution and distributions cannot be restricted to surfaces. Another version of this idea is obtained by taking $f \in L^p(\mathbb{R}^n)$, $1 \leq p \leq 2$; in this case $\hat{f}$ belongs to $L^q(\mathbb{R}^n)$, where q is the dual exponent: $\frac{1}{p} + \frac{1}{q} = 1$. If $q < \infty$ then the elements of L^q are measurable functions, defined only as almost everywhere and, in general, cannot be restricted to surfaces like Σ. However, if the index p satisfies $1 \leq p < \frac{2n}{n+1}$ then Φ belongs to $L^q(\mathbb{R}^n)$ and (4.5.27) shows that $\hat{f}$ can be restricted to Σ.

Another interesting concept related to these ideas is the notion of the wave front set of a distribution. If F belongs to $\mathcal{E}'(\mathbb{R}^n)$ then the set of singular frequencies of F is the set $S(F)$ of all $\xi \in \mathbb{R}^n$, $|\xi| = 1$ having no neighborhood V such that $\hat{F}(\lambda \boldsymbol{n}) = O(\lambda^{-\infty})$ as $\lambda \to \infty$ for each $\boldsymbol{n} \in V$. If $\boldsymbol{x}$ belongs to the singular support of $F \in \mathcal{D}'(\mathbb{R}^n)$ we introduce the set $S_{\boldsymbol{x}}$ of singular frequencies of F at $\boldsymbol{x}$ as

$$S_{\boldsymbol{x}} = \bigcap S(\phi F), \tag{4.5.28}$$

the intersection taken among all $\phi \in \mathcal{D}(\mathbb{R}^n)$ with $\phi(x) \neq 0$. The *wave front set* of the distribution F is the set

$$Wf(F) = \{(x, \xi) \in \mathbb{R}^n \times S_1 : \xi \in S_x(F)\}. \tag{4.5.29}$$

This set contains the exact information about the location of the singularities of F and the frequencies of such singularities.

Clearly, distributions whose singular supports do not meet can be multiplied in a natural way. A much improved version of this result says that two distributions whose wave front sets do not meet can be multiplied. In particular, if Σ is a regular surface, then the wave front set of $\delta(\Sigma)$ is the set $\{(x, n) : x \in \Sigma, n \text{ normal to } \Sigma \text{ at } x\}$ and thus a distribution $F \in \mathcal{D}'(\mathbb{R}^n)$ can be restricted to Σ if at each $x \in \Sigma$ the normal vector n is not a singular frequency of F at x.

We end this section with an example from the field of dispersive wave propagation.

Example 101. Let us consider the Klein-Gordon equation

$$c^2 \nabla^2 u - u_{tt} - b^2 u = 0. \tag{4.5.30}$$

In general, the simple exponentials $e^{i(k \cdot x - wt)}$ do not satisfy the equation. However, substitution into the equation shows that $e^{i(k \cdot x - wt)}$ is a solution of (4.5.30) if the exponents satisfy the *dispersion relation*

$$w^2 = c^2 |k|^2 + b^2 \tag{4.5.31}$$

which relates the wave vector k and the frequency w.

It follows that integrals of the form

$$\int_{\mathbb{R}^n} A_{\pm}(k) e^{i(k \cdot x \pm \sqrt{c^2 |k|^2 + b^2}\, t)} dk, \tag{4.5.32}$$

where $A_{\pm}(k)$ are arbitrary functions of k for which the integrals make sense, are solutions of the Klein–Gordon equation (4.5.30). In particular, the solution of the initial value problem

$$u(x, 0) = f(x), \qquad u_t(x, 0) = g(x),$$

associated to (4.5.30) is obtained by adding (4.5.32) with

$$A_{\pm}(k) = \frac{1}{2(2\pi)^n} \int (f(x) \pm iw^{-1} g(x)) e^{-ik \cdot x} dx. \tag{4.5.33}$$

More generally, the solution of any energy-conserving second order dispersive hyperbolic equation can be represented as the sum of integrals of the form

$$u(\boldsymbol{x}, t) = \int_{\mathbb{R}^n} A(\boldsymbol{k}) e^{i(\boldsymbol{k} \cdot \boldsymbol{x} - w(\boldsymbol{k})t)} \, d\boldsymbol{k}, \qquad (4.5.34)$$

where $w(\boldsymbol{k})$ satisfies a dispersion relation corresponding to the equation.

The analysis of the large time behavior of the solutions of those equations can thus be reduced to the analysis of integrals like (4.5.34).

If we apply (4.5.6), the asymptotic expansion of $u(\boldsymbol{x}, t) = u(t\boldsymbol{x}/t, t)$ for large t is obtained as

$$u(\boldsymbol{x}, t) \sim \frac{A(\boldsymbol{k}) \exp[i(\boldsymbol{k} \cdot \boldsymbol{x} - w(\boldsymbol{k})t) - \frac{i\pi}{4} sig\,(\boldsymbol{D}^2 w)]}{t^{3/2} |\det \boldsymbol{D}^2 w|^{\frac{1}{2}}}, \qquad (4.5.35)$$

where the condition for stationarity is

$$\boldsymbol{x} = \boldsymbol{V}(\boldsymbol{k})t, \qquad (4.5.36)$$

$$\boldsymbol{V}(\boldsymbol{k}) = \boldsymbol{D}w. \qquad (4.5.37)$$

For each $\boldsymbol{k}$ the relation (4.5.36) defines a *ray* in space-time, which corresponds to a point moving in space with the *group velocity* $\boldsymbol{V}(\boldsymbol{k})$. Along a ray the expansion of $u(\boldsymbol{x}, t)$ is given by (4.5.35), which is a plane wave with constant wave vector $\boldsymbol{k}$ and frequency $w(\boldsymbol{k})$ and whose amplitude decays as $t^{\frac{-3}{2}}$. When more than one ray passes though $(\boldsymbol{x}, t)$ then the expansion of $u(\boldsymbol{x}, t)$ is obtained by summing (4.5.35) over all the values of $\boldsymbol{k}$ with $\boldsymbol{x} = \boldsymbol{V}(\boldsymbol{k})t$. Observe that (4.5.35) cannot be applied if the Hessian matrix $\boldsymbol{D}^2 w$ is singular. The points in space time correponding to the rays where $\boldsymbol{D}^2 w$ is singular are called *caustics*: on them the expansion depends on the contribution from degenerate critical points.

4.6 Further Examples

As is the case with the expansion of generalized functions in one variable, the moment asymptotic expansion does not hold in spaces such as $\mathcal{D}'(\mathbb{R}^n)$ nor $\mathcal{S}'(\mathbb{R}^n)$. Our analysis in Section 3.8 suggests that in such cases the expansion of $f(\lambda \boldsymbol{x})$ as $\lambda \to \infty$ contains two types of terms: those arising from the ordinary expansion of $f(\boldsymbol{x})$ as $\mid \boldsymbol{x} \mid \to \infty$ and those arising from the moment expansion. We now present two interesting expansions from the field of potential theory.

Example 102. Let us consider the asymptotic development as $\lambda \to \infty$ of the distribution $f(\boldsymbol{x}) = \mathcal{P}f\,(\mid \boldsymbol{x} + \boldsymbol{y} \mid^q)$, where $\boldsymbol{y} \in \mathbb{R}^n$ is fixed. As is to be expected, two cases arise depending on whether $q + n$ is an even integer or not.

Let us first consider the case when $q \neq -n - 2m$, $m = 0, \pm 1, \pm 2, \cdots$. Then

$$f(\lambda \boldsymbol{x}) = \mathcal{P}f\left(\mid \lambda \boldsymbol{x} + \boldsymbol{y} \mid^q\right) = \lambda^q \mathcal{P}f(\mid \boldsymbol{x} + \frac{1}{\lambda}\boldsymbol{y} \mid^q)$$

$$\sim \lambda^q \sum_{N=0}^{\infty} \frac{1}{N!}(\boldsymbol{D}^N \mathcal{P}f\left(\mid \boldsymbol{x} \mid^q\right)\|\boldsymbol{y}^N)\lambda^{-N}, \quad \text{as } \lambda \to \infty.$$

But the formula for the derivative $\boldsymbol{D}^N \mathcal{P}f\left(\mid \boldsymbol{x} \mid^q\right)$ is given by

$$\boldsymbol{D}^N \mathcal{P}f(\mid \boldsymbol{x} \mid^q) =$$

$$\sum_{j=0}^{[N/2]} \frac{2^{N-2j}\Gamma\left(\frac{q}{2}+1\right)N!}{\Gamma\left(\frac{q}{2}-N+j+1\right)j!(N-2j)!}\boldsymbol{x}^{N-2j}\boldsymbol{\Delta}^j\,\mathcal{P}f\left(\mid \boldsymbol{x} \mid^{q-2N+2j}\right),$$

$$(4.6.1)$$

where the symmetric product is used, and thus

$$\mid \lambda \boldsymbol{x} + \boldsymbol{y} \mid^q \sim \mathcal{P}f\left(\mid \boldsymbol{x} \mid^q\right)\lambda^q + q(\boldsymbol{x} \cdot \boldsymbol{y})\mathcal{P}f(\mid \boldsymbol{x} \mid^{q-2})\lambda^{q-1}$$

$$+[q(q-2)(\boldsymbol{x} \cdot \boldsymbol{y})^2\,\mathcal{P}f(\mid \boldsymbol{x} \mid^{q-4}) + q \mid \boldsymbol{y} \mid^2\,\mathcal{P}f\left(\mid \boldsymbol{x} \mid^{q-2}\right)]\lambda^{q-2}$$

$$+[q(q-2)(q-4)(\boldsymbol{x} \cdot \boldsymbol{y})^3\,\mathcal{P}f\left(\mid \boldsymbol{x} \mid^{q-6}\right)+3q(q-2)(\boldsymbol{x} \cdot \boldsymbol{y}) \mid \boldsymbol{y} \mid^2\,\mathcal{P}f\left(\mid \boldsymbol{x} \mid^{q-4}\right)]\lambda^{q-3}$$

$$+\cdots, \quad \text{as } \lambda \to \infty, \qquad (4.6.2)$$

so that if $\phi \in \mathcal{S}(\mathbb{R}^n)$,

$$F.p. \int_{\mathbb{R}^n} \phi(\boldsymbol{x}) \mid \boldsymbol{x} + \lambda \boldsymbol{y} \mid^q\,d\boldsymbol{x} \sim \lambda^q F.p. \int_{\mathbb{R}^n} \phi(\boldsymbol{x}) \mid \boldsymbol{x} \mid^q\,d\boldsymbol{x}$$

$$+ q\lambda^{q-1} F.p. \int_{\mathbb{R}^n} (\boldsymbol{x} \cdot \boldsymbol{y}) \mid \boldsymbol{x} \mid^{q-2}\phi(\boldsymbol{x})d\boldsymbol{x} + \lambda^{q-2}[q(q-2)$$

$$\times F.p. \int_{\mathbb{R}^n} (\boldsymbol{x} \cdot \boldsymbol{y})^2 \mid \boldsymbol{x} \mid^{q-4}\phi(\boldsymbol{x})d\boldsymbol{x}+q \mid \boldsymbol{y} \mid^2\,F.p. \int_{\mathbb{R}^n} \mid \boldsymbol{x} \mid^{q-2}\phi(\boldsymbol{x})d\boldsymbol{x}]+\cdots$$

$$(4.6.3)$$

When $q = -n - 2m$, $m = 0, \pm 1, \pm 2, \cdots$, the formulas become more complicated, since the derivative $\boldsymbol{D}^N\,\mathcal{P}f\left(\mid \boldsymbol{x} \mid^{-n-2m}\right)$ contains some extra delta function terms:

$$\boldsymbol{D}^N \mathcal{P}f(\mid \boldsymbol{x} \mid^{-n-2m})$$

$$= \sum_{j=0}^{[N/2]} \frac{(-1)^{N-j}2^{N-2j}\Gamma\left(\frac{n}{2}+m+N-j\right)N!}{\Gamma(\frac{n}{2}+m)(N-2j)!j!}$$

$$\boldsymbol{\Delta}^j \boldsymbol{x}^{N-2j}\,\mathcal{P}f\left(\mid \boldsymbol{x} \mid^{-n-2m-2N+2j}\right)$$

$$-\sum_{j=\frac{|m|-m}{2}}^{[N/2]} \frac{2^j \Gamma\left(\frac{n}{2}+m+j\right) c_{m+j,n}\beta_{N,j}}{\Gamma(\frac{n}{2}+m)(N-2j)!j!} \Delta^j D^{N-2j}\nabla^{2m+2j}\delta(\boldsymbol{x}), \quad (4.6.4)$$

where the $\beta_{N,j}$ are the constants given by $\beta_{0,0}=0$,

$$\beta_{q,0}=\frac{1}{k}+\frac{1}{k+2}+\cdots+\frac{1}{k+2q-2}, \quad q\geq 1,$$

while $\beta_{0,p}=0$, $p\geq 1$, and

$$\beta_{q,p}=\sum_{j=0}^{p-1}\binom{p-1}{j}\frac{(-1)^j}{k+2q-2j-2}, \quad q\geq 1, p\geq 1.$$

Thus, we obtain

$$f(\lambda\boldsymbol{x})=\mathcal{P}f(\mid \lambda\boldsymbol{x}+\boldsymbol{y}\mid^{-n-2m})$$

$$=\lambda^{-n-2m}\mathcal{P}f(\mid \boldsymbol{x}+\frac{1}{\lambda}\boldsymbol{y}\mid^{-n-2m})-\frac{c_{m,n}\ln\lambda}{(2m)!\lambda^{n+2m}}\nabla^{2m}\delta\left(\boldsymbol{x}+\frac{1}{\lambda}\boldsymbol{y}\right)$$

$$\sim\left\{-\frac{c_{m,n}}{(2m)!}\nabla^{2m}\delta(\boldsymbol{x})\right\}\frac{\ln\lambda}{\lambda^{n+2m}}+\mathcal{P}f\left(\frac{1}{\mid \boldsymbol{x}\mid^{n+2m}}\right)\frac{1}{\lambda^{n+2m}}$$

$$+\left\{-\frac{c_{m,n}}{(2m)!}D_y\nabla^{2m}\delta(\boldsymbol{x})\right\}\frac{\ln\lambda}{\lambda^{n+2m+1}}$$

$$+\left\{-(n+2m)\boldsymbol{x}\cdot\boldsymbol{y}\mathcal{P}f\left(\frac{1}{\mid \boldsymbol{x}\mid^{n+2m+2}}\right)+\frac{c_{m,n}}{(2m)!(n+2m)}D_y\nabla^{2m}\delta(\boldsymbol{x})\right\}$$

$$\frac{1}{\lambda^{n+2m+1}}$$

$$+\left\{-\frac{c_{m,n}}{2!(2m)!}D_y^2\nabla^{2m}\delta(\boldsymbol{x})\right\}\frac{\ln\lambda}{\lambda^{n+2m+2}}$$

$$+\left\{(n+2m)(n+2m+2)(\boldsymbol{x}\cdot\boldsymbol{y})^2\mathcal{P}f\left(\frac{1}{\mid \boldsymbol{x}\mid^{n+2m+4}}\right)\right.$$

$$-(n+2m)\mid \boldsymbol{y}\mid^2\mathcal{P}f\left(\frac{1}{\mid \boldsymbol{x}\mid^{n+2m+2}}\right)$$

$$-\frac{(n+2m)c_{m+1,n}}{(2m+2)!(n+2m+2)}\mid \boldsymbol{y}\mid^2\nabla^{2m+2}\delta(\boldsymbol{x})$$

$$\left.-\frac{c_{m,n}}{(2m)!}\left(\frac{1}{n+2m}+\frac{1}{n+2m+2}\right)D_y^2\nabla^{2m}\delta(\boldsymbol{x})\right\}\times\frac{1}{\lambda^{n+2m+2}}+\cdots,$$

$$\text{as } \lambda\to\infty \quad (4.6.5)$$

in the space $\mathcal{S}'(\mathbb{R}^n)$, where $D_y = y \cdot D$ is the derivative in the y direction. Observe that $\nabla^{2j}\delta(x) = 0$ if $j = -1, -2, -3, \cdots$.

Example 103. We shall consider the asymptotic development of the integral

$$\psi(x) = \int_{\mathbb{R}^3} \frac{\phi(y)}{|x - y|} dy. \tag{4.6.6}$$

The Taylor expansion of $|x - y|^{-1}$ for $s << r$ where $r = |x|$, $s = |y|$ is

$$\frac{1}{|x - y|} \sim \frac{1}{r} - y_i \frac{\bar{\partial}}{\partial x_i}\left(\frac{1}{r}\right) + \frac{1}{2!}y_i y_j \frac{\bar{\partial}^2}{\partial x_i \partial x_j}\left(\frac{1}{r}\right) \tag{4.6.7}$$

$$- \frac{1}{3!}y_i y_j y_k \frac{\bar{\partial}^3}{\partial x_i \partial x_j \partial x_k}\left(\frac{1}{r}\right) + \cdots,$$

where the summation convention is used. Thus,

$$\psi(x) \sim \int \phi(y)\left[\frac{1}{r} - y_i \frac{\bar{\partial}}{\partial x_i}\frac{1}{r} + \frac{1}{2!}y_i y_j \frac{\bar{\partial}^2}{\partial x_i \partial x_j}\left(\frac{1}{r}\right)\right.$$

$$\left. - \frac{1}{3!}y_i y_j y_k \frac{\bar{\partial}^3}{\partial x_i \partial x_j \partial x_k}\left(\frac{1}{r}\right) + \cdots\right] dy.$$

If we now use (4.6.4) with $n = 3$, $m = -1$, that is

$$D^N \mathcal{P}f\left(\frac{1}{r}\right) = \sum_{j=0}^{[N/2]} \frac{(-1)^{N-j}2^{N-2j}\Gamma(N - j + 1/2)N!}{\sqrt{\pi}(N - 2j)!j!}$$

$$\Delta^j x^{N-2j}\mathcal{P}f\left(\frac{1}{r^{1+2N-2j}}\right)$$

$$- \sum_{j=1}^{[N/2]} \frac{N!\pi}{2^{2j-3}j!(j - 1)!(N - 2j)!} \sum_{q=0}^{j-1} \binom{j - 1}{q} \frac{(-1)^q}{2N - 2q - 1}\Delta^j$$

$$D^{N-2j}\nabla^{2j-2}\delta(x), \tag{4.6.8}$$

we obtain

$$\psi(\lambda x) \sim \left(\int \phi(y)dy\right)\frac{1}{|\lambda x|} + \left(\int \phi(y)y_i dy\right)\lambda x_i \mathcal{P}f\left(\frac{1}{|\lambda x|^3}\right)$$

$$- \left(\frac{2\pi}{3}\int \phi(y)s^2 dy\right) + \left(\frac{3}{2}\int \phi(y)y_i y_j dy\right)\lambda^2 x_i x_j \mathcal{P}f\left(\frac{1}{|\lambda x|^5}\right)$$

$$-\left(\frac{3}{2}\int \phi(\boldsymbol{y})s^2 d\boldsymbol{y}\right) \mathcal{P}f\left(\frac{1}{|\lambda \boldsymbol{x}|^3}\right) + \left(\frac{2\pi}{5}\int \phi(\boldsymbol{y})s^2 y_i d\boldsymbol{y}\right)\frac{\bar{\partial}}{\partial x_i}\delta(\lambda \boldsymbol{x})$$

$$-\left(\frac{3}{2}\int \phi(\boldsymbol{y})s^2 y_i d\boldsymbol{y}\right)\lambda x_i\, \mathcal{P}f\left(\frac{1}{|\lambda \boldsymbol{x}|^5}\right) + \cdots. \qquad (4.6.9)$$

Formula (4.6.9) contains two types of terms: the pseudofunction terms of the type $\mathcal{P}f\left(\frac{1}{|\lambda \boldsymbol{x}|^{2k+1}}\right)$ and the delta function terms. The pseudofunction terms arise from the classical expansion of $\psi(\boldsymbol{x})$ as $|\boldsymbol{x}| \to \infty$ while the delta function terms arise from the moment asymptotic expansion. Observe that this expansion contains logarithmic terms since the generalized function $\mathcal{P}f(r^{-3-2m}), m : 0, 1, 2, \cdots$ is not a homogeneous function but an associated homogeneous function.

4.7 Tensor Products and Partial Asymptotic Expansions

In this section we extend the results of previous sections to parametric and general multi-dimensional expansions by using the tools from the theory of topological tensor products [52], [110]. The basic idea is very simple: since the moment asymptotic expansion proved pivotal for scalar valued generalized functions it seems plausible that some form of it should remain valid for vector valued generalized functions as well. In fact, that is the case.

Our aim is to give partial expansions of the type $f(\lambda \boldsymbol{x}, \boldsymbol{y})$ as $\lambda \to \infty$ where $f(\boldsymbol{x}, \boldsymbol{y})$ is a generalized function of the variables $\boldsymbol{x} \in \mathbb{R}^n$ and $\boldsymbol{y} \in V$, where V is a smooth differentiable manifold. Of special interest is the expansion of $f(\lambda r, \boldsymbol{w}) = f(\lambda \boldsymbol{x})$, where $r = |\boldsymbol{x}|$, $\boldsymbol{w} \in S_1, (\boldsymbol{x} = r\boldsymbol{w})$ are polar coordinates in $\mathbb{R}^n$.

If X and Y are locally convex topological vector spaces, it is possible to construct several useful topologies in the tensor product $X \otimes Y$, such as the π-topology, the ε-topology and the σ- or Schmidt topology. We shall only work with the π-topology, but most of the results remain true for other topologies. Actually, if one of the factors is nuclear then all the topologies mentioned coincide. This applies in our situation since most spaces of generalized functions are nuclear [110].

If $\|\ \|_X$ and $\|\ \|_Y$ are seminorms, respectively, on the spaces X and Y then the π-norm $\|\ \|_\pi$ is defined on $X \otimes Y$ as

$$\|z\|_\pi = \inf\left\{\sum_{i=1}^n \|x_i\|_X \|y_i\|_Y : z = \sum_{i=1}^n x_i \otimes y_i\right\}. \qquad (4.7.1)$$

By allowing $\|\ \|_X$ and $\|\ \|_Y$ to vary among all continuous seminorms on X and Y we obtain a family of seminorms $\|\ \|_\pi$ on $X \otimes Y$ that generate the π-topology, denoted as $X \otimes_\pi Y$. Its completion is denoted as $X \hat{\otimes}_\pi Y$.

We remark that each $T \in (X \hat{\otimes} Y)'$ induces an operator $T_1 : X \to Y'$ given by $< T_1(x), y > = T(x \otimes y)$ and conversely.

Let $\mathcal{A}(\mathbb{R}^n)$ be any of the test function spaces $\mathcal{E}(\mathbb{R}^n), \mathcal{P}(\mathbb{R}^n), \mathcal{O}_C(\mathbb{R}^n)$ or $\mathcal{O}_\gamma(\mathbb{R}^n)$ (for suitable γ) where the moment asymptotic expansion holds. Let $\mathcal{B}(V)$ be any space of functions or generalized functions over the manifold V. The elements of $\mathcal{A}(\mathbb{R}^n) \otimes \mathcal{B}(V)$ are functions $\rho(x, y)$ that can be written as $\sum_{i=1}^m \phi_i(x)\psi_i(y)$, where $\phi_i \in \mathcal{A}(\mathbb{R}^n)$, $\psi_i \in \mathcal{B}(V)$; the elements of $\mathcal{A}(\mathbb{R}^n) \hat{\otimes}_\pi \mathcal{B}(V)$ are functions $\rho(x, y)$ that can be approximated, in the π topology, by such degenerate kernels.

Let $C_q = \{\rho \in \mathcal{A}(\mathbb{R}^n) \hat{\otimes}_\pi \mathcal{B}(V) : D_x^k \rho(0, y) = 0 \,\forall\, y \in V, |\, k\, | < q\}$. Then we have.

Lemma 9. *Let $\|\ \|_1$ be any continuous seminorm on $\mathcal{A}(\mathbb{R}^n) \otimes \mathcal{B}(V)$. Then for any q there exists another continuous seminorm $\|\ \|_2$ such that*

$$\|\rho(x/\lambda, y)\|_1 \le \lambda^{-q}\|\rho(x, y)\|_2, \quad \lambda \ge 1, \quad \rho \in C_q. \tag{4.7.2}$$

Proof. It suffices to prove the result if $\|\ \|_1$ is the π tensor product of seminorms $\|\ \|_1$ on $\mathcal{A}(\mathbb{R}^n)$ and $\|\ \|$ on $\mathcal{B}(V)$. It is also enough to establish (4.7.2) if $\rho \in \mathcal{A}_q \otimes \mathcal{B}$, where $\mathcal{A}_q = \{\phi \in \mathcal{A} : D^k\phi(0) = 0, |\, k\, | < q\}$, since $\mathcal{A}_q \otimes \mathcal{B}$ is dense in C_q.

Let $\|\ \|_2$ be a seminorm on $\mathcal{A}(\mathbb{R}^n)$ such that $\|\phi(x/\lambda)\|_1 \le \lambda^{-q}\|\phi(x)\|_2$ for $\lambda \ge 1$ if $\phi \in \mathcal{A}_q$. It is not true, in general, that the restriction of the π tensor of the norms $\|\ \|_2$ in $\mathcal{A}$ and $\|\ \|$ in $\mathcal{B}$ to $\mathcal{A}_q \otimes \mathcal{B}$ (denoted as $\|\ \|_2$) is equal to the π product of the restriction of $\|\ \|_2$ to $\mathcal{A}_q$ and $\|\ \|$ in $\mathcal{B}$ (denoted as $\|\ \|_3$). However, since $\mathcal{A}_q$ has a topological complement in $\mathcal{A}$, we can find a constant $K \ge 1$ such that $\|\ \|_3 \le K\|\ \|_2$.

Hence for $\rho \in \mathcal{A}_q \otimes \mathcal{B}$ we can find $\phi_1, \cdots, \phi_m \in \mathcal{A}_q$, $\psi_1, \cdots, \psi_m \in \mathcal{B}$ with $\rho = \sum_{i=1}^m \phi_i \otimes \psi_i$. Thus

$$\|\rho(x/\lambda, y)\|_1 \le \sum_{i=1}^m \|\phi_i(x/\lambda)\|_1 \|\psi_i(y)\|$$

$$\le \sum_{i=1}^m \lambda^{-q}\|\phi_i(x)\|_2 \|\psi_i(y)\|,$$

and taking the infimum over such representations,

$$\|\rho(x/\lambda, y)\|_1 \le \lambda^{-q}\|\rho(x, y)\|_3 \le \lambda^{-q}K\|\rho(x, y)\|_2. \qquad \blacksquare$$

This gives us the following partial moment expansion.

Theorem 40. *Let $f \in (\mathcal{A}(\mathbb{R}^n) \otimes_\pi \mathcal{B}(V))'$. Then*

$$f(\lambda x, y) = \sum_{|k|=0}^{N} \frac{(-1)^{|k|} \mu_k(y) D^k \delta(x)}{k! \lambda^{|k|+n}} + O\left(\frac{1}{\lambda^{N+n+1}}\right), \quad as \ \lambda \to \infty,$$

(4.7.3)

where the moments $\mu_k(y)$ are the distributions of $\mathcal{B}'(V)$ given by

$$< \mu_k(y), \psi(y) > = < f(x, y), x^k \psi(y) >,$$

(4.7.4)

for $\psi \in \mathcal{B}(V)$.

Proof. Let $\rho \in \mathcal{A}(\mathbb{R}^n) \otimes_\pi \mathcal{B}(V)$. Write $\rho = P + \rho_1$, where P is the partial Taylor polynomial of order N,

$$P(x, y) = \sum_{|k|=0}^{N} \frac{D^k \rho(0, y) x^k}{k!},$$

and where ρ_1 is the rest. If $f \in (\mathcal{A}(\mathbb{R}^n) \otimes_\pi \mathcal{B}(V))'$ then

$$< f(\lambda x, y), \rho(x, y) > = < f(\lambda x, y), P(x, y) + \rho_1(x, y) >$$

$$= \sum_{|k|=0}^{N} \frac{(-1)^{|k|} < \mu_k(y), D^k \rho(0, y) >}{k! \lambda^{|k|+n}} + R_N(\lambda),$$

where $R_N(\lambda) = \lambda^{-n} < f(x, y), \rho_1(x/\lambda, y) > = O(\lambda^{-n-N-1})$ as $\lambda \to \infty$ since $\rho_1 \in \mathcal{C}_{N+1}$. ∎

Let us now denote by $\mathcal{A}_p\{x^{\alpha_n}\}$ any of the spaces $\mathcal{E}_p, \mathcal{P}_p$ or $\mathcal{O}_{C,p}$, where the generalized moment expansion holds.

Theorem 41. *Let $f \in (\mathcal{A}_p\{x^{\alpha_n}\} \otimes \mathcal{B}(V))'$, where $\alpha_n \nearrow \infty$. Then*

$$f(\lambda x, y) \sim \sum_{n=0}^{\infty} \frac{\mu_{\alpha_n}(y) \delta_n(x)}{\lambda^{\alpha_n+1}},$$

(4.7.5)

where

$$< \mu_{\alpha_n}(y), \psi(y) > = < f(x, y), \psi(y) x^{\alpha_n} >$$

(4.7.6)

and where $\delta_n(x) = x^{-\alpha_n} \delta(x)$.

We do not give the result for logarithmic scales, but the corresponding expansion is clearly valid.

We remark that the functions $\rho \in \mathcal{A}_p\{x^{\alpha_n}\}\hat{\otimes}\mathcal{B}(V)$ are, generally speaking, those that belong to $\mathcal{A}_p\{x^{\alpha_n}\}$ for each y fixed (in a uniform way that depends on $\mathcal{B}$). Thus,

$$\rho(x,y) \sim x^{\alpha_1}a_1(y) + x^{\alpha_2}a_2(y) + \cdots \quad \text{as } x \to 0,$$

where $a_k(y) = \,<\delta_k(x), \rho(x,y)>$.

Example 104. Let $f(x,y) = e^{ix\cdot y}$ be the Fourier kernel. Its partial moments are given by

$$<\mu_k(y), \psi(y)> \,=\, <e^{ix\cdot y}, \psi(y)x^k> \,=\, <<e^{ix\cdot y}, x^k>, \psi(y)>$$
$$= \,<(2\pi)^n(-i)^{|k|}D^k\delta(y), \psi(y)>,$$

that is,

$$\mu_k(y) = (2\pi)^n(-i)^{|k|}D^k\delta(y). \tag{4.7.7}$$

Hence

$$e^{i\lambda x\cdot y} \sim (2\pi)^n \sum_{|k|=0}^{\infty} \frac{i^{|k|}D^k\delta(x)D^k\delta(y)}{k!\lambda^{|k|+n}}, \quad \text{as } \lambda \to \infty, \tag{4.7.8}$$

the expansion being valid in the spaces $\mathcal{A}(\mathbb{R}^n)\hat{\otimes}_\pi\mathcal{F}(\mathcal{A}'(\mathbb{R}^n))$, where $\mathcal{A}(\mathbb{R}^n)$ is a test function space for which the moment asymptotic holds, in particular, in the spaces $\mathcal{E}(\mathbb{R}^n)\hat{\otimes}\mathcal{O}_{exp}(\mathbb{R}^n)$ and $\mathcal{O}_C(\mathbb{R}^n)\hat{\otimes}\mathcal{O}_M(\mathbb{R}^n)$.

A particularly interesting case is the expansion of $f(\lambda r, w) = f(\lambda x)$, where $r = |x|$, $w \in S_1$, $x = rw$, are polar coodinates in $\mathbb{R}^n$. We can consider $f \in (\mathcal{A}\{r^{\alpha_n}\}\hat{\otimes}\mathcal{B}(S_1))'$, where $\Re e\,\alpha_n \nearrow \infty$. The expansion takes the form

$$f(\lambda r, w) \sim \sum_{j=0}^{\infty} \frac{\mu_{\alpha_j}(w)\delta_j(r)}{\lambda^{\alpha_j+1}}, \quad \text{as } \lambda \to \infty, \tag{4.7.9}$$

where $\delta_j(r) = r^{-\alpha_j}\delta(r)$. Observe that while the moment asymptotic expansion permits us to develop $<f(\lambda x), \phi(x)>$ when ϕ is smooth, expansion (4.7.9) permits us to consider ϕ to be smooth except at $x = 0$, where it has an expansion of the form

$$\phi(rw) \sim r^{\alpha_1}a_1(w) + r^{\alpha_2}a_2(w) + r^{\alpha_3}a_3(w) + \cdots, \quad \text{as } r \to 0, \tag{4.7.10}$$

and where $a_j(w) = \,<\delta_j(r), \phi(rw)>$ belongs to $\mathcal{B}(S_1)$.

Example 105. Let V be an open subset of the unit sphere S_1 of $\mathbb{R}^n$ and let $V_c = \{rw : w \in V, r > 0\}$ be the associated cone. Let $\gamma(w)$ be a continuous function on V that satisfies $\gamma(w) \geq m > 0$, $w \in V$. Then if $q \geq 1$, the kernel

$$F(r, w) = e^{-r^q \gamma(w)} \tag{4.7.11}$$

belongs to the space $(\mathcal{P}\{r^{\alpha_n}\} \hat{\otimes}_\pi C(\overline{V}))'$. In fact, what is required to be verifed is that for any $\phi \in C(V)$ the function

$$\Phi(r) = \int_V e^{-r^q \gamma(w)} \phi(w) d\sigma(w), \tag{4.7.12}$$

belongs to $\mathcal{P}'\{r^{\alpha_n}\}$. But $\Phi(r)$ is clearly continuous for $r \in \mathbb{R}$, while $|\Phi(r)| \leq e^{-mr^q} \int_V |\phi(w)| d\sigma(w)$ and thus $\Phi(r)$ can be regularized in $\mathcal{P}\{r^{\alpha_n}\}$.

The moments are

$$\mu_{\alpha_j}(w) = \int_0^\infty e^{-r^q \gamma(w)} r^{\alpha_j} dr = \frac{\Gamma(\alpha_j + 1)q^{-1}}{q\gamma(w)^{(\alpha_j+1)q^{-1}}},$$

and thus

$$e^{-\lambda r^q} \gamma(w) \sim \sum_{j=0}^\infty \frac{\Gamma\left(\frac{\alpha_j+1}{q}\right) \delta_j(r)}{q\gamma(w)^{\frac{\alpha_j+1}{q}} \lambda^{\frac{\alpha_j+1}{q}}}, \quad \text{as } \lambda \to \infty, \tag{4.7.13}$$

in the space $\mathcal{P}\{r^{\alpha_n}\} \hat{\otimes}_\pi C(\overline{V})$. Therefore if $\phi(x)$, $x \in V_c$ decreases exponentially as $|x| \to \infty$ while

$$\phi(rw) \sim r^{\alpha_1} a_1(w) + r^{\alpha_2} a_2(w) + \cdots, \quad \text{as } r \to 0,$$

where $a_i \in C(\overline{V})$, then we have

$$\int_{V_c} e^{-\lambda \gamma(x)} \phi(x) dx$$

$$\sim \sum_{j=0}^\infty \frac{1}{q} \Gamma\left(\frac{\alpha_j + n}{q}\right) \left[\int_V a_j(w)\gamma(w)^{-\frac{\alpha_j+n}{q}} d\sigma(w) \right] \lambda^{-\frac{\alpha_j+n}{q}}, \tag{4.7.14}$$

where we have denoted by $\gamma(x)$ the extension of $\gamma(w)$ given by $\gamma(x) = |x|^q \gamma(x/|x|)$.

In particular if $\gamma(x) = x \cdot y$, where $y \in \mathbb{R}^n$ is such that $x \cdot y > 0$, $y \in \overline{V}$, then we obtain the expansion of the following multi-dimensional Laplace transform

$$\int_{V_c} e^{-\lambda x \cdot y} \phi(x) dx \sim \sum_{j=0}^\infty \Gamma(\alpha_j + n) \int_V a_j(w)(x \cdot y)^{-\alpha_j - n} d\sigma(w) \lambda^{-\alpha_j - n},$$

$$\text{as } \lambda \to \infty. \tag{4.7.15}$$

Example 106. Let us now consider the expansion of Fourier transforms

$$\hat{\rho}(y) = \int_{\mathbb{R}^n} e^{ix \cdot y} \rho(x) dx, \tag{4.7.16}$$

as $| \, y \, | \to \infty$ for functions $\rho(x)$ that are smooth for $| \, x \, | \neq 0$ and admit the expansion

$$\rho(rw) \sim r^{\alpha_1} a_1(w) + r^{\alpha_2} a_2(w) + \cdots, \qquad \text{as } r \to 0, \tag{4.7.17}$$

where the a_i are smooth on S_1, i.e. $a_i \in \mathcal{D}(S_1)$. We assume polynomial behavior of the type of the space $\mathcal{K}$ as $r \to \infty$.

Writing (4.7.16) in polar coordinates yields

$$\hat{\rho}(\lambda y) = \int_0^\infty \int_{S_1} e^{i\lambda rw \cdot y} r^{n-1} \phi(rw) d\sigma(w) dr, \tag{4.7.18}$$

and it follows that we need to study the expansion of the kernel $f(\lambda r, w) = e^{i\lambda rw \cdot y}$ as $\lambda \to \infty$. In order to do so, we are going to show that $f \in (\mathcal{K}\{r^{\alpha_j}\} \hat{\otimes}_\pi \mathcal{D}(S))'$. This can be seen in two ways. On the one hand, if $\psi \in \mathcal{D}(S_1)$, the asymptotic behavior of $\Psi(r) = \int_{S_1} e^{irw \cdot y} \psi(w) d\sigma(w)$ as $r \to \infty$ shows that $\Psi \in \mathcal{K}'$ and thus it can be regularized in $\mathcal{K}\{r^{\alpha_j}\}$. Alternatively, if $\phi \in \mathcal{K}\{r^{\alpha_j}\}$, then the generalized function $\Phi(x) = \int_0^\infty e^{irx \cdot y} \phi(r) dr$ depends only on $x \cdot y$, i.e., $\Phi(x) = W(x \cdot y)$, where $W(t)$ is smooth for $t \neq 0$. The wave front set of $\Phi(x)$ is thus the plane $x \cdot y = 0$ with the constant direction given by y and thus the restriction results for generalized functions in Section 4.5 show that $\Phi(x)$ can'be restricted to S_1.

The partial moments of f are given by

$$< \mu_\alpha(w), \psi(w) > = < e^{irw \cdot y}, r^\alpha \psi(w) >,$$

and thus

$$\mu_\alpha(w) = \frac{\Gamma(\alpha+1) e^{i\pi(\alpha+1) \operatorname{sgn}(w \cdot y)/2}}{| \, w \cdot y \, |^{\alpha+1}}, \qquad \alpha \neq 0, 1, 2, \cdots, \tag{4.7.19a}$$

$$\mu_k(w) = 2\pi(-1)^k \delta^{+(k)}(w \cdot y), \quad k = 0, 1, 2, \cdots, \tag{4.7.19b}$$

where $\delta^+(x) = \frac{1}{2}\delta(x) - \frac{1}{2\pi i} \mathcal{P}f\left(\frac{1}{x}\right)$ is the Heisenberg delta distribution.

It follows that

$$\hat{\rho}(\lambda y) \sim \frac{< \mu_{\alpha_1}(w), a_1(w) >}{\lambda^{\alpha_1+n-1}} + \frac{< \mu_{\alpha_2}(w), a_2(w) >}{\lambda^{\alpha_2+n-1}} + \cdots, \quad \lambda \to \infty, \tag{4.7.20}$$

where the $a_j(w)$ are given in (4.7.17).

It should be clear that the smoothness of the $a_i(w)$ could be relaxed if one is interested in the asymptotic behavior of $\hat{\rho}(\lambda y)$ for a fixed direction y. In that case it is enough to ask $a_i(w)$ to be smooth near the intersection of the plane $w \cdot y = 0$ with the unit sphere.

4.8 An Application in Quantum Mechanics

In this section we apply our methods to the study of the quantum mechanical *twisted product*. Twisted products play a central role in the self-contained approach to Quantum Mechanics in the flat phase space $\mathbb{R}^{2n}$. Paradoxically, however, two different definitions of twisted products are currently used in the literature [2], [3], [5], [6], [50], [93].

Let q and p be the generalized coordinates and momenta, respectively, in the $2n$-dimensional phase space. Set $u = (q, p) \in \mathbb{R}^{2n}$. If $k = (k_1, \cdots, k_{2n}) \in \mathbb{N}^{2n}$ is a multi-index, the operator $\hat{D}^k$ is defined as

$$\hat{D}^k = (-1)^{k_{n+1}+\cdots+k_{2n}} \frac{\partial^{k_1}}{\partial u_{n+1}^{k_1}} \cdots \frac{\partial^{k_n}}{\partial u_{2n}^{k_n}} \frac{\partial^{k_{n+1}}}{\partial u_1^{k_{n+1}}} \cdots \frac{\partial^{k_{2n}}}{\partial u_n^{k_{2n}}}. \qquad (4.8.1)$$

Then the first definition of the quantum mechanical twisted product is

$$f \times_\hbar g = \sum_{|k|=0}^{\infty} \left(\frac{i\hbar}{2} \right)^{|k|} \frac{1}{k!} D^k f \hat{D}^k g, \qquad (4.8.2)$$

where $\hbar$ is Planck's constant.

To appreciate this definition we recall the notion of the Poisson bracket or Poisson commutator of the functions $f(q, p)$ and $g(q, p)$ defined in the $2n$-dimensional phase space with metric

$$(\alpha_{ij}) = J = \begin{pmatrix} O_n & I_n \\ -I_n & O_n \end{pmatrix}, \qquad (4.8.3)$$

where I_n and O_n are, respectively, the $n \times n$ identity and zero matrices. The Poisson bracket is defined to be the scalar product of their gradients:

$$\{f, g\} = \alpha^{ij} \frac{\partial f}{\partial u^i} \frac{\partial g}{\partial u^j} = \sum_{i=1}^{n} \left[\frac{\partial f}{\partial q^i} \frac{\partial g}{\partial p^i} - \frac{\partial g}{\partial q^i} \frac{\partial f}{\partial p^i} \right]. \qquad (4.8.4)$$

Let us now try to define a quantum mechanical product using the ordinary product and the Poisson bracket as ingredients. A first approximation would be

$$f \odot_\hbar g = fg + \frac{i\hbar}{2} \{f, g\}. \qquad (4.8.5)$$

Since the bracket satisfies Jacobi's identity, it follows that this product fails to be associative by a term of order $\hbar^2$. The natural correction would be the third term in (4.8.2). The new product fails to be associative by a term of order $\hbar^3$ and so on. Consequently, we need the entire expression in (4.8.2) for this purpose.

The second definition of the twisted product is given in terms of integrals as

$$(f \times_\hbar g)(u) = (\pi\hbar)^{-2n} \int_{\mathbb{R}^{2n}} \int_{\mathbb{R}^{2n}} f(u+v)g(u+w)\exp\left(\frac{2i}{\hbar}\sigma[v,w]\right) dv dw,$$
$$\tag{4.8.6}$$

where σ is the symplectic form on $\mathbb{R}^{2n}$ given by

$$\sigma[u,u'] = \sigma[(q,p),(q',p')] = u \cdot Ju' = \sum_{i=1}^{n}(q_i p'_{i+n} - p'_i q_{i+n}). \tag{4.8.7}$$

The definition (4.8.6) can be easily extended to generalized functions by duality. Indeed, if $\phi, \psi \in \mathcal{S}(\mathbb{R}^{2n})$ then clearly $\phi \times_\hbar \psi \in \mathcal{S}(\mathbb{R}^{2n})$. Since

$$< \phi \times_\hbar \psi, \rho > = < \phi, \psi \times_\hbar \rho > = < \psi, \rho \times_\hbar \phi >$$

$$= \int_{\mathbb{R}^{2n}} (\phi \times_\hbar \psi \times_\hbar \rho)(u) du, \tag{4.8.8}$$

the twisted product of $f \in \mathcal{S}'(\mathbb{R}^{2n})$ and $\phi \in \mathcal{S}(\mathbb{R}^{2n})$ can be defined as

$$< f \times_\hbar \phi, \psi > = < f, \phi \times_\hbar \psi > \tag{4.8.9a}$$

$$< \phi \times_\hbar f, \psi > = < f, \psi \times_\hbar \phi > . \tag{4.8.9b}$$

The products $f \times_\hbar \phi$ and $\phi \times_\hbar f$ turn out to be smooth functions of the space $\mathcal{O}_M$. The twisted product of two generalized functions f and g is defined as

$$< f \times_\hbar g, \phi > = < f, g \times_\hbar \phi > \tag{4.8.10}$$

if the right side can be evaluated for every $\phi \in \mathcal{S}$. Observe that like the ordinary and convolution products the twisted product of distributions is not always defined.

A case when $f \times_\hbar g$ is defined is when $f \in \mathcal{O}'_M$ and $g \in \mathcal{S}'$ or vice versa.

Example 107. The twisted product $\delta \times_\hbar g$ can be found from (4.8.6) as

$$(\delta \times_\hbar g)(u) = (\pi\hbar)^{-2n} \int_{\mathbb{R}^{2n}} g(u+w) \exp\left(\frac{-2i}{\hbar}\sigma[u,w]\right) dw.$$

If we now make a change of variables and observe that $\sigma[u, u] = 0$ we obtain

$$(\delta \times_\hbar g)(u) = (\pi\hbar)^{-2n} \hat{g}\left(\frac{2i}{\hbar} Ju\right), \qquad (4.8.11)$$

where $\hat{g}$ is the Fourier transform. In particular

$$\delta \times_\hbar \delta = (\pi\hbar)^{-2n}. \qquad (4.8.12)$$

Let us now consider the relationship between the two definitions of the twisted product. When one of the factors is a polynomial the two definitions coincide, as follows by using simple integration by parts. The general case is more complicated, however. Observe, for instance, that (4.8.2) only makes sense if one of the factors is smooth and thus (4.8.12) cannot be obtained from (4.8.2).

To reconcile the two definitions we return to formula (4.7.8), i.e.,

$$e^{i\lambda x \cdot y} \sim (2\pi)^n \sum_{|k|=0}^{\infty} \frac{i^{|k|} D^k \delta(x) D^k \delta(y)}{k! \lambda^{|k|+n}} \quad , \text{ as } \lambda \to \infty. \qquad (4.8.13)$$

When we replace y by Ay, where A is a non-singular matrix, we obtain the asymptotic expansion of $e^{i\lambda x \cdot Ay}$. For our problem we let $v, w \in \mathbb{R}^{2n}$ and let J be the matrix defined by (4.8.3). Then we obtain

$$e^{i\lambda v \cdot Jw} \sim (2\pi)^{2n} \sum_{|k|=0}^{\infty} \frac{i^{|k|} D^{|k|} \delta(v) \hat{D}^k \delta(w)}{k! \lambda^{|k|+2n}}. \qquad (4.8.14)$$

If $\phi \in \mathcal{A}(\mathbb{R}^{2n})$ and $\psi \in \mathcal{F}(\mathcal{A}(\mathbb{R}^{2n}))$, $\mathcal{F}$ being the Fourier transform and $\mathcal{A}'(\mathbb{R}^{2n})$ a generalized function space where the moment asymptotic expansion applies, then (4.8.6) with $\lambda = \frac{2}{\hbar}$ yields

$$(\phi \times_\hbar \psi)(u) = (\pi\hbar)^{-2n} < e^{i\lambda v \cdot Jw}, \phi(u+v)\psi(u+w) >$$

and thus

$$(\phi \times_\hbar \psi)(u) \sim \sum_{|k|=0}^{\infty} \left(\frac{i\hbar}{2}\right)^{|k|} \frac{D^k \phi(u) \hat{D}^k \psi(u)}{k!}, \qquad \text{as } \hbar \to 0.$$

We have proven the following result.

Theorem 42. *Let* $\phi \in \mathcal{A}(\mathbb{R}^{2n})$ *and* $\psi \in \mathcal{F}(\mathcal{A}(\mathbb{R}^{2n}))$, *the moment asymptotic expansion holding in* $\mathcal{A}'(\mathbb{R}^{2n})$. *Then for each* $u \in \mathbb{R}^{2n}$ *we have the asymptotic relation between the two definitions of the product,*

$$(\phi \times_\hbar \psi)(u) \sim \sum_{|k|=0}^{\infty} \left(\frac{i\hbar}{2}\right)^{|k|} \frac{D^k \phi(u) \hat{D}^k \psi(u)}{k!}, \qquad \text{as } \hbar \to 0. \qquad (4.8.15)$$

In particular, (4.8.15) holds if $\phi \in \mathcal{E}(\mathbb{R}^{2n})$ and $\psi \in \mathcal{O}_{exp}(\mathbb{R}^{2n}) = \mathcal{F}(\mathcal{E}(\mathbb{R}^{2n}))$ or if $\phi \in \mathcal{O}_C(\mathbb{R}^{2n})$ and $\psi \in \mathcal{O}_M(\mathbb{R}^{2n})$.

Clearly pointwise expansions of the twisted product can only hold if both factors are smooth. If we consider distributional asymptotic expansions then one of the factors can be a generalized function as long as the other remains smooth. That (4.8.15) need not hold in any sense when both factors are generalized functions is illustrated by the product $\delta \times_\hbar \delta = (\pi\hbar)^{-2n}$.

Theorem 43. *Let $\phi \in \mathcal{S}(\mathbb{R}^{2n})$ and $f \in \mathcal{S}'(\mathbb{R}^{2n})$. Then (4.8.15) holds in the distibutional sense, that is,*

$$< \phi \times_\hbar f, \rho > = \sum_{|\boldsymbol{k}|=0}^{N} \left(\frac{i\hbar}{2} \right)^{|\boldsymbol{k}|} \frac{< D^{\boldsymbol{k}}\phi \hat{D}^{\boldsymbol{k}} f, \rho >}{\boldsymbol{k}!} + O(\hbar^{N+1}) \qquad (4.8.16)$$

as $\hbar \to 0$, for any $\rho \in \mathcal{S}(\mathbb{R}^{2n})$.

Proof. Let R be the distribution of $\mathbb{R}^{4n}$ given by

$$R(\boldsymbol{v}, \boldsymbol{u}) = < f(\boldsymbol{u} + \boldsymbol{w}), e^{i\sigma[\boldsymbol{v}, \boldsymbol{w}]} >, \qquad (4.8.17)$$

the evaluation being with respect to $\boldsymbol{w}$.

The generalized function R belongs to $(\mathcal{O}_M(\mathbb{R}^{2n}) \hat{\otimes} \mathcal{S}'(\mathbb{R}^{2n}))'$ Indeed, if $\rho \in \mathcal{S}(\mathbb{R}^{2n})$ then

$$< R(\boldsymbol{v}, \boldsymbol{u}), \rho(\boldsymbol{u}) > = \mathcal{F}\{f * \check{\rho}; J\boldsymbol{v}\},$$

where $\check{\rho}(\boldsymbol{u}) = \rho(-\boldsymbol{u})$. But since $f * \check{\rho} \in \mathcal{O}_C$ it follows that $< R(\boldsymbol{v}, \boldsymbol{u}), \rho(\boldsymbol{u}) >$ belongs to $\mathcal{F}(\mathcal{O}_C) = \mathcal{O}'_M$. Hence the partial moment expansion of $R(\lambda\boldsymbol{v}, \boldsymbol{u})$ as $\lambda \to \infty$ holds for test functions of $\mathcal{O}_M(\mathbb{R}^{2n}) \hat{\otimes} \mathcal{S}(\mathbb{R}^{2n})$ and thus for test functions of $\mathcal{S}(\mathbb{R}^{2n}) \hat{\otimes} \mathcal{S}(\mathbb{R}^{2n}) = \mathcal{S}(\mathbb{R}^{4n})$. In particular, if we set $\psi(\boldsymbol{v}, \boldsymbol{u}) = \phi(\boldsymbol{u} + \boldsymbol{v})\rho(\boldsymbol{u})$ then $\psi \in \mathcal{S}(\mathbb{R}^{4n})$ and since the partial moments $\mu_{\boldsymbol{k}}(\boldsymbol{u})$ of $R(\boldsymbol{v}, \boldsymbol{u})$ are given as

$$\mu_{\boldsymbol{k}}(\boldsymbol{u}) = (2\pi)^n i^{|\boldsymbol{k}|} \hat{D}^{\boldsymbol{k}} f(\boldsymbol{u})$$

we obtain

$$< \phi \times_\hbar f, \rho > = (\pi\hbar)^{-2n} < R\left(\frac{2}{\hbar}\boldsymbol{v}, \boldsymbol{u} \right), \psi(\boldsymbol{v}, \boldsymbol{u}) >$$

$$\sim (\pi\hbar)^{-2n} \sum_{|\boldsymbol{k}|=0}^{\infty} \left(\frac{\hbar}{2} \right)^{|\boldsymbol{k}|+2n} \frac{(-1)^{|\boldsymbol{k}|}}{\boldsymbol{k}!} < D^{\boldsymbol{k}}\delta(\boldsymbol{v})\mu_{\boldsymbol{k}}(\boldsymbol{u}), \phi(\boldsymbol{u} + \boldsymbol{v})\rho(\boldsymbol{u}) >$$

$$\sim (2\pi)^{-2n} \sum_{|\boldsymbol{k}|=0}^{\infty} \left(\frac{\hbar}{2}\right)^{|\boldsymbol{k}|} \frac{1}{\boldsymbol{k}!} < D^{\boldsymbol{k}}\phi(\boldsymbol{u})\mu_{\boldsymbol{k}}(\boldsymbol{u}), \rho(\boldsymbol{u}) >$$

$$\sim \sum_{|\boldsymbol{k}|=0}^{\infty} \left(\frac{i\hbar}{2}\right)^{|\boldsymbol{k}|} \frac{1}{\boldsymbol{k}!} < D^{\boldsymbol{k}}\phi(\boldsymbol{u})\hat{D}^{\boldsymbol{k}}f(\boldsymbol{u}), \rho(\boldsymbol{u}) >,$$

as required. ∎

Finally, let us examine the question of the convergence of the twisted product expansion. Naturally, the series (4.8.2) may be divergent and even if convergent its sum is, in general, different from the twisted product. A striking example is the following. If ϕ and ψ both belong to $\mathcal{D}(\mathbb{R}^{2n})$ but their supports do not meet, then the twisted product $\phi \times_{\hbar} \psi$ extends to an entire analytic function in $\mathbb{C}^{2n}$ and thus is generically non-zero on $\mathbb{R}^{2n}$; however, $\phi \times_{\hbar} \psi \sim 0$ to all orders as $\hbar \to 0$.

Even so, the asymptotic development of the twisted product becomes a convergent series in some cases. The basic step for obtaining convergence results is the following. Suppose $\phi \in \mathcal{E}(\mathbb{R}^n)$ extends to an entire function in $\mathbb{C}^n$, then its Taylor series

$$\phi(\boldsymbol{x}) = \sum_{|\boldsymbol{k}|=0}^{\infty} \frac{D^{\boldsymbol{k}}\phi(0)}{\boldsymbol{k}!} \boldsymbol{x}^{\boldsymbol{k}}$$

converges in the topology of $\mathcal{E}(\mathbb{R}^n)$. Since $< f(\lambda\boldsymbol{x}), \boldsymbol{x}^{\boldsymbol{k}} >= \lambda^{-|\boldsymbol{k}|-n}\mu_{\boldsymbol{k}}$ we obtain the convergent series

$$< f(\lambda\boldsymbol{x}), \phi(\boldsymbol{x}) >= \sum_{|\boldsymbol{k}|=0}^{\infty} \frac{D^{\boldsymbol{k}}\phi(0)\mu_{\boldsymbol{k}}}{\boldsymbol{k}!\lambda^{|\boldsymbol{k}|+n}},$$

for every λ if $f \in \mathcal{E}'(\mathbb{R}^n)$. Use of this result immediately gives the following theorem.

Theorem 44. *Let $\phi \in \mathcal{E}(\mathbb{R}^{2n})$ be a function that extends to an entire function in $\mathbb{C}^{2n}$ and let $\psi \in \mathcal{O}_{exp}(\mathbb{R}^{2n}) = \mathcal{F}(\mathcal{E}(\mathbb{R}^n))$ or vice versa. Then*

$$(\phi \times_{\hbar} \psi)(\boldsymbol{u}) = \sum_{|\boldsymbol{k}|=0}^{\infty} \left(\frac{i\hbar}{2}\right)^{|\boldsymbol{k}|} \frac{1}{\boldsymbol{k}!} D^{\boldsymbol{k}}\phi(\boldsymbol{u})\hat{D}^{\boldsymbol{k}}\psi(\boldsymbol{u}), \qquad (4.8.18)$$

for every $\hbar > 0$, the convergence being uniform on compacts of $\mathbb{R}^{2n}$.

For further details and examples the reader is referred to Estrada, Gracia-Bondía and Várilly [32].

CHAPTER 5

The Asymptotic Expansion of Certain Series Considered by Ramanujan

5.1 Introduction

In this chapter we study the asymptotic behavior as $\varepsilon \to 0$ of series of the type

$$\sum_{n=1}^{\infty} a_n \phi(n\varepsilon). \tag{5.1.1}$$

Series of this kind were considered by Ramanujan in Chapter 15 of his second notebook [95]. It seems that he found his expansions by a formal application of the Euler–Maclaurin formula, but he provided no proofs. He started with an expansion of the sum $\sum_{1}^{\infty} \phi(n\varepsilon)$ for a function with an expansion of the form $\phi(x) \sim b_1 x^{\alpha_1} + b_2 x^{\alpha_2} + \cdots$, as $x \to 0$, where $-1 < \alpha_1 < \alpha_2 < \alpha_3 < \cdots$. In our notation his formula becomes

$$\sum_{n=1}^{\infty} \phi(n\varepsilon) \sim \frac{1}{\varepsilon} \int_0^{\infty} \phi(x)dx + \sum_{j=1}^{\infty} b_j \zeta(-\alpha_j)\varepsilon^{\alpha_j}, \qquad \text{as } \varepsilon \to 0, \tag{5.1.2}$$

where $\zeta(\alpha)$ is the Riemann zeta function. Later he derived many other interesting expansions, mostly in the case where $\phi(x) = e^{-x^p}$, $p > 0$.

In recent years, Berndt and Evans have undertaken the task of providing proofs of the results stated by Ramanujan in his notebooks. In their work [7] they study the first part of Chapter 15 of Ramanujan's second notebook and establish (5.1.2) and some of its generalizations in the case where $\phi(x) = e^{-x^p}$. Their proofs are based on the Mellin transform techniques.

A proof of expansion (5.1.2) for an extensive class of smooth functions was given by us using the distributional theory of asymptotic expansions [30], [43].

In this chapter the idea of studying (5.1.1) via the asymptotic expansion of series of delta functions is pursued further.

It is interesting to note that Ramanujan also included the asymptotic expansion of some arithmetical functions in this context. This suggests the possibility of using the asymptotic expansion of the series $\sum_{n=1}^{\infty} a_n \phi(n\varepsilon)$ to study the behavior of the sequence $\{a_n\}$. We explore this idea and show that much information about the sequence $\{a_n\}$ can indeed be obtained from a knowledge of the development of the sum $\sum_{n=1}^{\infty} a_n \phi(n\varepsilon)$. It could be that Ramanujan adopted a similar point of view.

Section 5.2 considers (5.1.2) as well as many related expansions. The notion of Lambert type series is considered in Section 5.3. As we show, the manipulations needed to handle the Lambert series become rather simple within the present context.

In Section 5.4 expansions are studied with regard to the notion of "smallness." The concept of a distributionally small sequence is introduced and it is shown that many famous residual terms in number theory are indeed distributionally small. The fact that distributionally small sequences are actually small in many ways, particularly in the Cesàro sense, is established.

In the last section the results are extended to multiple series. These, in turn, find application in the study of the number of solutions of Diophantine equations and in the study of finite partitions.

5.2 Basic Formulas

In this section we shall apply the moment asymptotic expansion to series of Dirac delta functions in order to obtain the asymptotic expansion of certain important series.

Let us start with the distribution $g(x) = \sum_{-\infty}^{\infty}(-1)^n\delta(x-n)$. Since $g(x)$ is periodic of period 2 and since its Fourier series has no constant term, it follows that $g \in \mathcal{K}'$. Its moments are

$$\mu_k = \sum_{-\infty}^{\infty}(-1)^n n^k = 0, \qquad k = 0, 1, 2, \cdots \tag{5.2.1}$$

It follows that if $\phi \in \mathcal{K}$ then

$$\sum_{-\infty}^{\infty}(-1)^n\phi(n\varepsilon) = o(\varepsilon^{\infty}), \qquad \text{as } \varepsilon \to 0, \tag{5.2.2}$$

where the notation $o(\varepsilon^{\infty})$ means a function that is $o(\varepsilon^n)$ for every n. Observe that the alternating series $\sum_{n=-\infty}^{\infty}(-1)^n\phi(n)$ may be divergent when $\phi \in \mathcal{K}$, but since $g \in \mathcal{K}'$, it can be regularized to give a canonical finite value.

The generalized function $f(x) = \sum_{n=1}^{\infty}(-1)^n\delta(x-n)$ also belongs to $\mathcal{K}'$ since we can write $f(x) = \theta(x)g(x)$, where $\theta(x)$ is a smooth cut-off function that satisfies $\theta(x) = 0$, $x < 1/4$, $\theta(x) = 1$, $x > 3/4$. Actually, since 0 does not belong to the support of f, the generalized function $f(x)$ belongs to $\mathcal{K}'\{x^{\alpha_n}\}$ for each sequence $\alpha_n \nearrow \infty$ and its moment function $\mu(\alpha)$ is entire. Here we have

$$\mu(\alpha) = \sum_{n=1}^{\infty}(-1)^n n^{\alpha} = (2^{\alpha+1} - 1)\zeta(-\alpha), \tag{5.2.3}$$

where $\zeta(s)$ is the zeta function.

It follows that if $\phi \in \mathcal{K}\{x^{\alpha_n}\}$, with expansion $\phi(x) \sim a_1 x^{\alpha_1} + a_2 x^{\alpha_2} + \cdots$, as $x \to 0$ then

$$\sum_{n=1}^{\infty}(-1)^n\phi(n\varepsilon) \sim (2^{\alpha_1+1} - 1)\zeta(-\alpha_1)a_1\varepsilon^{\alpha_1}$$

$$+ (2^{\alpha_2+1} - 1)\zeta(-\alpha_2)a_2\varepsilon^{\alpha_2} + \cdots, \quad \varepsilon \to 0. \qquad (5.2.4)$$

In particular if $\phi \in \mathcal{K}$ then

$$\sum_{n=1}^{\infty}(-1)^n\phi(n\varepsilon) \sim \sum_{n=0}^{\infty}\frac{(2^{n+1} - 1)\zeta(-n)\phi^{(n)}(0)\varepsilon^n}{n!} \qquad (5.2.5)$$

$$\sim \frac{-\phi(0)}{2} + \sum_{m=0}^{\infty}\frac{(2^{2m+1} - 1)\zeta(-2m + 1)\phi^{(2m+1)}(0)\varepsilon^{2m+1}}{(2m + 1)!},$$

where we have used the fact that $\zeta(0) = \frac{-1}{2}$ and $\zeta(-2k) = 0$, $k = 1, 2, 3, \cdots$.

We remark that the alternating series $\sum_{n=1}^{\infty}(-1)^n\phi(n)$ could be divergent when $\phi \in \mathcal{K}$ and in such cases its value is defined as $< f(x), \phi(x) >$. Actually it will follow from the results of Section 5.4 that the series is always Cesàro summable to the value $< f(x), \phi(x) >$.

Taking $\phi = e^{-x}$ we obtain in particular

$$\frac{1}{1 + e^{-x}} = \sum_{n=0}^{\infty}(-1)^n e^{-nx} = \frac{1}{2} + \sum_{m=0}^{\infty}\frac{(2^{2m+1} - 1)B_{2m+2}x^{2m+1}}{(2m + 2)!}, \quad |x| < 1,$$

where $B_n = n\zeta(1 - n)$ are the Bernoulli numbers.

The expansion

$$\sum_{n=1}^{\infty}(-1)^n n^q\phi(n\varepsilon) \sim \sum_{j=1}^{\infty}(2^{\alpha_j+q+1} - 1)\zeta(-\alpha_j - q)a_j\varepsilon^{\alpha_j}, \quad \varepsilon \to 0, \qquad (5.2.6)$$

follows by applying (5.2.4) to $\tau(x) = x^q\phi(x)$ and simplifying.

Since $\sum_{n=1}^{\infty}(-1)^n n^q \ln n\delta(x - n)$ also belongs to $\mathcal{K}'\{x^{\alpha_n}\}$ we can differentiate both sides of (5.2.6) with respect to q to obtain

$$\sum_{n=1}^{\infty}(-1)^n n^q \ln n\phi(n\varepsilon) \sim \sum_{j=1}^{\infty}\{(\ln 2)2^{\alpha_j+q+1}\zeta(-\alpha_j - q)$$

$$- (2^{\alpha_j+q+1} - 1)\zeta'(-\alpha_j - q)\}a_j\varepsilon^j \quad \text{as } \varepsilon \to 0. \qquad (5.2.7)$$

Next, let us consider the generalized function $h(x) = \sum_{n=-\infty}^{\infty} \delta(x-n)$. It does not belong to $\mathcal{K}'$ since the constant in its Fourier expansion is 1, but we can write $h(x) = 1 + h_1(x)$, where $h_1 \in \mathcal{K}'$. The moments of h_1 vanish and we obtain

$$\sum_{n=-\infty}^{\infty} \phi(n\varepsilon) = \frac{1}{\varepsilon} \int_{-\infty}^{\infty} \phi(x)dx + o(\varepsilon^{\infty}), \quad \text{as } \varepsilon \to 0, \qquad (5.2.8)$$

whenever $\phi \in L^1(\mathbb{R}) \cap \mathcal{K}(\mathbb{R})$.

The distribution $f(x) = \sum_{n=1}^{\infty} \delta(x-n)$ can be written as $f(x) = H(x) + f_1(x)$ where $H(x)$ is the Heaviside function and where $f_1 \in \mathcal{K}'$. The moment function of f_1 is given as

$$< f_1(x), x^{\alpha} > = \zeta(-\alpha), \quad \alpha \neq -1, \qquad (5.2.9)$$

while at $\alpha = -1$ we use the finite part value $< f_1(x), x^{-1} > = \gamma$, the Euler constant.

It follows that if $\phi \in \mathcal{K} \cap L^1(1, \infty)$ then

$$\sum_{n=1}^{\infty} \phi(n\varepsilon) \sim \frac{1}{\varepsilon} \int_0^{\infty} \phi(x)dx + \sum_{n=0}^{\infty} \frac{\zeta(-n)\phi^{(n)}(0)\varepsilon^n}{n!}, \quad \text{as } \varepsilon \to 0. \quad (5.2.10)$$

More generally, let $\phi \in \mathcal{K}\{x^{\alpha_n}\} \cap L^1(1, \infty)$ with development $\phi(x) \sim \sum_{j=1}^{\infty} a_j x^{\alpha_j}$ as $x \to 0$. Let $\alpha_k = -1$ so that $\Re e\alpha_1 \leq \cdots \leq \Re e\alpha_k \leq -1 \leq \Re e\alpha_{k+1} \leq \cdots$. Since $H(\lambda x) = H(x) + \ln \lambda \delta_k(x)$ in $\mathcal{K}\{x^{\alpha_n}\}$ we obtain

$$\sum_{n=1}^{\infty} \phi(n\varepsilon) = \frac{1}{\varepsilon} < H(x/\varepsilon) + f_1(x/\varepsilon), \phi(x) >$$

$$= \frac{1}{\varepsilon} < H(x) - \ln\varepsilon\delta_k(x) + f_1(x/\varepsilon), \phi(x) >$$

$$\sim \frac{1}{\varepsilon} \text{ F.p.} \int_0^{\infty} \phi(x)dx - \frac{\ln\varepsilon}{\varepsilon}a_k + \frac{\gamma}{\varepsilon}a_k + \sum_{j=1}^{\infty} \zeta(-\alpha_j)a_j\varepsilon^{\alpha_j}. \quad (5.2.11)$$

Similarly, by considering $\tau(x) = x^q\phi(x)$ we obtain the development of $\sum_{n=1}^{\infty} n^q\phi(n\varepsilon)$. If $\alpha_k = -(1+q)$ then

$$\sum_{n=1}^{\infty} n^q\phi(n\varepsilon) \sim \sum_{j=1}^{k-1} \zeta(-\alpha_j - q)a_j\varepsilon^{\alpha_j} - \frac{a_k \ln\varepsilon}{\varepsilon^{q+1}}$$

$$+ \left[\text{F.p.} \int_0^{\infty} x^q\phi(x)dx + \gamma a_k \right] \varepsilon^{-q-1}$$

$$+ \sum_{j=k+1}^{\infty} \zeta(-\alpha_j - q)a_j \varepsilon^{\alpha_j}. \tag{5.2.12}$$

In particular, if $\phi(x) = e^{-x^p}$, $p > 0$, we obtain the expansions

$$\sum_{n=1}^{\infty} n^q e^{-\varepsilon n^p} \sim p^{-1}\Gamma\left(\frac{q+1}{p}\right)\varepsilon^{-(q+1)/p} + \sum_{n=0}^{\infty}\frac{(-1)^n \zeta(-pn-q)\varepsilon^n}{n!} \tag{5.2.13}$$

as $\varepsilon \to 0$ if $(q+1)/p \neq 0, -1, -2, \cdots$ while if $k = 0, 1, 2, \cdots$

$$\sum_{n=1}^{\infty}\frac{e^{-\varepsilon n^p}}{n^{1+kp}} \sim \sum_{n=0}^{k-1}\frac{(-1)^n \zeta(1+(k-n)p)\varepsilon^n}{n!}$$

$$-\frac{(-1)^k}{k!}\varepsilon^k \ln\varepsilon + \varepsilon^k\left[\frac{(-1)^k\psi(k+1)}{pk!} + \frac{\gamma(-1)^k}{k!}\right]$$

$$+ \sum_{n=k+1}^{\infty}\frac{(-1)^n \zeta(1+(k-n)p)\varepsilon^n}{n!}, \quad \text{as } \varepsilon \to 0, \tag{5.2.14}$$

where $\psi(s)$ is the digamma function, so that $\psi(k) = 1 + 1/2 + \cdots + 1/(k-1) - \gamma$.

Observe that if we set $p = q = 1$ in (5.2.13) and replace ε by $-x$ we recover the well-known relation

$$\frac{1}{e^x - 1} = \frac{1}{x} + \frac{1}{2} + \sum_{n=1}^{\infty}\frac{(-1)^n \zeta(-n)x^n}{n!}, \quad |x| < 1. \tag{5.2.15}$$

Another useful expansion is the following. Let $\phi \in \mathcal{K}\{x^{\alpha_n}\}$ with $x^q \ln x\phi(x) \in L^1(1,\infty)$. Let $\alpha_k = -(1+q)$. Then

$$\sum_{n=1}^{\infty} n^q \ln n\phi(n\varepsilon) \sim \frac{(\ln\varepsilon)^2 a_k}{2\varepsilon^{q+1}} - \frac{\ln\varepsilon}{\varepsilon^{q+1}} \text{F.p.} \int_0^{\infty} x^q\phi(x)dx$$

$$+ \frac{1}{\varepsilon^{q+1}}\left[\text{F.p.} \int_0^{\infty} (\ln x)x^q\phi(x)dx + c_1 a_k\right]$$

$$- \sum_{\substack{j=1 \\ j \neq k}}^{\infty} \zeta'(-\alpha_j - q)a_j\varepsilon^{\alpha_j}, \quad \varepsilon \to 0, \tag{5.2.16}$$

where c_1 is the finite part of $-\zeta'(-s)$ at $s = -1$, so that

$$c_1 = \lim_{s \to -1}\left[-\zeta'(-s) - \left(\frac{1}{s+1}\right)^2\right] = \lim_{n \to \infty}\left[\sum_{j=1}^{n}\frac{\ln j}{j} - \frac{\ln^2 n}{2}\right]. \tag{5.2.17}$$

The expansion (5.2.16) is obtained by writing $\sum_{n=1}^{\infty} n^q \ln n \delta(x - n) = x^q \ln x H(x) + h(x)$ where $h \in \mathcal{K}'$ and using the identity

$$(\lambda x)^q \ln(\lambda x) H(\lambda x) = \lambda^q \ln \lambda x^q H(x) + \lambda^q x^q \ln x H(x) + \frac{\lambda^q (\ln \lambda)^2}{2} \delta_k(x).$$
(5.2.18)

Let us now consider a periodic sequence $\{b_n\}$, specifically $b_{k+n} = b_k$, where k is the period. Let $c = 1/k(b_1 + \cdots + b_k)$. The distribution $f(x) = \sum_{n=1}^{\infty} b_n \delta(x - n)$ can be decomposed as $cH(x) + f_1(x)$, where $f_1 \in \mathcal{K}'$. The moment function $\mu(\alpha)$ of $f_1(x)$ is the analytic continuation of

$$\mu(\alpha) = < f_1(x), x^\alpha > = \sum_{n=1}^{\infty} b_n n^\alpha, \qquad \Re \alpha < -1.$$
(5.2.19)

If $c = 0$ then 0 does not belong to the support of f_1 and thus $\mu(\alpha)$ is entire. When $c \neq 0$ then $f_2(x) = f_1(x) + cH(x)H(1 - x)$ belongs to $\mathcal{K}'$ and is supported in $[1, \infty)$. It follows that the moment function of f_2 is entire and consequently $\mu(\alpha)$ is analytic in $\mathbb{C} \backslash \{-1\}$, the point $\alpha = -1$ being a simple pole with residue $-c$. We denote by $\mu(-1) = \lim_{\alpha \to -1} \left[\mu(\alpha) - \frac{c}{\alpha+1} \right]$ the finite part of $\mu(\alpha)$ at $\alpha = -1$.

In particular, if $a_n = 1$ for $n \equiv r (\mathrm{mod}\ k)$ while $a_n = 0$ otherwise, and $1 \leq r \leq k$, then the moment function is given by

$$\mu(\alpha) = \mu(r, k; \alpha) = \sum_{\substack{n=1 \\ n \equiv r (\mathrm{mod}\ k)}}^{\infty} n^\alpha = k^\alpha \zeta(-\alpha, r/k),$$
(5.2.20)

where $\zeta(s, a)$ is the Hurwitz zeta function given by $\zeta(s, a) = \sum_{n=0}^{\infty} (n + a)^{-s}$ for $\Re s > 1$.

Thus, if $\phi \in \mathcal{K}\{x^{\alpha_n}\} \cap L^1(1, \infty)$ with expansion $\phi(x) \sim a_1 x^{\alpha_1} + a_2 x^{\alpha_2} + \cdots$, as $x \to 0$ and with $\alpha_k = -1$, then

$$\sum_{\substack{n=1 \\ n \equiv r (\mathrm{mod}\ k)}}^{\infty} \phi(n\varepsilon) \sim \frac{\ln \varepsilon}{k\varepsilon} a_k + \left[\frac{1}{k} \int_0^{\infty} \phi(x) dx + \mu(-1) a_k \right] \frac{1}{\varepsilon}$$

$$+ \sum_{\substack{j=1 \\ j \neq k}}^{\infty} \zeta(-\alpha_j, r/k) a_j (k\varepsilon)^{\alpha_j}.$$
(5.2.21)

Also, if $\beta(n)$ is a non-principal character modulo k [4], then

$$\sum_{n=1}^{\infty} \beta(n) \phi(n\varepsilon) \sim \sum_{j=1}^{\infty} L(-\alpha_j; \beta) a_j \varepsilon^{\alpha_j},$$
(5.2.22)

where $L(s; \beta)$ is the Dirichlet L-function associated with β, which is the entire function given by the series $\sum_{n=1}^{\infty} \beta(n) n^{-s}$ for $\Re e\, s > -1$.

The preceding analysis can be generalized to smooth functions whose expansions near $x = 0$ contain logarithmic kernels. In particular, let $\phi \in \mathcal{K}\{x^{\alpha_n} \ln x, x^{\alpha_n}\}$ with expansion $\phi(x) \sim \sum_{j=1}^{\infty} (a'_j \ln x + a_j) x^{\alpha_j}$, as $x \to 0$. Let $\alpha_k = -(1 + q)$ and suppose $x^q \phi(x) \in L^1(1, \infty)$. Then

$$
\sum_{n=1}^{\infty} n^q \phi(n\varepsilon) \sim \frac{-(\ln \varepsilon)^2 a'_k}{2\varepsilon^{q+1}} + [-a_k + \gamma a'_k] \frac{\ln \varepsilon}{\varepsilon^{q+1}}
$$

$$
+ \left[F.p. \int_0^{\infty} x^q \phi(x) dx + \gamma a_k + c_1 a'_k \right] \frac{1}{\varepsilon^{q+1}}
$$

$$
+ \sum_{\substack{j=1 \\ j \neq k}}^{\infty} \{\zeta(-\alpha_j - q) a'_j \ln \varepsilon
$$

$$
+ [\zeta(-\alpha_j - q) a_j - \zeta'(-\alpha_j - q) a'_j]\} \varepsilon^{\alpha_j}, \tag{5.2.23}
$$

where c_1 is given by (5.2.17).

5.3 Lambert Type Series

In this section we shall consider the asymptotic development of divisor type series such as

$$
F(\varepsilon) = \sum_{k=1}^{\infty} \sum_{j=1}^{\infty} k^r j^s \phi(k^p j^q \varepsilon), \tag{5.3.1}
$$

where $p, q > 0$. In the case where $\phi(x) = e^{-x}, p = q = 1$, setting $z = e^{-\varepsilon}$ yields the Lambert series

$$
\sum_{k=1}^{\infty} \sum_{j=1}^{\infty} k^r j^s z^{kj}, \tag{5.3.2}
$$

and that is why we designated series like (5.3.1) and its generalizations considered below as Lambert type series.

The asymptotic expansion of a Lambert type series can be found by iteration. Let us illustrate this procedure by considering the development of (5.3.1). Suppose $\phi \in \mathcal{S}\{x^{\alpha_n}\}$ with development $\phi(x) \sim a_1 x^{\alpha_1} + a_2 x^{\alpha_2} + \cdots +$, as $x \to 0$ and set

$$
\tau(x) = \sum_{j=1}^{\infty} j^s \phi(xj^q). \tag{5.3.3}
$$

Using (5.2.12) we get

$$\tau(x) \sim -q^{-1}a_n x^{\alpha_n} \ln x$$
$$+ x^{\alpha_n} \left(q^{-1} F.p. \int_0^\infty x^{\alpha_n-1}\phi(x)dx + \gamma a_n \right)$$
$$+ \sum_{j \neq n} \zeta(-\alpha_j q - s)a_j x^{\alpha_j}, \tag{5.3.4}$$

as $x \to 0$, where $\alpha_n = -(1+s)/q$. Therefore $\tau \in \mathcal{S}\{x^{\alpha_j} \ln x, x^{\alpha_j}\}$.

But it follows from (5.3.1) and (5.3.3) that

$$F(\varepsilon) = \sum_{k=1}^\infty k^r \tau(k^p \varepsilon),$$

and thus the expansion of $F(\varepsilon)$ can be obtained from the formula (5.2.23). Suppose first that $(1+s)/q \neq (1+r)/p$ and let n, m be the indices with $\alpha_n = -(1+s)/q$, $\alpha_m = -(1+r)/p$. Then

$$F(\varepsilon) \sim \left(q^{-1} F.p. \int_0^\infty x^{(s+1)/q-1}\phi(x)dx + \gamma a_n \right)\zeta(-\alpha_n p - r)\varepsilon^{\alpha_n}$$
$$+ q^{-1} p a_n \zeta'(-\alpha_n p - r)\varepsilon^{\alpha_n} - q^{-1}a_n\zeta(-\alpha_n p - r)\varepsilon^{\alpha_n} \ln \varepsilon$$
$$+ \left(p^{-1} F.p. \int_0^\infty x^{-\alpha_m-1}\tau(x)dx + \gamma\zeta(-\alpha_m q - s)a_m \right)\varepsilon^{\alpha_m}$$
$$-p^{-1}a_m\zeta(-\alpha_m q-s)\varepsilon^{\alpha_m} \ln \varepsilon \quad + \sum_{j \neq n,m} \zeta(-\alpha_j q-s)\zeta(-\alpha_j p-r)a_j \varepsilon^{\alpha_j},$$

$$\text{as } \varepsilon \to 0.$$

Observe now that

$$F.p. \int_0^\infty x^{-\alpha_m-1}\tau(x)dx = \sum_{j=1}^\infty j^s F.p. \int_0^\infty x^{(r+1)/p-1}\phi(xj^q)dx$$
$$= \sum_{j=1}^\infty j^{s+\alpha_m q} \left[F.p. \int_0^\infty x^{(r+1)/p-1}\phi(x)dx - (\ln j^q)a_m \right]$$
$$= \zeta(-\alpha_m q - s) F.p. \int_0^\infty x^{(r+1)/p-1}\phi(x)dx + qa_m\zeta'(-\alpha_m q - s),$$

and therefore the expansion of the Lambert type series (5.3.1) takes the form

$$\sum_{k=1}^{\infty}\sum_{j=1}^{\infty}k^r j^s \phi(k^p j^q \varepsilon) \sim -q^{-1}a_n\zeta(-\alpha_n p - r)\varepsilon^{\alpha_n}\ln\varepsilon$$

$$+\left\{-q^{-1}\zeta(-\alpha_n p - r)F.p\int_0^{\infty}x^{-\alpha_n-1}\phi(x)dx\right.$$

$$\left.+a_n[\gamma\zeta(-\alpha_n p - r) + pq^{-1}\zeta'(-\alpha_n p - r)]\right\}\varepsilon^{\alpha_n}$$

$$-p^{-1}a_m\zeta(-\alpha_m q - s)\varepsilon^{\alpha_m}\ln\varepsilon$$

$$+\left\{-p^{-1}\zeta(-\alpha_m q - s)F.p\int_0^{\infty}x^{-\alpha_m-1}\phi(x)dx\right.$$

$$\left.+a_m[\gamma\zeta(-\alpha_m q - s) + qp^{-1}\zeta'(-\alpha_n q - s)]\right\}\varepsilon^{\alpha_m}$$

$$\sum_{j\neq n,m}\zeta(-\alpha_j q - s)\zeta(-\alpha_j p - r)a_j\varepsilon^{\alpha_j}. \quad (5.3.5)$$

The case when $(1+s)/q = (1+r)/p$ is slightly more complicated. As in the previous case we need an expression for the finite part of the integral $\int_0^{\infty}x^{\beta}\tau(x)dx$ in terms of finite part integrals involving $\phi(x)$. Here $\beta = (1+r)/p - 1 = -(\alpha_m + 1) = (1+s)/q - 1 = -(\alpha_n + 1)$. We proceed as follows.

$$F.p\int_0^{\infty}x^{\beta}\tau(x)dx = F.p\lim_{\varepsilon\to 0}\int_{\varepsilon}^{\infty}x^{\beta}\tau(x)dx$$

$$= F.p\lim_{\varepsilon\to 0}\sum_{j=1}^{\infty}j^s\int_{\varepsilon}^{\infty}x^{\beta}\phi(xj^q)dx$$

$$= F.p\lim_{\varepsilon\to 0}\sum_{j=1}^{\infty}\frac{1}{j}\int_{\varepsilon j^q}^{\infty}x^{\beta}\phi(x)dx$$

$$= F.p\lim_{\varepsilon\to 0}\sum_{j=1}^{\infty}\frac{\omega(j\varepsilon^{\frac{1}{q}})}{j},$$

where

$$\omega(x) = \int_{x^q}^{\infty}t^{\beta}\phi(t)dt.$$

The finite part of the limit of $\sum_{j=1}^{\infty} j^{-1} w(j \varepsilon^{1/q})$ as $\varepsilon \to 0$ can be found from the expansion (5.2.12)

$$
w(x) \sim F.p. \int_0^\infty x^\beta \phi(x) dx - q a_n \ln x
$$
$$
+ \sum_{j \neq n} (\alpha_j + \beta + 1)^{-1} a_j x^{q(\alpha_j + \beta + 1)}. \qquad (5.3.6)
$$

The result is

$$
F.p. \lim_{\varepsilon \to 0} \sum_{j=1}^{\infty} \frac{1}{j} w(j \varepsilon^{\frac{1}{q}}) = F.p. \int_0^\infty \frac{w(x)}{x} dx + \gamma F.p. \int_0^\infty x^\beta \phi(x) dx - c_1 q a_n
$$
$$
= \frac{F.p.}{q} \int_0^\infty x^\beta \ln x \phi(x) dx + \gamma F.p. \int_0^\infty x^\beta \phi(x) dx - c_1 q a_n. \qquad (5.3.7)
$$

If we now use (5.3.7) and (5.2.23) we obtain that when $(1+r)/p = (1+s)/q = -\alpha_n$ then

$$
\sum_{k=1}^{\infty} \sum_{j=1}^{\infty} k^r j^s \phi(k^p j^q \varepsilon) \sim \sum_{j \neq n} \zeta(-\alpha_j p - r) \zeta(-\alpha_j q - s) a_j \varepsilon^{\alpha_j} + (2pq)^{-1} a_n \varepsilon^{\alpha_n} (\ln \varepsilon)^2
$$
$$
- \left[\frac{1}{pq} F.p. \int_0^\infty x^{-\alpha_n - 1} \phi(x) dx + \gamma \left(\frac{1}{p} + \frac{1}{q} \right) a_n \right] \varepsilon^{\alpha_n} \ln \varepsilon
$$
$$
+ \left\{ F.p. \int_0^\infty x^{-\alpha_n - 1} \left[\frac{\ln x}{pq} + \gamma \left(\frac{1}{p} + \frac{1}{q} \right) \right] \phi(x) dx \right.
$$
$$
\left. + \left(\gamma^2 - \left(\frac{q}{p} + \frac{p}{q} \right) c_1 \right) a_n \right\} \varepsilon^{\alpha_n}. \qquad (5.3.8)
$$

Formulas (5.3.5) and (5.3.8) can be written in an alternative way by introducing the arithmetical functions $\sigma(r, s; p, q; m)$ given by

$$
\sigma(r, s; p, q; m) = \sum_{k^p j^q = m} k^r j^s. \qquad (5.3.9)
$$

We remark that in particular when $p = q = 1$, $s = 0$ then

$$
\sigma(r, 0; 1, 1; m) = \sigma_r(m) = \sum_{k | m} k^r, \qquad (5.3.10)
$$

the notation $d(m)$ being used when $r = 0$, i.e., for the number of divisors of m.

With this notation we could summarize our formulas as follows.

Theorem 45. *Let $p > 0$, $q > 0$, $r, s \in \mathbb{R}$ and let $\phi \in \mathcal{S}\{x^{\alpha_j}\}$ with the expansion $\phi(x) \sim \sum_{j=1}^{\infty} a_j x^{\alpha_j}$ as $x \to 0$. Denote by n and m the indices that satisfy $\alpha_n = -(1+r)/p$, $\alpha_m = -(1+s)/q$.*

(a) If $\alpha_n \neq \alpha_m$ then

$$\sum_{j=1}^{\infty} \sigma(r, s; p, q; j)\phi(j\varepsilon) = \sum_{k=1}^{\infty}\sum_{j=1}^{\infty} k^r j^s \phi(k^p j^q \varepsilon) \sim -q^{-1} a_n \zeta(-\alpha_m p - r)\varepsilon^{\alpha_n} \ln \varepsilon$$

$$- p^{-1} a_m \zeta(-\alpha_m q - s)\varepsilon^{\alpha_m} \ln \varepsilon$$

$$+ \left\{ q^{-1}\zeta(-\alpha_n p - r) F.p. \int_0^{\infty} x^{-\alpha_n - 1}\phi(x)dx \right.$$

$$\left. + a_n \left(\gamma \zeta(-\alpha_n p - r) + \frac{p}{q}\zeta'(-\alpha_n p - r) \right) \right\} \varepsilon^{\alpha_n}$$

$$+ \left\{ p^{-1}\zeta(-\alpha_m q - s) F.p. \int_0^{\infty} x^{-\alpha_m - 1}\phi(x)dx \right.$$

$$\left. + a_m \left(\gamma \zeta(-\alpha_m q - s) + \frac{q}{p}\zeta'(-\alpha_m q - s) \right) \right\} \varepsilon^{\alpha_m}$$

$$+ \sum_{j \neq n, m} \zeta(-\alpha_j q - s)\zeta(-\alpha_j p - r)\varepsilon^{\alpha_j} a_j, \quad \varepsilon \to 0. \qquad (5.3.11)$$

(b) If $\alpha_n = \alpha_m$ then

$$\sum_{j=1}^{\infty} \sigma(r, s; p, q; j)\phi(j\varepsilon) = \sum_{k=1}^{\infty}\sum_{j=1}^{\infty} k^r j^s \phi(k^p j^s \varepsilon) \sim \frac{1}{2qp} a_n \varepsilon^{\alpha_n} (\ln \varepsilon)^2$$

$$- \left(\frac{1}{pq} F.p. \int_0^{\infty} x^{-\alpha_n - 1}\phi(x)dx + \gamma \left(\frac{1}{p} + \frac{1}{q} \right) a_n \right) \varepsilon^{\alpha_n} \ln \varepsilon$$

$$+ \left\{ F.p. \int_0^{\infty} x^{-\alpha_n - 1} \left(\frac{\ln x}{pq} + \gamma \left(\frac{1}{p} + \frac{1}{q} \right) \right) \phi(x)dx \right.$$

$$\left. + \left(\gamma^2 - \left(\frac{q}{p} + \frac{p}{q} \right) c_1 \right) a_n \right\} \varepsilon^{\alpha_n}$$

$$+ \sum_{j \neq n, m} \zeta(-\alpha_j p - r)\zeta(-\alpha_j q - s)a_j \varepsilon^{\alpha_j}, \qquad \text{as } \varepsilon \to 0. \qquad (5.3.12)$$

In particular, if $r \neq 0$ then

$$\sum_{j=1}^{\infty} \sigma_r(j)\phi(j\varepsilon) \sim -a_n\zeta(1-r)\varepsilon^{-1}\ln\varepsilon + \left[\zeta(1-r)F.p.\int_0^{\infty}\phi(x)dx\right.$$

$$\left. +a_n(\gamma\zeta(1-r)+\zeta'(1-r))\right]\varepsilon^{-1} - a_m\zeta(1+r)\varepsilon^{-1-r}\ln\varepsilon$$

$$+ \left[\zeta(1+r)F.p.\int_0^{\infty}x^r\phi(x)dx + a_m(\gamma\zeta(1+r)+\zeta'(1+r))\right]\varepsilon^{-1-r}$$

$$+ \sum_{j\neq n,m}\zeta(-\alpha_j)\zeta(-\alpha_j-r)a_j\varepsilon^{\alpha_j}, \quad \text{as } \varepsilon\to 0, \tag{5.3.13}$$

where $\alpha_m = -1 - r$, $\alpha_n = -1$.

Similarly, if $\alpha_n = -1$,

$$\sum_{j=1}^{\infty} d(j)\phi(j\varepsilon) \sim a_n\frac{(\ln\varepsilon)^2}{2\varepsilon} - \left[F.p.\int_0^{\infty}\phi(x)dx + 2\gamma a_n\right]\frac{\ln\varepsilon}{\varepsilon}$$

$$+ \left\{F.p.\int_0^{\infty}(\ln x + 2\gamma)\phi(x)dx + (\gamma^2 - 2c_1)a_n\right\}\frac{1}{\varepsilon}$$

$$+ \sum_{j\neq n}\zeta(-\alpha_j)^2 a_j\varepsilon^{\alpha_j}, \quad \text{as } \varepsilon\to 0. \tag{5.3.14}$$

It is interesting to observe that when considering formulas of this kind in Chapter 15 of his second notebook [7], [95], Ramanujan gave the formula

$$\sum_{n\leq x} d(n) \sim x\ln x + (2\gamma - 1)x. \tag{5.3.15}$$

This is a well-known formula in elementary number theory, but what is interesting is where it is placed. In fact, it is possible that the development of $\sum_{n=1}^{\infty} b_n\phi(n\varepsilon)$ led Ramanujan to some of his celebrated asymptotic formulas for arithmetical functions [61]. Let us see how this could have come about.

The expansion for the series $\sum_{j=1}^{\infty} d(j)\phi(j\varepsilon)$ given in (5.3.14) is very similar to that of $\sum_{j=1}^{\infty}(\ln j)\phi(j\varepsilon)$ given in (5.2.16) since the leading terms are the same. Subtracting these two expressions we obtain

$$\sum_{j=1}^{\infty}(d(j) - \ln j)\phi(j\varepsilon) \sim -2\gamma a_n\varepsilon^{-1}\ln\varepsilon + 2\gamma\int_0^{\infty}\phi(x)dx + (\gamma^2 - 3c_1)a_n$$

$$+ \sum_{j \neq n} (\zeta(-\alpha_j)^2 + \zeta'(-\alpha_j)) a_j \varepsilon^{\alpha_j},$$

where $\alpha_n = -1$. The leading term in this expression is nothing but 2γ times the leading term in the formula (5.2.11). Therefore,

$$\sum_{j=1}^{\infty} (d(j) - \ln j - 2\gamma)\phi(j\varepsilon) \sim -(\gamma^2 + 3c_1)a_n \varepsilon^{-1}$$

$$+ \sum_{j \neq n} (\zeta(-\alpha_j)^2 + \zeta'(-\alpha_j) - 2\gamma\zeta(-\alpha_j)) a_j \varepsilon^{\alpha_j},$$

where we remark the fact that $\zeta(-\alpha)^2 + \zeta'(-\alpha) - 2\gamma\zeta(-\alpha)$ is entire and its value at $\alpha = -1$ is precisely $-\gamma^2 - 3c_1$.

Thus $d(n) = \ln n + 2\gamma + e(n)$, where $e(n)$ is an error term that is expected to be small in some sense. The fact that $\{e(n)\}$ and similar residual term are indeed small in several ways is established in Section 5.4. Presently we would like to observe that our analysis leads to the approximations

$$\sigma(r, s; p, q; n) \approx q^{-1} \zeta \left(\frac{p(1+s)}{q} - r \right) n^{(1+s)/q-1}$$

$$+ p^{-1} \zeta \left(\frac{q(1+r)}{p} - s \right) n^{(1+r)/p-1}, \quad \frac{1+r}{p} \neq \frac{1+s}{q}, \tag{5.3.16a}$$

$$\sigma_r(n) \approx \zeta(1 - r) + \zeta(1 + r)n^r, \quad r \neq 0, \tag{5.3.16b}$$

$$\sigma(r, s; p, q; n) \approx p^{-1} q^{-1} n^{(1+r)/p-1} \ln n + \gamma \left(\frac{1}{p} + \frac{1}{q} \right), \quad \frac{1+r}{p} = \frac{1+s}{q}, \tag{5.3.16c}$$

$$d(n) \approx \ln n + 2\gamma. \tag{5.3.16d}$$

Berndt and Evans [7] gave proofs of the results of Theorem 45 in the case where $\phi(x) = e^{-x}$, the case considered by Ramanujan. This choice of ϕ is useful enough, particularly because of its relation with Abel summability. The asymptotic formula for the series $\sum_{n=1}^{\infty} d(n)e^{-n\varepsilon}$ as $\varepsilon \to 0$ was obtained by Wigert [115] and plays a role in obtaining mean value theorems for the Riemann zeta function [109]. If we set

$$f(r, s; p, q; \varepsilon) = \sum_{k=1}^{\infty} \sum_{j=1}^{\infty} k^r j^s e^{-\varepsilon k^p j^q}, \tag{5.3.17}$$

and use the values

$$F.p. \int_0^\infty e^{-x} x^\alpha dx = \Gamma(\alpha + 1), \quad \alpha \neq -1, -2, -3, \cdots,$$

$$F.p. \int_0^\infty e^{-x} x^{-k} dx = \frac{(-1)^{k-1}\psi(k)}{(k-1)!}, \quad k = -1, -2, -3, \cdots,$$

where $\psi(k) = 1 + 1/2 + \cdots + 1/(k-1) - \gamma$, we obtain

I. $(1+r)/p \neq (1+s)/q$

I. (a) $-(1+r)/p, -(1+s)/q \notin \{1, 2, 3, \cdots\}$

$$f(r, s; p, q; \varepsilon) \sim \frac{\zeta(p/q(1+s) - r)}{q} \Gamma\left(\frac{1+s}{q}\right) \varepsilon^{-(1+s)/q}$$

$$+ \frac{\zeta(q/p(1+r) - s)}{p} \Gamma\left(\frac{1+r}{p}\right) \varepsilon^{-(1+r)/p}$$

$$+ \sum_{n=0}^\infty \frac{\zeta(-nq - s)\zeta(-np - r)}{n!} (-\varepsilon)^n. \tag{5.3.18a}$$

I. (b) $-(1+s)/q \notin \{1, 2, 3, \cdots\}, -(1+r)/p = k \in \mathbb{N}$,

$$f(r, s; p, q; \varepsilon) \sim q^{-1}\zeta(p/q(1+s) - r)\Gamma\left(\frac{1+s}{q}\right) \varepsilon^{-(1+s)/q}$$

$$\frac{(-1)^k}{k!} \left\{ \frac{\zeta(kq - s)}{p} (\psi(k+1) - \ln\varepsilon) + \gamma\zeta(kq - s) + \frac{q}{p}\zeta'(kq - s) \right\} \varepsilon^k$$

$$+ \sum_{\substack{n=0 \\ n \neq k}}^\infty \frac{\zeta(-nq - s)\zeta(-np - r)}{n!} (-\varepsilon)^n. \tag{5.3.18b}$$

I. (c) $-(1+r)/p = k, -(1+s)/q = j, k, j \in \mathbb{N}$,

$$f(r, s; p, q; \varepsilon) \sim$$

$$\frac{(-1)^j}{j!} \left\{ \frac{\zeta(jp - r)}{q} (\psi(j+1) - \ln\varepsilon) + \gamma\zeta(jp - r) + \frac{p}{q}\zeta'(jp - r) \right\} \varepsilon^j$$

$$\frac{(-1)^k}{k!} \left\{ \frac{\zeta(kp - s)}{p} (\psi(k+1) - \ln\varepsilon) + \gamma\zeta(kq - s) + \frac{q}{p}\zeta'(kq - s) \right\} \varepsilon^k$$

$$+ \sum_{\substack{n=0 \\ n \neq k, j}}^\infty \frac{\zeta(-nq - s)\zeta(-np - r)}{n!} (-\varepsilon)^n. \tag{5.3.18c}$$

II. $(1+r)/p = (1+s)/q = -\alpha$

II. (a) $\alpha \notin \mathbb{N}$

$$f(r, s; p, q; \varepsilon) \sim -p^{-1}q^{-1}\Gamma \left(\frac{1+r}{p}\right) \varepsilon^{-(1+r)/p} \ln \varepsilon$$

$$+ \left(p^{-1}q^{-1}\Gamma' \left(\frac{1+r}{p}\right) + \gamma(p^{-1} + q^{-1})\Gamma \left(\frac{1+r}{p}\right)\right) \varepsilon^{-(1+r)/p}$$

$$+ \sum_{n=0}^{\infty} \frac{\zeta(-nq - s)\zeta(-np - r)}{n!}(-\varepsilon)^n. \tag{5.3.18d}$$

II. (b) $\alpha = k \in \mathbb{N}$

$$f(r, s; p, q; \varepsilon) \sim \frac{(-1)^j}{2pqk!}\varepsilon^k (\ln \varepsilon)^2$$

$$- \left(\frac{(-1)^k \psi(k+1)}{pqk!} + \left(\frac{1}{p} + \frac{1}{q}\right)\frac{(-1)^k}{k!}\right) \varepsilon^k \ln \varepsilon$$

$$+ \left\{z_k + \gamma \left(\frac{1}{p} + \frac{1}{q}\right)\frac{(-1)^k \psi(k+1)}{k!} + \left(\gamma^2 - \left(\frac{p}{q} + \frac{q}{p}\right)c_1\right)\frac{(-1)^k}{k!}\right\} \varepsilon^k$$

$$+ \sum_{\substack{n=0 \\ n \neq k}}^{\infty} \frac{\zeta(-nq - s)\zeta(-np - r)}{n!}(-\varepsilon)^n, \tag{5.3.18e}$$

where $z_k = F.p. \int_0^{\infty} x^{-k-1}e^{-x} \ln x dx$.

Our analysis can be easily generalized to n-dimensional series. Let $K = \{(k_1, \cdots, k_n) \in \mathbb{Z}^n : k_i \geq 0\}$ and let us use the standard notation $\mathbf{k}^{\mathbf{p}} = k_1^{p_1} \cdots k_n^{p_n}$, if $\mathbf{k} = (k_1 \cdots k_n)$ and $\mathbf{p} = (p_1 \cdots p_n)$. Let

$$F(\varepsilon) = \sum_{\mathbf{k} \in K} \mathbf{k}^{\mathbf{r}} \phi(\varepsilon \mathbf{k}^{\mathbf{p}}), \tag{5.3.19}$$

where $\mathbf{r}, \mathbf{p} \in \mathbb{R}^n$, $p_i > 0$.

In order to give the asymptotic development of $F(\varepsilon)$ as $\varepsilon \to 0$, it is convenient to introduce the following notation. If $\mathbf{x} \in \mathbb{R}^n$ then

$$\zeta_n(\mathbf{x}) = \zeta(x_1) \cdots \zeta(x_n),$$

where the finite part value $\zeta(1) = \gamma$ is used if some x_i are equal to 1. For $i = 1, 2, \cdots, n$, $\mathbf{x}(i)$ is the $(n-1)$-dimensional vector obtained by deleting the i-th entry x_i.

Theorem 46. *Let $\mathbf{p}, \mathbf{r} \in \mathbb{R}^n$, $p_i > 0$. Let $\beta_i = (1 + r_i)/p_i$, $1 \leq i \leq n$ and asume $\beta_i \neq \beta_j$ for $i \neq j$. Then if $\phi \in S\{x^{\alpha n}\}$ with expansion $\phi(x) \sim$*

$\sum_{j=1}^{\infty} a_j x^{\alpha_j}$, $x \to 0$,

$$\sum_{k \in K} k^r \phi(\varepsilon k^p) \sim \sum_{\substack{j=1 \\ \alpha_j \neq \beta_i}}^{\infty} \zeta_n(-\alpha_j p - r) a_j \varepsilon^{\alpha_j}$$

$$+ \sum_{i=1}^{n} \zeta_{n-1}(-\beta_i p(i) - r(i)) \left[-p_i^{-1} a_{\beta_i} \ln \varepsilon + p_i^{-1} F.p. \int_0^{\infty} x^{-\beta_i - 1} \phi(x) dx + \gamma a_{\beta_i} \right.$$

$$\left. + \sum_{\substack{k=1 \\ k \neq i}}^{n} \frac{p_k \zeta'(-\beta_i p_k - r_k)}{p_i \zeta(-\beta_i p_k - r_k)} a_{\beta_i} \right] \varepsilon^{\beta_i}. \qquad (5.3.20)$$

In particular, if $\phi(x) = e^{-x}$, *let* $B_1 = \{\beta_i : \beta_i \in \mathbb{N}\}$, $B_2 = \{\beta_i : \beta_i \notin \mathbb{N}\}$. *Then*

$$\sum_{k \in K} k^r e^{-\varepsilon k^p} \sim \sum_{\beta_i \in B_2} \frac{\zeta_{n-1}(-\beta_i p(i) - r(i))}{p_i} \Gamma(-\beta_i) \varepsilon^{\beta_i} + \sum_{\beta_i = j \in B_1} \zeta_{n-1}(jp(i)$$

$$- r(i)) \frac{(-1)^j}{j!} \left[\frac{-\ln \varepsilon}{p_i} + \frac{\psi(j+1)}{p_i} + \gamma + \sum_{\beta_k \neq \beta_i} \frac{p_k \zeta'(-jp_k - r_k)}{p_i \zeta(-jp_k - r_k)} \right] \varepsilon^j$$

$$+ \sum_{n \in \mathbb{N} - B_1} \frac{\zeta_n(-np - r)}{n!} (-\varepsilon)^n, \qquad \text{as } \varepsilon \to 0. \qquad (5.3.21)$$

Proof. Formula (5.3.20) follows by iteration of the results of Theorem 45. ∎

Introducing the arithmetical function

$$\sigma(r; p; m) = \sum_{k^p = m} k^r, \qquad (5.3.22)$$

for $p \in \mathbb{N}^n$, then (5.3.20) suggests the approximation

$$\sigma(r; p; m) \sim \sum_{i=1}^{n} \frac{\zeta_{n-1}(-\beta_i p(i) - r(i))}{p_i} m^{\beta_i - 1}. \qquad (5.3.23)$$

The case when some of the β_i coincide can be handled in a similar way, but the notation becomes very complicated.

As should be clear, this method can also be applied to the study of the asymptotic development of a series of the type $\sum_{n=1}^{\infty} c_n \phi(n\varepsilon)$, where

$c_n = \sum_{k^p j^q = n} a_k b_j$ if the development of the series $\sum_{n=1}^{\infty} a_n \phi(n\varepsilon)$ and $\sum_{n=1}^{\infty} b_n \phi(n\varepsilon)$ are known.

5.4 Distributionally Small Sequences

In Section 5.3 we obtained the asymptotic expansion of the series $\sum_{n=1}^{\infty} d(n)\phi(n\varepsilon)$ and this suggests that we could write $d(n) = \ln n + 2\gamma + e(n)$, where $e(n)$ is a remainder term. A similar analysis yields $\sigma_r(n) = \zeta(1-r) + \zeta(1+r)n^r + e_r(n)$ when $r \neq 0$.

In this section we introduce the notion of a distributionally small sequence. The remainders $\{e_r(n)\}$ turn out to be distributionally small. Distributionally small sequences do not have to be small in the ordinary sense, but as we show they are indeed small in the Cesàro sense, i.e., after they are averaged enough times. The formula $\sum_{n=1}^{N} d(n) \sim (\ln n + 2\gamma)$ is just an indication of the behavior of the distributionally small sequences; in fact, much more is true.

Definition. *Let $\{b_n\}$ be a sequence of complex numbers. We say that $\{b_n\}$ is distributionally small and write $\{b_n\} \in \mathcal{DS}$ if the moment asymptotic expansion holds for the distribution $f(x) = \sum_{n=1}^{\infty} b_n \delta(x - n)$ in the space $\mathcal{D}(\mathbb{R})$. This means that there are constants $\mu_0, \mu_1, \mu_2, \cdots$, such that for each $\phi \in \mathcal{D}(\mathbb{R})$ we have*

$$\sum_{n=1}^{\infty} b_n \phi(n\varepsilon) = < f(x), \phi(\varepsilon x) > = \sum_{j=0}^{N} \frac{\mu_j \phi^{(j)}(0)\varepsilon^j}{j!} + O(\varepsilon^{N+1}), \qquad as \ \varepsilon \to 0.$$
$$(5.4.1)$$

The idea behind this definition is the following. The distributions for which the moment asymptotic expansion holds are those of "rapid decay at infinity," in the distributional sense. Therefore, if the moment asymptotic expansion holds for $\sum_{n=1}^{\infty} b_n \delta(x - n)$, it means that the sequence $\{b_n\}$ should be of "rapid decay" in some sense.

We remark that if $f(x) = \sum_{n=1}^{\infty} b_n \delta(x - n)$ belongs to $\mathcal{O}'_C$ then $\{b_n\}$ is distributionally small. An identical conclusion is obtained if $f \in \mathcal{K}'$ or more generally if $f(x^p) \in \mathcal{O}'_C$ for some $p > 0$.

As a special case, suppose $\{b_n\}$ is of *rapid decay*, that is, $\lim_{n \to \infty} n^\alpha b_n = 0$ for each $\alpha > 0$. Then $\sum_{n=1}^{\infty} b_n \delta(x - n)$ belongs to $\mathcal{O}'_C$ and thus $\{b_n\}$ is distributionally small. Conversely, if $\{b_n\}$ is a *positive* distributionally small sequence then $\{b_n\}$ should be of rapid decay. To see it, choose a test function $\phi \in \mathcal{D}(\mathbb{R})$ that satisfies the following conditions:

(a) $\phi \geq 0$,

(b) $\phi^{(j)}(0) = 0, \quad j = 0, 1, 2, \cdots ,$

(c) $\phi(1) = 1.$

Then since $\{b_n\} \in \mathcal{DS},$

$$F(\varepsilon) = \sum_{n=1}^{\infty} b_n \phi(n\varepsilon) = o(\varepsilon^{\alpha}), \quad \text{as } \varepsilon \to 0, \tag{5.4.2}$$

for each $\alpha > 0$. But since $\{b_n\}$ is positive,

$$0 \leq \varliminf_{n \to \infty} n^{\alpha} b_n \leq \varlimsup_{n \to \infty} n^{\alpha} b_n \leq \varlimsup_{n \to \infty} n^{\alpha} F(1/n) \leq \varlimsup_{\varepsilon \to 0} \varepsilon^{-\alpha} F(\varepsilon) = 0,$$

and the desired result $\lim_{n \to \infty} n^{\alpha} b_n = 0$ follows. Summarizing,

Theorem 47. *Any sequence of rapid decay is distributionally small. Any positive distributionally small sequence is of rapid decay.*

Nevertheless, distributionally small sequences do not have to tend to zero; they do not even have to be bounded. An example is provided by the sequence $\{(-1)^n n^q\}$, which is distributionally small for any $q \in \mathbb{C}$ since $\sum_{n=1}^{\infty} (-1)^n n^q \delta(x - n)$ belongs to $\mathcal{K}'$. The sequence $\{(-1)^n n^q (\ln n)^r\}$ is another example.

However, when distributionally small sequences are analyzed by using the summability theory of divergent series they do appear to be of rapid decay. Take the sequence $\{(-1)^n\}$ for instance. It does not tend to zero, but its average $1/n \sum_{j=1}^{n} (-1)^j = (1 + (-1)^{n+1})/n$ does. For the sequence $\{(-1)^n n\}$ we need to average twice to obtain a null sequence. Actually if any distributionally small sequence is averaged enough times a null sequence is eventually obtained.

In order to prove these results we need to give a short summary of some concepts of the theory of summability of divergent series. A complete account can be found in [60].

Let $\{b_n\}$ be a sequence. If b_n tends to L then so does $(b_1 + \cdots + b_n)/n$. Therefore we can extend the notation of limit of certain non-convergent sequences by setting

$$\lim_{n \to \infty} b_n = L \quad (C, 1) \quad \text{if} \quad \lim_{n \to \infty} \frac{1}{n} \sum_{j=1}^{n} b_j = L. \tag{5.4.3}$$

The C refers to Cesàro. Convergence in the (C, k) sense $k = 2, 3, \cdots$ is defined inductively as follows:

$$\lim_{n \to \infty} b_n = L \quad (C, k) \quad \text{if} \quad \lim_{n \to \infty} \frac{1}{n} \sum_{j=1}^{n} b_j = L \quad (C, k - 1). \tag{5.4.4}$$

For a series $\sum_{n=1}^{\infty} b_n$ summability in the Cesàro sense means (C, k) convergence of its partial sums $B_n = b_1 + \cdots + b_n$, that is,

$$\sum_{n=1}^{\infty} b_n = S \quad (C, k) \quad \text{if} \quad \lim_{n \to \infty} B_n = S \quad (C, k). \tag{5.4.5}$$

Alternatively, let us define inductively the sequence $\{B_n^k\}$, $k = 0, 1, 2, \cdots$, as follows: $B_n^0 = B_n$, $B_n^{k+1} = \sum_{j=1}^{n} B_j^k$. Then $\sum_{n=1}^{\infty} b_n$ is (C, k) summable to S if and only if $\lim_{n \to \infty} k! n^{-k} B_n^k = S$. We use the notation $\sum b_n = S(C)$ to mean that the series is (C, k) summable for some k.

Another important summation method is that of Abel. If $\lim_{x \to 1} \sum_{n=1}^{\infty} b_n x^n = S$ we say that $\sum_{n=1}^{\infty} b_n$ is Abel summable to S and write $\sum_{n=1}^{\infty} b_n = S(A)$. Both Cesàro and Abel methods are *regular*, that is, convergent series are summable to their ordinary sum. Any Cesàro summable series is Abel summable, but the converse is not true.

Another useful summation method is the following. Let $f(x)$ be a function defined for $x \geq 0$, smooth for x near 0 and $f(0) = 1$. Then if $\{b_n\}$ is a sequence we set

$$F(\varepsilon) = \sum_{n=1}^{\infty} b_n f(n\varepsilon) \tag{5.4.6}$$

and say that $\sum_{n=1}^{\infty} b_n$ is (f) summable to S and write $\sum_{n=1}^{\infty} b_n = S \quad (f)$ provided $\lim_{\varepsilon \to 0} F(\varepsilon) = S$.

In order to have a regular summation method we need to ask for some conditions of f. In particular, if $f \in \mathcal{D}(\mathbb{R})$ then the method is regular. When $f(x) = e^{-x}$ we recover Abel summability while if

$$f_k(x) = \begin{cases} (1 - x)^k, & x \leq 1, \\ 0, & x \geq 1, \end{cases} \tag{5.4.7}$$

we obtain the Riesz typical means

$$\sum_{n=1}^{\infty} b_n f_k(n\varepsilon) = \varepsilon^k \sum_{n \leq 1/\varepsilon} b_n \left(\frac{1}{\varepsilon} - n \right)^k$$

that are equivalent to (C, k) summability.

We can also define order relations with respect to a summation method. For instance, if $\alpha > -1$ we say that $b_n = o(n^\alpha) \quad (f)$ whenever $F(\varepsilon) = o(\varepsilon^{-\alpha-1})$ as $\varepsilon \to 0$, and similarly with the big O symbol. In the case where $f = f_k$ the order relations $b_n = O(n^\alpha) \quad (f_k)$ or $b_n = o(n^\alpha) \quad (f_k)$ are also written as $b_n = O(n^\alpha) \quad (C, k)$ or $b_n = o(n^\alpha) \quad (C, k)$ and they are equivalent to the order relations $B_n^k = O(n^{\alpha+k+1})$ or $B_n^k = o(n^{\alpha+k+1})$. These definitions are consistent with the ordinary relations. Suppose for instance that $a_n = O(n^\alpha)$ while $f \in \mathcal{D}(\mathbb{R})$. Then we can find $M > 0$ such that $\mid b_n \mid \leq M n^\alpha$ and $g \in \mathcal{D}(\mathbb{R})$ such that $\mid f \mid \leq g$. Use of (5.2.12) yields

$$\mid F(\varepsilon) \mid \leq \sum_{n=1}^{\infty} \mid b_n \mid \mid f(n\varepsilon) \mid \leq M \sum_{n=1}^{\infty} n^\alpha g(n\varepsilon) = \frac{M}{\varepsilon^{\alpha+1}} \int_0^{\infty} x^\alpha g(x) dx + O(1)$$

$$= O(\varepsilon^{-\alpha-1})$$

since $\alpha > -1$.

Lemma 10. *Let $\{b_n\}$ be a distributionally small sequence. Then $\{nb_n\}$ is also distributionally small.*

Proof. If $\phi \in \mathcal{D}$ we have

$$\sum_{n=1}^{\infty} b_n \phi(n\varepsilon) \sim \sum_{j=0}^{\infty} \mu_j \frac{\phi^{(j)}(0)}{j!} \varepsilon^j$$

Replacing $\phi(x)$ by $\tau(x) = x\phi(x)$ we obtain

$$\sum_{n=1}^{\infty} n b_n \phi(n\varepsilon) = \frac{1}{\varepsilon} \sum_{n=1}^{\infty} b_n \tau(n\varepsilon) \sim \frac{1}{\varepsilon} \sum_{j=1}^{\infty} \frac{\mu_j \tau^{(j)}(0) \varepsilon^j}{j!}$$

$$= \sum_{j=0}^{\infty} \frac{\mu_{j+1} \phi^{(j)}(0) \varepsilon^j}{j!}. \qquad \blacksquare$$

Use of this lemma and the definition at the beginning of this section immediately give

Theorem 48. *Let $\{b_n\}$ be a distributionally small sequence. Let $\phi \in \mathcal{D}(\mathbb{R})$ with $\phi(0) = 1$. Then for each $k = 0, 1, 2, \cdots$ the series $\sum_{n=1}^{\infty} n^k b_n$ is (ϕ) summable and we have*

$$\sum_{n=1}^{\infty} n^k b_n = \mu_k \qquad (\phi). \qquad (5.4.8)$$

Also $n^k b_n = o(1)$ (ϕ) for every $k \in \mathbb{N}$.

A deeper result is true.

Theorem 49. *Let $\{b_n\}$ be a distributionally small sequence. Then the series $\sum_{n=1}^{\infty} n^\alpha b_n$ is Cesàro summable for each $\alpha \in \mathbb{C}$. In particular $n^\alpha b_n = o(1)$ (C) for each complex number α.*

Proof. Using Lemma 10 and the fact that the Cesàro summability of $\sum_{n=1}^{\infty} n^\beta b_n$ implies that $\sum_{n=1}^{\infty} n^\alpha b_n$ for $\Re e\, \alpha < \Re e\beta$, all that is required to be established is that $\sum_{n=1}^{\infty} b_n$ is (C) summable. Actually we shall prove a little more.

Let $g(x) = \sum_{n=1}^{\infty} b_n \delta(x-n)$, $R_N(x) = g(x) - \sum_{j=0}^{N-1}(-1)^j \mu_j \delta^{(j)}(x)/j!$. The moment asymptotic expansion holds for $g(x)$ in $\mathcal{D}(\mathbb{R})$ and thus $< R_N(\lambda x),\, \phi(x) > = o(\lambda^{-\beta})$ as $\lambda \to \infty$ provided $\beta < N+1$ and $\phi \in \mathcal{D}$. But the weak distributional convergence of $\lambda^\beta R_N(\lambda x)$ to zero implies its strong convergence. Therefore, given $\varepsilon > 0$ there exists $m \in \mathbb{N}$ such that if $\phi \in \mathcal{D}$ with supp $\phi \subset [-1, 1]$ then

$$| \lambda^\beta < R_N(\lambda x), \phi(x) >| \le \varepsilon \sum_{j=0}^{m} \sup \left\{| \phi^{(j)}(x) |: 0 \le x \le 1\right\}, \qquad (5.4.9)$$

for $\lambda \ge \lambda_0$. Observe that the values of $\phi(x)$ for $x < 0$ are irrelevant since supp $R_\lambda \subset [0, \infty)$.

Let $\mathcal{N}$ be the normed space formed by those functions ϕ defined in $[0, \infty)$ of class C^m there (even at $x = 0!$) and whose support is contained in $[0, 1]$. The norm is given by

$$\|\phi\| = \sum_{j=0}^{m} \sup \left\{| \phi^{(j)}(x) |: 0 \le x \le 1\right\}. \qquad (5.4.10)$$

Then the set of restrictions to $[0, \infty)$ of elements of $\mathcal{D}(\mathbb{R})$ with support in $[-1, 1]$ is dense in $\mathcal{N}$ and thus $\lambda^\beta R_N(\lambda x)$ can be extended to $\mathcal{N}$ in a way that $| \lambda^\beta < R_N(\lambda x), \phi(x) >| \le \varepsilon\|\phi\|$, $\phi \in \mathcal{N}$.

If $k > m$ then $f_k(x)$ defined by (5.4.7) belongs to $\mathcal{N}$ and thus we obtain

$$\frac{1}{\lambda^k} \sum_{n \le \lambda} b_n(\lambda - n)^k = < \lambda g(\lambda x), f_k(x) >$$

$$= \sum_{j=0}^{N-1} \frac{\mu_j f_k^{(j)}(0)\lambda^{-j}}{j!} + < \lambda R_\lambda(x), f_k(x) >$$

and since $< \lambda R_\lambda(x), f_k(x) > = O(\lambda^{1-\beta})$ while $f_k(x) \sim 1 - \binom{k}{1}x + \binom{k}{2}x^2 - \cdots$, as $x \to 0$ we obtain

$$\frac{1}{\lambda^k} \sum_{n \le \lambda} b_n(\lambda - n)^k = \sum_{j=0}^{N-1}(-1)^j \binom{k}{j}\frac{\mu_j}{\lambda^j} + o\left(\frac{1}{\lambda^\alpha}\right), \qquad \text{as } \lambda \to \infty, \quad (5.4.11)$$

for $\alpha < N$.

In particular, $(1/\lambda^k)\sum_{n\leq\lambda} b_n(\lambda - n)^k \to \mu_0$ when $\lambda \to \infty$ and thus $\sum_{n=1}^{\infty} b_n$ is (C,k) summable to μ_0. ∎

The converse result also holds.

Theorem 50. *Let $\{b_n\}$ be a sequence such that $\sum_{n=1}^{\infty} b_n n^k$ is Cesàro summable for each $k \in \mathbb{N}$. Then $\{b_n\}$ is distributionally small.*

Proof. Let $\mu_k = \sum_{n=1}^{\infty} n^k b_n$ (C). We want to show that if $\phi \in \mathcal{D}$ then

$$F(\varepsilon) = \sum_{n=1}^{\infty} b_n\phi(n\varepsilon) \sim \sum_{k=0}^{\infty} \frac{\mu_k\phi^{(k)}(0)}{k!}\varepsilon^k, \quad \text{as } \varepsilon \to 0. \qquad (5.4.12)$$

In order to establish (5.4.12) we first show that $F(\varepsilon) = O(1)$ as $\varepsilon \to 0$ provided $\phi \in C_0^m(\mathbb{R})$, where m is large enough. Indeed, let supp $\phi \subset (-\infty, c]$ and let $k \in \mathbb{N}$ be such that $n^{-k}B_n^k \to k!\,\mu_0$, where $\{B_n^k\}$ is the k-th order Cesàro mean of $\{b_n\}$. Applying summation by parts k times we obtain

$$F(\varepsilon) = \sum_{n=1}^{\infty} b_n\phi(n\varepsilon) = \sum_{n=1}^{\infty} B_n^k(-1)^{k+1}\triangle^{k+1}\phi(n\varepsilon)$$

$$= \sum_{n=1}^{N} B_n^k(-1)^{k+1}\triangle^{k+1}\phi(n\varepsilon), \qquad (5.4.13)$$

where $\triangle$ is the forward difference operator $\triangle a_n = a_{n+1} - a_n$ and $N = [\![c/\varepsilon]\!]$. But if m is large enough we can find a constant $M > 0$ such that $|\triangle^{k+1}\phi(n\varepsilon)| \leq M\varepsilon^{k+1}$ while $|B_n^k| \leq Mn^k$ and thus

$$|F(\varepsilon)| \leq \sum_{n=1}^{N} |B_n^k\triangle^{k+1}\phi(n\varepsilon)| \leq M^2\varepsilon^{k+1}\sum_{n=1}^{N} n^k \leq M^2\varepsilon^{k+1}N^{k+1} \leq M^2 c^{k+1},$$

and the relation $F(\varepsilon) = O(1)$ as $\varepsilon \to 0$ follows.

Next we choose q large enough so that $\lim_{\varepsilon\to 0}\sum_{n=1}^{\infty} b_n f_q(n\varepsilon) = \mu_0$, where $f_q(x)$ is defined by (5.4.7), and such that $\sum_{n=1}^{\infty} nb_n\tau(n\varepsilon) = O(1)$ as $\varepsilon \to 0$ whenever τ is of class C_0^{q-1}. Thus if ϕ is of class C_0^q then by writing $\phi(x) = \phi(0)f_q(x) + x\tau(x)$ we obtain

$$\lim_{\varepsilon\to 0}\sum_{n=1}^{\infty} b_n\phi(n\varepsilon) = \lim_{\varepsilon\to 0}\left[\phi(0)\sum_{n=1}^{\infty} b_n f_q(n\varepsilon) + \varepsilon\sum_{n=1}^{\infty} nb_n\tau(n\varepsilon)\right] = \mu_0\phi(0).$$

It follows that, given any $k \in \mathbb{N}$, there exist $m \in \mathbb{N}$ such that if $\phi \in C_0^m$ then

$$\lim_{\varepsilon\to 0}\sum_{n=1}^{\infty} n^j b_n\phi(n\varepsilon) = \mu_j\phi(0), \quad j = 0, 1, \cdots, k. \qquad (5.4.14)$$

Therefore, if $\phi \in \mathcal{D}(\mathbb{R})$ we have

$$\lim_{\varepsilon \to 0} \frac{\sum_{n=1}^{\infty} b_n \phi(n\varepsilon) - \mu_0 \phi(0)}{\varepsilon} = \lim_{\varepsilon \to 0} \sum_{n=1}^{\infty} n b_n \phi'(n\varepsilon) = \mu_1 \phi'(0),$$

and more generally

$$\lim_{\varepsilon \to 0} \frac{\sum_{n=1}^{\infty} b_n \phi(n\varepsilon) - \sum_{j=0}^{k-1} \frac{\mu_j \phi^{(j)}(0)\varepsilon^j}{j!}}{\varepsilon^k} =$$

$$\lim_{\varepsilon \to 0} \frac{\sum_{n=1}^{\infty} n^k b_n \phi^{(k)}(n\varepsilon)}{k!} = \frac{\mu_k \phi^{(k)}(0)}{k!},$$

which is precisely Poincaré's definition of the asymptotic expansion (5.4.12).
∎

These results permit us to derive some properties of distributionally small sequences.

Corollary. *Let $\{b_n\}$ be a distributionally small sequence. Then*

(a) $b_n = O(n^k)$ for some $k \in \mathbb{N}$.

(b) $\{n^\alpha b_n\}$ is also distributionally small for each $\alpha \in \mathbb{C}$.

(c) the moment function $\mu(\alpha) = \sum_{n=1}^{\infty} n^\alpha b_n \quad (C)$, is entire.

It is worthwhile pointing out that corresponding results for Abel summability do not hold. Actually, if $b_n = (-1)^n e^{\sqrt{n}}$ then $\sum_{n=1}^{\infty} n^k b_n$ is Abel summable for $k = 1, 2, 3, \cdots$ but the sequence $\{b_n\}$ is not distributionally small.

If $\{b_n\}$ is a distributionally small sequence then the distribution $G(x) = \sum_{n=1}^{\infty} b_n \delta(x - n)$ can be extended to larger spaces than $\mathcal{D}$. Actually, since $b_n = O(n^k)$ for some k, G can be extended to $\mathcal{S}$ and in many cases to even larger spaces such as $\mathcal{O}_C$ or $\mathcal{K}$. On the other hand, since 0 does not belong to the support of G, it can be extended to $\mathcal{S}\{x^{\alpha_n}\}$ for each sequence $\alpha_n \nearrow \infty$.

Theorem 51. *Let $\{b_n\}$ be a distributionally small sequence and let $\phi \in \mathcal{S}\{x^{\alpha_n}\}$ with expansion $\phi(x) \sim a_1 x^{\alpha_1} + a_2 x^{\alpha_2} + \cdots$, as $x \to 0$. Then*

$$\sum_{n=1}^{\infty} b_n \phi(n\varepsilon) \sim \mu(\alpha_1) a_1 \varepsilon^{\alpha_1} + \mu(\alpha_2) a_2 \varepsilon^{\alpha_2} + \cdots, \quad \text{as } \varepsilon \to 0. \quad (5.4.15)$$

Proof. Suppose first that $\phi \in \mathcal{D}\{x^{\alpha_n}\}$. Let $\theta \in \mathcal{D}$ be such that $\theta(x) = 1$, $0 \leq x \leq 1$ and let m be large enough so that $\sum_{n=1}^{\infty} n^{\alpha_p} b_n \phi(n\varepsilon) = O(1)$ as

$\varepsilon \to 0$ if $\phi \in C_0^m(\mathbb{R})$. Let $\alpha_{q+1} > \alpha_p + m$. Writing $\phi(x) = \theta(x)[a_1 x^{\alpha_1} + \cdots + a_q x^{\alpha_q}] + x^{\alpha_p}\tau(x)$, then $\tau \in C_0^m(\mathbb{R})$. Thus

$$\sum_{n=1}^{\infty} b_n \phi(n\varepsilon) = \sum_{j=1}^{q} a_j \varepsilon^{\alpha_j} \sum_{n=1}^{\infty} n^{\alpha_j} b_n \theta(n\varepsilon) + \varepsilon^{\alpha_p} \sum_{n=1}^{\infty} n^{\alpha_p} b_n \tau(n\varepsilon)$$

$$= \sum_{n=1}^{p-1} a_j \mu(\alpha_j) \varepsilon^{\alpha_j} + O(\varepsilon^{\alpha_p}),$$

since the hypothesis implies that $\{n^\alpha b_n\}$ is distributionally small for any α and thus

$$\sum_{n=1}^{\infty} n^\alpha b_n \theta(n\varepsilon) = \mu(\alpha) + o(\varepsilon^\infty), \qquad \text{as } \varepsilon \to \infty.$$

To complete the proof it is enough to show that if $g(x) = \sum_{n=1}^{\infty} b_n \delta(x-n)$ then $< g(x), \phi(\varepsilon x) > = o(\varepsilon^\infty)$ as $\varepsilon \to 0$ for any $\phi \in \mathcal{S}$ with $\phi^{(j)}(0) = 0, j = 0, 1, 2, \cdots$. Actually, we could further assume that $\int_0^\infty \phi(x)dx = 0$ since if that is not the case we can find $\phi_1 \in \mathcal{D}$ with $\phi_1^{(j)}(0) = 0, j = 0, 1, 2, \cdots$ and with $\int_0^\infty \phi_1(x)dx = \int_0^\infty \phi(x)dx$ and we would have $< g(x), \phi(\varepsilon x) > = < g(x), \phi(\varepsilon x) - \phi_1(\varepsilon x) > + o(\varepsilon^\infty)$ as $\varepsilon \to 0$. Observe that under these conditions $\int_0^\infty x^k \phi^{(n)}(x)dx = 0$ for $k = 0, 1, 2, \cdots, n$.

Let $G_n(x) = 1/n! \sum_{j \leq x} b_j (x - j)^n = x^n/n! < g(y), f_n(y/x) >$, where f_n is given by (5.4.7). Then $G_n^{(n)}(x) = g(x)$ while as we showed in Theorem 49, for each $k \in \mathbb{N}$ we can find n large enough such that

$$G_n(x) = \sum_{j=0}^{k-1} \frac{(-1)^j \mu_j x^{n-j}}{j!(n-j)!} + O(x^{n-k}), \qquad \text{as } x \to 0. \tag{5.4.16}$$

Therefore

$$< g(x), \phi(\varepsilon x) > = \varepsilon^n < G_n(x), \phi^{(n)}(\varepsilon x) >$$

$$= \varepsilon^n \left[\sum_{j=0}^{k-1} \frac{(-1)^j \mu_j}{j!(n-j)!} \int_0^\infty x^{n-j} \phi(x)dx + O(\varepsilon^{k-n-1}) \right]$$

$$= O(\varepsilon^{k-1}). \qquad \blacksquare$$

Other useful properties of distributionally small sequences are given in the next theorem.

Theorem 52. *Let $\{b_n\}$ and $\{c_n\}$ be two distributionally small sequences with moment functions $B(\alpha) = \sum_{n=1}^{\infty} b_n n^\alpha$, $C(\alpha) = \sum_{n=1}^{\infty} c_n n^\alpha$. Let $p, q \in \mathbb{N}$, $r, s \in \mathbb{R}$.*

(a) The sequence

$$d_n = \sum_{k^p j^q = n} k^r j^s b_k c_j \tag{5.4.17}$$

is distributionally small and its moment function is $D(\alpha) = B(\alpha p + r)C(\alpha q + s)$.

(b) The sequence

$$e_n = \sum_{k^p j^q = n} k^r j^s b_k - \frac{B\left(\frac{-p(1+s)}{q} - r\right) n^{(s+1)/q-1}}{q} \tag{5.4.18}$$

is distributionally small and its moment function is the entire function

$$E(\alpha) = B(\alpha p + r)\zeta(\alpha q - s) - \frac{B\left(\frac{-p(1+s)}{q} - r\right)\zeta\left(-\alpha - \frac{r+1}{q} + 1\right)}{q}$$

(c) The sequence

$$f_n = \sum_{k=1}^{n} k^r b_k - B(r) \tag{5.4.19}$$

is distributionally small.

Proof. Both (a) and (b) follow easily by using the methods of Section 5.3 while (c) can be obtained by summation by parts. ∎

It is convenient to introduce the following notation: $b_n \approx c_n$ if $\{b_n - c_n\}$ is distributionally small. Observe that the above theorem yields $b_n * n^s \approx B(-(1+s))n^s$ if $\{b_n\} \in \mathcal{DS}$, where $*$ denotes the Dirichlet multiplication: $(b * c)_n = \sum_{kj=n} b_k c_j$.

Example 108. Let us consider the arithmetical function $d(n)$, the number of divisors of n. The asymptotic formula

$$\sum_{n \leq x} d(n) \sim x \ln x + (2\gamma - 1)x,$$

goes back to Dirichlet [4], [17], [66], [109]. The problem of determining the exact order of the error term $E(x) = \sum_{n \leq x} d(n) - x \ln x - (2\gamma - 1)x$ is one of the most famous problems in number theory. It is easy to show that $E(x) = O(x^{1/2})$. Many authors such as Voronoi, Van der Corput and Richert have improved this result. The best available O estimate is $E(x) = O(x^{35/108+\delta})$, for each $\delta > 0$, which was obtained by Kolesnik [77]. On the other hand, the Ω results $\overline{\lim}_{x\to\infty} x^{-1/4}E(x) = \infty$ and $\underline{\lim}_{x\to\infty} x^{-1/4}E(x) = -\infty$ were established by Hardy [59] and Ingham [65], respectively. Improved Ω results are given in Hafner [54]. It is believed that $E(x) = O(x^{1/4+\delta})$ for each $\delta > 0$.

Many of these results have very general versions. Motivated by the celebrated Erdös–Fuchs theorem [29], Richert [96] gave an interesting Ω result valid for generalizations of $d(n)$. A comprehensive account can be found in [55]. Other approaches to this problem have been devised. We indicate for instance the mean value result [109]

$$\frac{1}{x}\int_0^x E(x)^2 dx = O(x^{1/2+\delta}). \tag{5.4.20}$$

Our ideas provide some related information. Since $d(n) \approx \ln n + 2\gamma$ then the series $\sum_{n=1}^{\infty}(d(n) - \ln n - 2\gamma)n^\alpha$ is Cesàro summable for each $\alpha \in \mathbb{C}$ to the value $F(\alpha) = \zeta(-\alpha)^2 + \zeta'(-\alpha) - 2\gamma\zeta(-\alpha)$. For each $k \in \mathbb{N}$ and $\alpha \in \mathbb{C}$ we can find $\beta \in \mathbb{R}$ such that

$$x^{-\beta}\sum_{n\leq x}(d(n) - \ln n - 2\gamma)(x - n)^\beta n^\alpha = F(\alpha) - \binom{\beta}{1}\frac{F(\alpha+1)}{x} + \cdots$$

$$\cdots + (-1)^k\binom{\beta}{k}\frac{F(\alpha+k)}{x^k} + O(x^{-k-1}). \tag{5.4.21}$$

Also, it follows that

$$\sum_{j=1}^{N} d(j) = \ln N! + 2\gamma N + \frac{1}{4} + \frac{1}{2}\ln 2\pi + \gamma + r_n, \tag{5.4.22}$$

where $\{r_n\}$ is distributionally small. Alternatively

$$\sum_{j\leq x} d(j) = \ln[\![x]\!]! + 2\gamma[\![x]\!] + \frac{1}{4} + \frac{1}{2}\ln 2\pi + \gamma + \tilde{E}(x), \tag{5.4.23}$$

where $\tilde{E}(x)$ is "distributionally small" since it satisfies the moment asymptotic expansion in the space $\mathcal{S}'$. The approximations $\ln[\![x]\!]! + 2\gamma[\![x]\!] + 1/4 + 1/2\ln 2\pi + \gamma$ and $x\ln x + (2\gamma - 1)x$ are different but closely related as Stirling's formula shows.

Similar results can be obtained for the arithmetical function $\sigma_r(n)$.

5.5 Multiple Series

We shall now study the asymptotic behavior of multiple series of the type

$$\sum_{k \in K} a_k \phi(\varepsilon k), \qquad (5.5.1)$$

where K is a subset of $\mathbb{Z}^n$ and ϕ belongs to a suitable space of n-dimensional test functions.

We start with the series $\sum_{k \in \mathbb{Z}^n} \phi(\varepsilon k)$.

Theorem 53. *Let $\phi \in L^1(\mathbb{R}^n) \cap \mathcal{K}(\mathbb{R}^n)$. Then*

$$\sum_{k \in \mathbb{Z}^n} \phi(\varepsilon k) = \frac{1}{\varepsilon^n} \int_{\mathbb{R}^n} \phi(x) dx + o(\varepsilon^\infty), \qquad as \ \varepsilon \to 0. \qquad (5.5.2)$$

Proof. Let $f(x) = \sum_{j \in \mathbb{Z}} \delta(x - j)$. We can write $f(x) = 1 + g(x)$ where $g \in \mathcal{K}'(\mathbb{R})$. Thus $f(x_1) \cdots f(x_n) = 1 + \sum g_I(x_1, \cdots, x_n)$, the sum taken over all non-empty subsets I of $\{1, \cdots, n\}$, where $g_I(x_1, \cdots, x_n) = \prod_{i \in I} g(x_i)$.

Observe now that

$$\sum_{k \in \mathbb{Z}^n} \phi(\varepsilon k) = < f(x_1) \cdots f(x_n), \phi(\varepsilon x_1 \cdots, \varepsilon x_n) >$$

$$= < 1 + \sum g_I(x_1, \cdots, x_n), \phi(\varepsilon x_1, \cdots, \varepsilon x_n) >$$

$$= \frac{1}{\varepsilon^n} \int_{\mathbb{R}^n} \phi(x) dx + \sum < g_I(x_1, \cdots, x_n), \phi(\varepsilon x_1, \cdots, \varepsilon x_n) > .$$

Recall now that moments of $g(x)$ vanish and therefore $< g(x), \tau(\varepsilon x) > = o(\varepsilon^\infty)$ provided $\tau \in \mathcal{K}(\mathbb{R})$. It follows that if $I = \{i_1, \cdots, i_p\}, \{1, \cdots, n\} \setminus I = \{j_1, \cdots, j_q\}$ then if $\phi \in L^1(\mathbb{R}^n) \cap \mathcal{K}(\mathbb{R}^n)$,

$$< g_I(x_1, \cdots, x_n), \phi(\varepsilon x_1, \cdots, \varepsilon x_n) > = \varepsilon^{-n} < g(\varepsilon^{-1} x_{i_1}) \cdots g(\varepsilon^{-1} x_{i_p}),$$

$$\int_{\mathbb{R}^n} \phi(x_1, \cdots, x_n) dx_{j_1} \cdots dx_{j_q} >$$

$$= o(\varepsilon^\infty)$$

and (5.5.2) follows. ∎

Example 109. Let $S_{n,2}(k)$ denote the number of solutions of the diophantine equation

$$k = j_1^2 + \cdots j_n^2, \qquad j_1, \cdots j_n \in \mathbb{Z}. \qquad (5.5.3)$$

Then if $\phi \in S(\mathbb{R})$, formula (5.5.2) yields

$$\sum_{k=0}^{\infty} S_{n,2}(k)\phi(\varepsilon k) = \sum_{j \in \mathbb{Z}^n} \phi(\varepsilon \mid j \mid^2) = \frac{1}{\varepsilon^{\frac{1}{2}n}} \int_{\mathbb{R}^n} \phi(\mid x \mid^2)dx + o(\varepsilon^{\infty}).$$

But

$$\int_{\mathbb{R}^n} \phi(\mid x \mid^2)dx = \frac{\pi^{\frac{1}{2}n}}{\Gamma(n/2)} \int_0^{\infty} x^{\frac{1}{2}n-1}\phi(x)dx,$$

and thus

$$\sum_{k=0}^{\infty} S_{n,2}(k)\phi(k\varepsilon) = \frac{\pi^{\frac{1}{2}n}}{\Gamma(n/2)} \int_0^{\infty} x^{\frac{1}{2}n-1}\phi(x)dx + o(\varepsilon^{\infty}), \qquad \text{as } \varepsilon \to 0.$$

$$(5.5.4)$$

This result implies that the sequence $S_{n,2}(k) - \pi^{\frac{1}{2}n}/\Gamma(n/2)k^{\frac{1}{2}n-1}$ is distributionally small. Therefore for each $\alpha \in \mathbb{C}$ the series

$$F(\alpha) = \sum_{k=1}^{\infty} \left[S_{n,2}(k) - \frac{\pi^{\frac{1}{2}n}}{\Gamma(n/2)} k^{\frac{1}{2}n-1} \right] k^{\alpha}$$

is Cesàro summable. It follows that the moment function $U_{n,2}(\alpha)$ initially defined for $\Re e\ \alpha < -n/2$ by

$$U_{n,2}(\alpha) = \sum_{k=1}^{\infty} S_{n,2}(k)k^{\alpha} = \sum_{j \in \mathbb{Z}^n} \mid j \mid^{2\alpha} \qquad (5.5.5)$$

can be extended to $\mathbb{C}\backslash\{-n/2\}$. Actually,

$$U_{n,2}(\alpha) = \frac{\pi^{\frac{1}{2}n}}{\Gamma(n/2)}\zeta(1 - \frac{1}{2}n - \alpha) + F(\alpha),$$

where $F(\alpha)$ is entire. Observe the special values $U_{n,2}(0) = -1$, $U_{n,2}(k) = 0$, $k = 1, 2, 3, \cdots$.

If $\phi \in S\{x^{\alpha_j}\}$, with expansion $\phi(x) \sim a_1 x^{\alpha_1} + a_2 x^{\alpha_2} \cdots$ as $x \to 0$ and with $\alpha_q = -n/2$ then

$$\sum_{k=1}^{\infty} S_{n,2}(k)\phi(k\varepsilon) \sim \sum_{j \neq q} a_j U_{n,2}(\alpha_j)\varepsilon^{\alpha_j}$$

$$+ \frac{1}{\varepsilon^{\frac{1}{2}n}} \left\{ \frac{\pi^{\frac{1}{2}n}}{\Gamma(n/2)} F.p. \int_0^{\infty} x^{\frac{1}{2}n-1}\phi(x)dx \right.$$

$$+ \left(\frac{\gamma\pi^{\frac{1}{2}n}}{\Gamma(n/2)} + F(-n/2) \right) a_q - \left. \frac{\pi^{\frac{1}{2}n}a_q}{\Gamma(n/2)} \ln\varepsilon \right\} \qquad (5.5.6)$$

More generally, let $c = (c_1, \cdots, c_n) \in \mathbb{N}^n$ and let $S_{n,2p}(c; k)$ denote the number of solutions of the diophantine equation

$$k = c_1 j_1^{2p} + \cdots c_n j_n^{2p}. \tag{5.5.7}$$

Then for each $\phi \in \mathcal{S}(\mathbb{R})$ we have

$$\sum_{k=0}^{\infty} S_{n,2p}(c, k)\phi(k\varepsilon) = \sum_{j \in \mathbb{Z}^n} \phi(\varepsilon(c_1 j_1^{2p} + \cdots + c_n j_n^{2p}))$$

$$= \frac{\Gamma(1/2p)^n (c_1 \cdots c_n)^{-1/2p} \varepsilon^{-n/2p}}{p^n \Gamma(n/2p)} \int_0^{\infty} x^{n/2p-1}\phi(x)dx + o(\varepsilon^{\infty}), \quad (5.5.8)$$

as $\varepsilon \to \infty$. Therefore

$$S_{n,2p}(c, k) \approx \frac{\Gamma(1/2p)^n k^{n/2p-1}}{(c_1 \cdots c_n)^{1/2p} p^n \Gamma(n/2p)}. \tag{5.5.9}$$

Observe also that the moment function given for $\Re e \; \alpha < -n/2p$ by

$$U_{n,2p}(\alpha) = \sum_{k=1}^{\infty} S_{n,2p}(c; k)k^{\alpha} = \sum_{j \in \mathbb{Z}^n} (c_1 j_1^{2p} + \cdots + c_n j_n^{2p})^{\alpha}, \tag{5.5.10}$$

can be extended to an analytic function in $\mathbb{C}\backslash\{-n/2p\}$.

Formula (5.2.2) can also be generalized to multiple series. A similar analysis to that of Theorem 53 yields

Theorem 54. *Let $\phi \in \mathcal{S}(\mathbb{R}^n \times \mathbb{R}^m)$ where $m \geq 1$. Then*

$$\sum_{j \in \mathbb{Z}^n} \sum_{k \in \mathbb{Z}^m} (-1)^{|k|}\phi(\varepsilon j, \varepsilon k) = o(\varepsilon^{\infty}), \quad \text{as } \varepsilon \to 0. \tag{5.5.11}$$

Example 110. Since $j^2 \equiv j \pmod 2$ it follows that if at least one c_i is odd then the sequence $(-1)^k S_{n,2p}(c; k)$ is distributionally small. Indeed, suppose the first $n_1 c_i$'s to be even, the remaining n_2 odd. Then if $\phi \in S(\mathbb{R})$,

$$\sum_{k=0}^{\infty} (-1)^k S_{n,2p}(c; k)\phi(k\varepsilon) = \sum_{j \in \mathbb{Z}^n} (-1)^{c_1 j_1^{2p} + \cdots + c_n j_n^{2p}} \phi(\varepsilon(c_1 j_1^{2p} + \cdots + c_n j_n^{2p}))$$

$$= \sum_{j \in \mathbb{Z}^{n_1}} \sum_{k \in \mathbb{Z}^{n_2}} (-1)^{|k|}\phi(\varepsilon(c_1 j_1^{2p} + \cdots + c_{n_1} j_{n_1}^{2p} + c_{n_1+1} k_1^{2p} + \cdots + c_n k_{n_2}^{2p}))$$

$$= o(\varepsilon^\infty)$$

Thus, in the Cesàro sense the Diophantine equation $c_1 j_1^{2p} + \cdots + c_n j_n^{2p} = k$ has as many solutions for even k as for odd k provided at least one c_i is odd. Actually,

$$\lim_{N \to \infty} \frac{\sum_{k=1}^{N} S_{n,2p}(c; 2k)}{\sum_{k=1}^{N} S_{n,2p}(c; 2k-1)} = 1. \qquad (5.5.12)$$

The asymptotic behavior of the alternating series $\sum_{\substack{k \in \mathbb{Z}^n \\ k_i \geq 1}} (-1)^{|k|} \phi(\varepsilon k)$ can be analyzed by using tensor product considerations. In fact, since

$$\sum_{k=1}^{\infty} (-1)^k k^q \delta(x - k\varepsilon) \sim \sum_{k=0}^{\infty} \frac{(-1)^k \theta(k+q) \delta^{(k)}(x)}{k!} \varepsilon^k, \qquad \text{as } \varepsilon \to 0,$$

$$\qquad (5.5.13)$$

where $\theta(x) = (2^{x+1} - 1)\zeta(-x)$, it follows that

$$\sum_{\substack{k \in \mathbb{Z}^n \\ k_i \geq 1}} (-1)^{|k|} k^q \delta(x - \varepsilon k) \sim \sum_{k \in \mathbb{N}^n} \frac{(-1)^{|k|} \theta_n(k+q)}{k!} D^k \delta(x) \varepsilon^{|k|}, \qquad (5.5.14)$$

for any $q \in \mathbb{C}^n$ in the space $\mathcal{K}(\mathbb{R}^n)$. Here $\theta_n(x) = \theta(x_1) \cdots \theta(x_n)$.

Theorem 55. *Let $\phi \in \mathcal{K}(\mathbb{R}^n)$. Then*

$$\sum_{\substack{k \in \mathbb{Z}^n \\ k_i \geq 1}} (-1)^k k^q \phi(\varepsilon k) \sim \sum_{k \in \mathbb{N}^n} \frac{\theta_n(k+q)}{k!} D^k \phi(0) \varepsilon^{|k|}. \qquad (5.5.15)$$

Recently, D. Borwein and J.M. Borwein [11] have given the generalization to multiple series of the well-known Leibniz test for alternating series according to which $\sum_{n=1}^{\infty} (-1)^n a_n$ converges if $a_n \searrow 0$. See also [86], [88].

In [43] it was proven that if $a_n = \tau(n)$, where $\tau \in \mathcal{K}$ then the alternating series $\sum_{n=1}^{\infty} (-1)^n a_n$ is Abel summable to the value $< \sum_{n=1}^{\infty} (-1)^n \delta(x - n), \tau(x) >$. Actually, if $\tau \in \mathcal{K}$ then $\tau(x) \sum_{n=1}^{\infty} (-1)^n \delta(x-n) = \sum_{n=1}^{\infty} (-1)^n a_n \delta(x-n)$ belongs to $\mathcal{K}'$, therefore $\{(-1)^n a_n\}$ is distributionally small and thus $\sum_{n=1}^{\infty} (-1)^n a_n$ is Cesàro summable.

Let now $\tau \in \mathcal{K}(\mathbb{R}^n)$ and set $a_k = \tau(k)$, $k \in \mathbb{Z}^n$.

Then $\tau(x) \sum_{\substack{k \in \mathbb{Z}^n \\ k_i \geq 1}} (-1)^{|k|} \delta(x - k) = \sum_{\substack{k \in \mathbb{Z}^n \\ k_i \geq 1}} (-1)^{|k|} a_k \delta(x - k)$ belongs to $\mathcal{K}'(\mathbb{R}^n)$. It follows that the alternating series $\sum_{\substack{k \in \mathbb{Z}^n \\ k_i \geq 1}} (-1)^{|k|} a_k$ is summable to the value $L = < \sum_{\substack{k \in \mathbb{Z}^n \\ k_i \geq 1}} (-1)^{|k|} \delta(x - k), \tau(x) >$ in the sense that for every $\phi \in \mathcal{K}(\mathbb{R})$ with $\phi(0) = 1$ we have

$$\sum_{\substack{k \in \mathbb{Z}^n \\ k_i \geq 1}} (-1)^{|k|} a_k \phi(\varepsilon k) = L + O(\varepsilon), \qquad \text{as } \varepsilon \to 0. \qquad (5.5.16)$$

In particular, if $\phi(x) = e^{-|x|}$ we obtain Abel type summability:

$$\lim_{\varepsilon \to 0} \sum_{\substack{k \in \mathbb{Z}^n \\ k_i \geq 1}} (-1)^{|k|} a_k e^{-\varepsilon |k|} = L. \tag{5.5.17}$$

Next, let us consider the asymptotic development of the multiple series $\sum_{\substack{k \in \mathbb{Z}^n \\ k_i \geq 1}} \phi(\varepsilon k)$. We start with the formula

$$f(\lambda x) = \sum_{k=1}^{\infty} \delta(\lambda x - k) \sim H(x) + \sum_{k=0}^{\infty} \frac{(-1)^k \zeta(-k) \delta^{(k)}(x)}{k! \lambda^{k+1}}, \quad \text{as } \lambda \to \infty, \tag{5.5.18}$$

which holds in the space $\mathcal{S}(\mathbb{R})$. Then the asymptotic expansion of the tensor product $f(\lambda x_1) \cdots f(\lambda x_n)$ in the space $\mathcal{S}(\mathbb{R}^n)$ follows from (5.5.18) as

$$\sum_{\substack{k \in \mathbb{Z}^n \\ k_i \geq 1}} \delta(\lambda x - k) \sim \prod_{i=1}^{n} \left\{ H(x_i) + \sum_{k=0}^{\infty} \frac{(-1)^k \zeta(-k) \delta^{(k)}(x_i)}{k! \lambda^{k+1}} \right\}$$

$$\sim H(x_1) \cdots H(x_n) + \left(\sum_{i=1}^{n} \frac{H(x_1) \cdots H(x_n)}{H(x_i)} \zeta(0) \delta(x_i) \right) \lambda^{-1}$$

$$+ \left[\sum_{i \neq j} \frac{H(x_1) \cdots H(x_n)}{H(x_i) H(x_j)} \zeta(0)^2 \delta(x_i) \delta(x_j) \right.$$

$$\left. - \sum_{i=1}^{n} \frac{H(x_1) \cdots H(x_n)}{H(x_i)} \zeta(-1) \delta'(x_i) \right] \lambda^{-2} + \cdots \tag{5.5.19}$$

Use of (5.5.19) gives the expansion

$$\sum_{\substack{k \in \mathbb{Z}^n \\ k_i \geq 1}} \phi(\varepsilon k) \sim \frac{1}{\varepsilon^n} \int_0^{\infty} \cdots \int_0^{\infty} \phi(x_1, \cdots, x_n) dx_1 \cdots dx_n$$

$$- \frac{1}{2\varepsilon^{n-1}} \sum_{i=1}^{n} \int_0^{\infty} \cdots \int_0^{\infty} \phi(x_1, \cdots, x_{i-1}, 0, x_{i+1}, \cdots, x_n) dx_1 \cdots d\hat{x}_i \cdots dx_n$$

$$+ \left[\frac{1}{4} \sum_{i \neq j} \int_0^{\infty} \cdots \int_0^{\infty} \phi(x_1, \cdots, x_{i-1}, 0, \cdots, x_{j-1}, 0, \cdots x_n) dx_1 \cdots d\hat{x}_i \cdots d\hat{x}_j \cdots dx_n \right.$$

$$\left. - \frac{1}{12} \sum_{i=1}^{n} \int_0^{\infty} \cdots \int_0^{\infty} \frac{\partial \phi}{\partial x_i}(x_1, \cdots, x_{i-1}, 0, x_{i+1}, \cdots, x_n) dx_1 \cdots d\hat{x}_i \cdots dx_n \right] \frac{1}{\varepsilon^{n-2}} + \cdots$$

$$\tag{5.5.20}$$

The asymptotic expansion of the series $\sum_{i=1}^{n} \sum_{k_i=1}^{\infty} k_i^{q_i} \phi(\varepsilon k_1, \cdots, \varepsilon k_n)$ can also be obtained by taking the tensor product of the expansions of $\sum_{k_i=1}^{\infty} k_i^{q_i} \delta(\lambda x_i - k_i)$ as $\lambda \to \infty$. Although for a given set of q_i''s it is an easy matter to give the first few terms of the expansion, the notation becomes very complicated in the general case. We prefer to illustrate the ideas with the following particular cases.

(a) If $q_i > -1$ for each i then

$$\sum_{i=1}^{n} \sum_{k_i=1}^{\infty} k_i^{q_i} \phi(\varepsilon k_1, \cdots, \varepsilon k_n) \sim$$

$$\frac{1}{\varepsilon^{q_1 + \cdots + q_n + n}} \int_0^\infty \cdots \int_0^\infty x_1^{q_1} \cdots x_n^{q_n} \phi(x_1, \cdots, x_n) dx_1 \cdots dx_n. \quad (5.5.21)$$

(b) If $q > p > -1$ set $m = [\![q - p]\!]$. Then

$$\sum_{k=1}^{\infty} \sum_{j=1}^{\infty} k^q j^p \phi(\varepsilon k, \varepsilon j) = \frac{1}{\varepsilon^{p+q+2}} \int_0^\infty \int_0^\infty \phi(x, y) x^p y^q dx dy$$

$$+ \frac{\zeta(-p)}{\varepsilon^{q+1}} \int_0^\infty x^q \phi(x, 0) dx$$

$$+ \frac{\zeta(-p-1)}{\varepsilon^q} \int_0^\infty x^q \frac{\partial \phi}{\partial y}(x, 0) dx + \cdots$$

$$\cdots + \frac{\zeta(-p-m)}{m! \varepsilon^{q-p+1}} \int_0^\infty x^q \frac{\partial^m \phi}{\partial y^m}(x, 0) dx$$

$$+ \frac{\zeta(-q)}{\varepsilon^{p+1}} \int_0^\infty y^q \phi(0, y) dy + O(\varepsilon^\alpha), \quad (5.5.22)$$

where $\alpha = \min\{m - q, 0\}$.

(c)

$$\sum_{k=1}^{\infty} \sum_{j=1}^{\infty} \frac{\phi(\varepsilon k, \varepsilon j)}{k} = \left(-\int_0^\infty \phi(0, y) dy \right) \frac{\ln \varepsilon}{\varepsilon}$$

$$+ \left[F.p. \int_0^\infty \int_0^\infty \frac{\phi(x, y)}{x} dx dy + \gamma \int_0^\infty \phi(0, y) dy \right] \frac{1}{\varepsilon}$$

$$+ \frac{1}{2} \phi(0, 0) \ln \varepsilon - \frac{1}{2} \left[F.p. \int_0^\infty \frac{\phi(x, 0)}{x} + \gamma \phi(0, 0) \right] + O(\varepsilon \ln \varepsilon). \quad (5.5.23)$$

(d)

$$\sum_{k=1}^{\infty}\sum_{j=1}^{\infty}\frac{\phi(\varepsilon k,\varepsilon j)}{kj}=\phi(0,0)(\ln\varepsilon)^2-\left[F.p.\int_0^{\infty}\frac{\phi(0,y)}{y}dy\right.$$

$$\left.+F.p.\int_0^{\infty}\frac{\phi(x,0)}{x}dx+2\gamma\phi(0,0)\right]\ln\varepsilon$$

$$+F.p.\int_0^{\infty}\int_0^{\infty}\frac{\phi(x,y)}{xy}dxdy+\gamma F.p.\int_0^{\infty}\frac{\phi(0,y)}{y}dy$$

$$+\gamma F.p.\int_0^{\infty}\frac{\phi(x,0)}{x}dx+\gamma^2\phi(0,0)+O(\varepsilon\ln\varepsilon). \tag{5.5.24}$$

We shall now consider several applications of these ideas.

Example 111. Let b_n be the number of solutions of the equation

$$k+j^2=n,\qquad k,j\in\{0,1,\cdots,n\}.$$

Then if $\phi\in S(\mathbb{R})$ we have

$$\sum_{n=0}^{\infty}b_n\phi(n\varepsilon)=\sum_{j=0}^{\infty}\sum_{k=0}^{\infty}\phi(\varepsilon j^2+\varepsilon k)$$

$$=\left\langle\sum_{j=0}^{\infty}\sum_{k=0}^{\infty}\delta(x-\varepsilon^{1/2}j)\delta(y-\varepsilon k),\phi(x^2+y)\right\rangle$$

$$=\frac{1}{\varepsilon^{3/2}}\int_0^{\infty}x^{1/2}\phi(x)dx+\frac{1}{\varepsilon}\int_0^{\infty}\phi(x)dx$$

$$+\frac{1}{4\varepsilon^{1/2}}\int_0^{\infty}x^{-1/2}\phi(x)dx+\frac{\phi(0)}{4}$$

$$+\frac{\varepsilon^{1/2}}{48}F.p.\int_0^{\infty}x^{-3/2}\phi(x)dx+O(\varepsilon). \tag{5.5.25}$$

In particular, of course, $\sum_{n=0}^{N}b_n\sim 2N^{\frac{3}{2}}/3$ as $N\to\infty$.

Example 112. Let $\phi\in S(\mathbb{R})$. Then

$$\sum_{n=1}^{\infty}n\phi(n\varepsilon)\sim\left\langle\frac{xH(x)}{\varepsilon^2}+\sum_{n=0}^{\infty}\frac{(-1)^n\zeta(-n-1)\delta^{(n)}(x)}{n!}\varepsilon^n,\phi(x)\right\rangle.$$

But also

$$\sum_{n=1}^{\infty} n\phi(n\varepsilon) = \sum_{j=0}^{\infty}\sum_{k=1}^{\infty} \phi(\varepsilon(k+j))$$

$$\sim \left\langle \left\{ \frac{H(x)}{\varepsilon} + \frac{1}{2}\delta(x) + \sum_{n=1}^{\infty} \frac{(-1)^n \zeta(-n)\delta^{(n)}(x)}{n!}\varepsilon^n \right\} \right.$$

$$\left. \left\{ \frac{H(y)}{\varepsilon} + \sum_{n=0}^{\infty} \frac{(-1)^n \zeta(-n)\delta^{(n)}(y)}{n!}\varepsilon^n \right\}, \phi(x+y) \right\rangle,$$

and thus using the definition of the convolution of distributions,

$$\frac{xH(x)}{\varepsilon^2} + \sum_{n=0}^{\infty} \frac{(-1)^n \zeta(-n-1)\delta^{(n)}(x)}{n!}\varepsilon^n =$$

$$\left(\frac{H(x)}{\varepsilon} + \frac{\delta(x)}{2} + \sum_{n=1}^{\infty} \frac{(-1)^n \zeta(-n)\delta^{(n)}(x)}{n!}\varepsilon^n \right)$$

$$* \left(\frac{H(x)}{\varepsilon} + \sum_{n=0}^{\infty} \frac{(-1)^n \zeta(-n)\delta^{(n)}(x)}{n!}\varepsilon^n \right).$$

Since $\delta^{(n)}(x) * \delta^{(m)}(x) = \delta^{(n+m)}(x)$, this yields the recursion relation

$$\zeta(-n-1) = \frac{n+1}{n+3}\sum_{k=1}^{n-1}\binom{n}{k}\zeta(-k)\zeta(k-n). \qquad (5.5.26)$$

Let us consider the asymptotic development of the series

$$F(\varepsilon) = \sum_{k_1=0}^{\infty}\sum_{k_2=0}^{\infty}\cdots\sum_{k_N=0}^{\infty}\phi(\varepsilon(a_1 k_1 + \cdots a_N k_N)), \qquad (5.5.27)$$

where $0 \le a_1 \le a_2 \le \cdots \le a_N$ where $\phi \in \mathcal{S}(\mathbb{R})$. In the case where $a_1, \cdots, a_N \in \mathbb{N}$ then

$$F(\varepsilon) = \sum_{n=0}^{\infty} A(n)\phi(n\varepsilon), \qquad (5.5.28)$$

where $A(n)$ is the number of solutions of the equation

$$a_1 k_1 + a_2 k_2 + \cdots + a_N k_N = n, \qquad k_i \in \{0, 1, \cdots, n\}. \qquad (5.5.29)$$

Alternatively, $A(n)$ can be realized as a partition function. Actually, if $a_i = i$ then $A(n) = p_N(n)$, the number of partitions of the number n into N integral parts,

$$n = x_1 + \cdots + x_N, \qquad 0 \le x_i \le n, \qquad (5.5.30)$$

without regard to the order of the factors and with repetitions allowed [1], [17], [94].

In the sequel we shall often need the identity

$$\int_0^\infty \cdots \int_0^\infty \phi(a_1 x_1 + \cdots + a_N x_N) dx_1 \cdots dx_N$$

$$= \frac{1}{(a_1 \cdots a_N)(N-1)!} \int_0^\infty x^{N-1} \phi(x) dx, \qquad (5.5.31)$$

which follows by a simple change of variables.

The asymptotic development of $F(\varepsilon)$ has the form

$$F(\varepsilon) \sim L_N(\phi)\varepsilon^{-N} + L_{N-1}(\phi)\varepsilon^{-N+1} + \cdots + L_0(\phi) + \cdots, \qquad \text{as } \varepsilon \to 0,$$
$$(5.5.32)$$

for certain functionals $L_i(\phi)$. The formula for $L_i(\phi)$ can be obtained from (5.5.19).

The leading order term is

$$L_N(\phi) = < H(x_1) \cdots H(x_N), \phi(a_1 x_1 + \cdots + a_N x_N) >$$

$$= \frac{1}{a_1 \cdots a_N} \int_0^\infty \frac{x^{N-1}}{(N-1)!} \phi(x) dx. \qquad (5.5.33)$$

The next term is obtained as

$$L_{N-1}(\phi) = \left\langle \left(\frac{1}{2}\right) N H^{N-1}\delta, \phi(a_1 x_1 + \cdots a_N x_N) \right\rangle$$

$$= \frac{1}{2a_1 \cdots a_N} \sum_{i=1}^n a_i \int_0^\infty \frac{x^{N-2}}{(N-2)!} \phi(x) dx, \qquad (5.5.34)$$

while

$$L_{N-2}(\phi) =$$

$$\left\langle \left(\frac{1}{2}\right)^2 \frac{N}{2} H^{N-2}\delta^2 + \left(\frac{1}{12}\right) N H^{N-1}\delta', \phi(a_1 x_1 + \cdots + a_N x_N) \right\rangle$$

$$= \frac{1}{a_1 \cdots a_N} \left[\left(\frac{1}{2}\right)^2 \sum_{i \ne j} a_i a_j + \frac{1}{12} \sum_i a_i^2 \right] \int_0^\infty \frac{x^{N-3}}{(N-3)!} \phi(x) dx. \quad (5.5.35)$$

In order to write the general form of L_i we need to introduce some notation. If $p = (p_1, \cdots, p_k) \in \mathbb{N}^k$ we put

$$\sigma(p; a) = \sum_{\tau} a_{\tau(1)}^{p_1} \cdots a_{\tau(k)}^{p_k}, \tag{5.5.36}$$

the sum being taken over all injections τ from $\{1, \cdots, k\}$ to $\{1, \cdots, N\}$. The $\sigma(p; a)$ are basic symmetric functions of the a_i's. Observe in particular that

$$\sigma(1; a) = \sum_{i=1}^{n} a_i, \qquad \sigma(2; a) = \sum_{i=1}^{n} a_i^2,$$

$$\sigma((1,1); a) = \sum_{i \neq j}^{n} a_i a_j.$$

If $Q \in \mathbb{Z}$ we denote by $W(Q)$ the set of sequences $(k_{-1}, k_0, k_1, k_2, \cdots), k_i \in \mathbb{N}$ that satisfy the two conditions $k_{-1} + k_0 + k_1 + \cdots = N$ and $k_{-1} - k_1 - 2k_2 - 3k_3 - \cdots = Q$. For instance, $W(N-1) = \{(N-1, 1, 0, 0, \cdots)\}$ while $W(N-2) = \{(N-2, 2, 0, 0, \cdots), (N-1, 0, 1, 0, 0, \cdots)\}$. If $k = (k_{-1}, k_0, k_1, \cdots)$ we set $p(k) = (1, \cdots, 1, 2, \cdots, 2, \cdots)$, where $|\{j : p_j = i\}| = k_{i-1}$. We also set $A = a_1 \cdots a_N$.

Then it is easy see that if $1 \leq Q \leq N$,

$$L_Q(\phi) = \frac{1}{A} \sum_{k \in W(Q)} \sigma(p(k); a) \left(\frac{1}{2}\right)^{k_0} \prod_{n \geq 1} \left(\frac{(-1)^n \zeta(-n)}{n!}\right)^{k_n}$$

$$\times \int_0^{\infty} \frac{x^{Q-1}}{(Q-1)} \phi(x) dx, \tag{5.5.37}$$

while if $Q \leq 0$,

$$L_Q(\phi) = \frac{(-1)^Q}{A} \sum_{k \in W(Q)} \sigma(p(k); a) \left(\frac{1}{2}\right)^{k_0}$$

$$\times \prod_{n \geq 1} \left(\frac{(-1)^n \zeta(-n)}{n!}\right)^{k_n} \phi^{(-Q)}(0). \tag{5.5.38}$$

Taking $\phi(x) = e^{-x}$ and putting $r = e^{-\varepsilon}$ we obtain

$$\sum_{n=0}^{\infty} A(n) r^n = \frac{1}{(1 - r^{a_1})(1 - r^{a_2}) \cdots (1 - r^{a_N})} \tag{5.5.39}$$

and writing this in partial fractions we immediately obtain

$$A(n) = \sum_\omega B^{(\omega)}(n)\omega^n, \tag{5.5.40}$$

the sum being taken over all complex numbers ω such that $\omega^{a_i} = 1$ for some i, where $B^{(\omega)}(n)$ are polynomials in n whose degree is one less than the number of i's with $\omega^{a_i} = 1$.

It follows from (5.5.40) that if $\phi \in S(\mathbb{R})$ then

$$\sum_{n=0}^\infty A(n)\phi(\varepsilon n) = \frac{1}{\varepsilon} \int_0^\infty B^{(1)}(x/\varepsilon)\phi(x)dx + O(1), \qquad \text{as } \varepsilon \to 0, \quad (5.5.41)$$

and a comparison of (5.5.32), (5.5.38) and (5.5.41) yields

$$B^{(1)}(x) = \frac{1}{A} \sum_{Q=1}^N \left[\sum_{k \in W(Q)} \sigma(p(k); a) \left(\frac{1}{2}\right)^{k_0} \right.$$

$$\left. \times \prod_{n \geq 1} \left(\frac{(-1)^n \zeta(-n)}{n!}\right)^{k_n} \right] \frac{x^{Q-1}}{(Q-1)!}. \tag{5.5.42}$$

Observe now that

$$A(n) \approx B^{(1)}(n), \tag{5.5.43}$$

while

$$A(n) = B^{(1)}(n) + O(n^r), \tag{5.5.44}$$

where r is the maximum of the degrees of $B^{(\omega)}(n)$ for $\omega \neq 1$.

In particular, if the a_i are relative primes, $(a_i, a_j) = 1$, $i \neq j$, then

$$A(n) = B^{(1)}(n) + O(1), \tag{5.5.45}$$

while if the a_i do not have a common factor, we have

$$A(n) = \frac{n^{N-1}}{A(N-1)!} + O(n^{N-2}). \tag{5.5.46}$$

In case no subset of $N-1$ elements of $a_1, \cdots, a_N$ has a common factor then

$$A(n) = \frac{1}{A} \left[\frac{n^{N-1}}{(N-1)!} + \frac{1}{2} \left(\sum a_i\right) \frac{n^{N-2}}{(N-2)!} \right] + O(n^{N-3}), \tag{5.5.47}$$

and

$$A(n) = \frac{1}{A}\left[\frac{n^{N-1}}{(N-1)!} + \frac{1}{2}\left(\sum a_i\right)\frac{n^{N-2}}{(N-2)!} \right.$$

$$\left. + \left(\frac{1}{4}\sum_{i\neq j} a_i a_j + \frac{1}{12}\sum a_i^2\right)\frac{n^{N-3}}{(N-3)!}\right] + O(n^{N-4}), \qquad (5.5.48)$$

if no subset of $N-2$ elements of $a_1, \cdots, a_N$ has a common factor.

When $a_i = i$ and $A(n) = p_N(n)$ we obtain

$$p_N(n) = B^{(1)}(n) + O(n^{[\![N/2]\!]}), \qquad (5.5.49)$$

so that if $N \geq 8$

$$p_N(n) = \frac{1}{N!}\left[\frac{n^{N-1}}{(N-1)!} + \frac{N(N+1)n^{N-2}}{4(N-2)!} \right.$$

$$\left. + \left(\frac{N^2(N+1)^2}{16} - \frac{N(N+1)(2N+1)}{36}\right)\frac{n^{N-3}}{(N-3)!}\right]$$

$$+ O(n^{N-4}). \qquad (5.5.50)$$

This formula is an extension of the Erdös–Lehner asymptotic formula [1]

$$p_N(n) \sim \frac{1}{N!}\binom{n-1}{N-1}, \qquad \text{as } n \to \infty. \qquad (5.5.51)$$

In fact (5.5.42) and (5.5.49) provide an even sharper estimate.

Finally, when $a_1 = \cdots = a_N = 1$ then it is well known that $A(n) = \binom{n+N-1}{N-1}$ and thus we obtain the identity

$$\binom{n+N-1}{N-1} = \sum_{Q=1}^{N}\left[\sum_{k\in W(Q)} \binom{N}{k-1}\left(\frac{1}{2}\right)^{k_0} \right.$$

$$\left. \times \prod_{n\geq 1}\left(\frac{(-1)^n \zeta(-n)}{n!}\right)^{k_n} \right]\frac{n^{Q-1}}{(Q-1)!}. \qquad (5.5.52)$$

CHAPTER 6

Series of Dirac Delta Functions

6.1 Introduction

In this chapter we discuss several properties of series of Dirac delta functions [31] of the type

$$\sum_{n=0}^{\infty} a_n \delta^{(n)}(x). \tag{6.1.1}$$

As we have seen in the previous chapters, not only do such series form the building blocks in the asymptotic expansion of distributions, but they also arise in other contexts.

Despite the fact that series as (6.1.1) diverge in the distributional sense unless only finitely many a_n do not vanish, they have been used as formal tools in several areas of applied mathematics, including orthogonal polynomials [73], [74], [75], [89] and in the solution of differential and functional equations [18], [62], [81], [99], [114].

In Section 6.2 we give the basic definitions needed to handle such divergent series. Several problems leading to series of Dirac delta functions are presented in Section 6.3. Section 6.4 gives the relationship between formal solutions and the asymptotics of classical and distributional solutions of ordinary differential equations. Section 6.5 considers some singular perturbation problems using this perspective.

6.2 Basic Notions

In this section we give the basic definitions associated with series of Dirac delta functions of the type

$$u(x) = \sum_{n=0}^{\infty} a_n \delta^{(n)}(x). \tag{6.2.1}$$

Recall that the delta functions $\delta^{(n)}(x)$, the distributional derivative of order n of $\delta(x)$, are distributions of $\mathcal{D}'(\mathbb{R})$ supported at $\{0\}$, whose action on a test function ϕ is given by

$$< \delta^{(n)}(x), \phi(x) > = (-1)^n \phi^{(n)}(0). \tag{6.2.2}$$

We remark the important orthogonality relation

$$< \delta^{(n)}(x), x^m > = \begin{cases} 0, & n \neq m, \\ (-1)^n n!, & n = m. \end{cases} \tag{6.2.3}$$

We can say that the sequences $\{x^n/n!\}$ and $\{(-1)^n \delta^{(n)}(x)\}$ are biorthogonal or the dual of one another.

We proved in Chapter 2 that (6.2.1) diverges in $\mathcal{D}'(\mathbb{R})$ unless only a finite number of a_n's do not vanish. Indeed, if we formally evaluate (6.2.1) at a test function $\phi \in \mathcal{D}$, we have

$$< u(x), \phi(x) > = \sum_{n=0}^{\infty} (-1)^n a_n \phi^{(n)}(0), \tag{6.2.4}$$

and since the $\phi^{(n)}(0)$ can be taken as arbitrary real or complex numbers according to Borel's theorem (Theorem 8), it follows that (6.2.4) is generally divergent. Therefore, in general, (6.2.1) diverges in $\mathcal{D}'(\mathbb{R})$: $u(x)$ is not a distribution. Actually, none of the standard summability methods produces a distribution out of the series $\sum_{n=0}^{\infty} a_n \delta^{(n)}(x)$ unless $a_n = 0$ for $n > N$.

One could attempt to solve this problem by trying to interpret the series as ultradistributions or, more generally, as hyperfunctions. However, very soon one finds problems with series of delta solutions not belonging to any of these spaces. For instance, it follows from the general theory of hyperfunctions [70] that if

$$\varlimsup_{n \to \infty} (| a_n | n!)^{\frac{1}{n}} > 0, \tag{6.2.5}$$

then (6.2.1) does not define a hyperfunction concentrated at $\{0\}$.

Consequently, in the general case, series of Dirac delta functions cannot be considered as distributions or ultradistributions, not even as hyperfunctions. Naturally, any of these series can be considered as a functional in the space of polynomials, but it is not clear what the relationship of such functionals is with classical analytical objects.

The term *dual Taylor series* can be used for them, since in a sense they are "dual" to the Taylor series $\sum_{n=0}^{\infty} b_n x^n$. But again, this term does not say very much about what they are.

Notice, however, that many algebraic and analytical operations, such as differentiation and multiplication with polynomials can be performed on dual Taylor series. Indeed,

$$\frac{d}{dx} \left(\sum_{n=0}^{\infty} a_n \delta^{(n)}(x) \right) = \sum_{n=1}^{\infty} a_{n-1} \delta^{(n)}(x), \tag{6.2.6}$$

$$x \left(\sum_{n=0}^{\infty} a_n \delta^{(n)}(x) \right) = \sum_{n=1}^{\infty} -(n+1)a_{n+1}\delta^{(n)}(x), \qquad (6.2.7)$$

the second formula coming from the fact that $x\delta^{(n)}(x) = -n\delta^{(n-1)}(x)$.

6.3 Several Problems that Lead to Series of Deltas

We shall now consider several problems where dual Taylor series appear naturally. Let us start with the *problem of moments*.

Let $\{\mu_n\}$ be a sequence of real or complex numbers. The problem of finding a function $f(x)$ that satisfies

$$< f(x), x^n > = \int_{-\infty}^{\infty} f(x)x^n dx = \mu_n, \qquad (6.3.1)$$

is called the problem of moments.

Classically (6.3.1) was studied under the restriction that $f(x)$ is a positive measure, supported in a given closed set [101]. Since positivity of f is not needed in many situations, the problem of moments was eventually studied for signed measures and more recently in spaces of distributions and ultradistributions.

Observe that the series of delta functions provides an immediate solution to the problem of moments (6.3.1). In fact, using (6.2.3) it follows that

$$u(x) = \sum_{n=0}^{\infty} \frac{(-1)^n \mu_n \delta^{(n)}(x)}{n!} \qquad (6.3.2)$$

is a solution of (6.3.1).

Writing (6.3.2) was very simple and natural, but can a classical solution be obtained from it? Or, put in a somewhat different way, if $f(x)$ is a solution of (6.3.1) in the ordinary sense, what is the relationship between $f(x)$ and $u(x)$?

The results of the previous chapters show that if $f(x)$ is a solution of (6.3.1) and if $f(x)$ belongs to $\mathcal{E}'(\mathbb{R}), \mathcal{O}'_C(\mathbb{R}), \mathcal{O}'_M(\mathbb{R}), \mathcal{K}'(\mathbb{R})$ or any other space where the moment asymptotic expansion holds, then f and u are related in an asymptotic way, namely,

$$f(\lambda x) \sim u(\lambda x) = \sum_{n=0}^{\infty} \frac{(-1)^n \mu_n \delta^{(n)}(x)}{n! \lambda^{n+1}}, \quad \text{as } \lambda \to \infty. \qquad (6.3.3)$$

Observe that this can be expressed equivalently as

$$f(\lambda x) = \sum_{n=0}^{N} \frac{(-1)^n \mu_n \delta^{(n)}(x)}{n! \lambda^{n+1}} + O\left(\frac{1}{\lambda^{N+2}}\right), \quad \text{as } \lambda \to \infty, \qquad (6.3.4)$$

a relation that involves no infinite series.

The problem of moments in the space $\mathcal{E}'(\mathbb{R})$ was studied by us [41]. The basic result is the following.

Theorem 56. *Let $\{\mu_n\}$ be a sequence of real or complex numbers. Then $\{\mu_n\}$ is the moment sequence of a distribution $f \in \mathcal{E}'[-a, a]$ if and only if there exist constants M and γ such that for every $n \in \mathbb{N}$,*

$$\left| \sum_{j=1}^{n} (-1)^{n+j} \binom{n+j-1}{2j-1} \mu_{2j-1} \left(\frac{2}{a} \right)^{2j} \right| \leq M n^{\gamma}, \qquad (6.3.5a)$$

$$\left| \sum_{j=0}^{n-1} (-1)^{n+j} \binom{n+j-1}{2j} \mu_{2j} \left(\frac{2}{a} \right)^{2j+1} \right| \leq M n^{\gamma}. \qquad (6.3.5b)$$

If a solution exists, it is unique.

Proof. The idea of the proof is to associate to each $f \in \mathcal{E}'[-a, a]$ its analytic or Cauchy representation

$$F(z) = \frac{1}{2\pi i} < f(x), \frac{1}{x - z} >, \qquad (6.3.6)$$

defined in the region $\overline{\mathbb{C}} \setminus [-a, a]$ of the complex Riemann sphere. If $\{\mu_n\}$ is the moment sequence of f, then the Taylor series of $F(z)$ for large z is

$$F(z) = -\frac{1}{2\pi i} \sum_{n=0}^{\infty} \frac{\mu_n}{z^{n+1}}. \qquad (6.3.7)$$

Next, we consider the conformal mapping

$$z = \frac{a}{2} \left(w + \frac{1}{w} \right), \qquad (6.3.8a)$$

$$w = \frac{z}{a} - \sqrt{\left(\frac{z}{a} \right)^2 - 1}, \qquad (6.3.8b)$$

between the unit disc of the w-plane and the region $\overline{\mathbb{C}} \setminus [-a, a]$ of the z-sphere. Define the analytic function $G(w)$ in the unit disc by

$$G(w) = F(z) = F\left(\frac{a}{2} \left(w + \frac{1}{w} \right) \right). \qquad (6.3.9)$$

Then f is a distribution if and only if the limit $\lim_{r\to 1} G(re^{i\theta})$ exists in the distributional sense. According to the results of [34] this happens if and only if $\dfrac{G^{(n)}(0)}{n!} = O(n^\gamma)$ for some γ: this is (6.3.5). ■

Observe that in the case where (6.3.5) is satisfied then the series

$$\sum_{n=0}^{\infty} \frac{(-1)^n \mu_n \delta^{(n)}(x)}{n!}$$

does not represent, in general, a distribution supported at $\{0\}$.

When the support of f is allowed to be an infinite interval then (6.3.1) has infinite solutions for any arbitrary sequence $\{\mu_n\}$, a result that goes back to Boas [9], [10]. Recently Durán [25] established this for the space $S(0,\infty) = \{\phi \in S(\mathbb{R}) : supp\ \phi \subseteq [0,\infty)\}$. We give the proof based on Ritt's theorem (Theorem 12) as explained in [24].

Theorem 57. *Let $\{\mu_n\}$ be an arbitrary sequence of real or complex numbers. Then there exists $\phi \in S(0,\infty)$ with*

$$\int_0^\infty x^n \phi(x)dx = \mu_n, \quad n \in \mathbb{N}. \tag{6.3.10}$$

Proof. Observe that a function $\psi \in S(\mathbb{R})$ is the Fourier transform

$$\psi(u) = \hat{\phi}(u) = \int_0^\infty e^{ixu}\phi(x)dx \tag{6.3.11}$$

of a function ϕ of the class $S(0,\infty)$ if and only if it can be extended to a bounded continuous function $\Psi(z)$ in the upper half plane $\Im m\ z \geq 0$, analytic in $\Im m\ z > 0$ and vanishing as $z \to \infty$.

Since

$$\frac{d^n \hat{\phi}(0)}{du^n} = i^n \int_0^\infty x^n \phi(x)dx, \tag{6.3.12}$$

the problem of (6.3.10) is equivalent to that of finding $\psi = \hat{\phi}$ in the class $\mathcal{F}(S(0,\infty))$ that satisfies

$$\psi^{(n)}(0) = i^n \mu_n, \quad n \in \mathbb{N}. \tag{6.3.13}$$

Let

$$G(z) = e^{(1-i)(z+i)^{\frac{1}{2}}}, \tag{6.3.14}$$

where the branch of the square root is chosen to ensure that $G(z) \to 0$ as $z \to \infty$ in the upper half plane $\Im m; z \geq 0$.

Let the sequence $a_0, a_1, a_2, \cdots$ be defined as

$$\frac{1}{G(z)} = a_0 + a_1 z + a_2 z^2 + \cdots, \quad | z | < 1. \tag{6.3.15}$$

Let $\delta \in (0, \frac{\pi}{2})$ and let $F(z)$ be a bounded analytic function in the sector $S : -\delta < \arg z < \pi + \delta$, with the asymptotic power series

$$F(z) \sim b_0 + b_1 z + b_2 z^2 + \cdots, \quad z \to 0, \ z \in S, \tag{6.3.16}$$

where

$$b_n = \sum_{k=0}^{n} \frac{i^k}{k!} \mu_k a_{n-k}. \tag{6.3.17}$$

Let $\Psi(z) = F(z)G(z)$. Then Ψ is analytic in the sector S and vanishes at infinity. Also

$$\Psi(z) \sim \sum_{n=0}^{\infty} \frac{i^n \mu_n z^n}{n!}, \quad z \to 0, \ z \in S. \tag{6.3.18}$$

It follows that if ψ is the restriction of Ψ to the real axis, then $\psi \in \mathcal{F}(S(0, \infty))$ and ψ satisfies (6.3.13), as required. ∎

The next problem we would like to mention is the *problem of weights for orthogonal polynomials*. Actually, this is one of the first problems where series of Dirac delta functions have been used. Motivated by the failure of classical methods for the study of the weights for the Bessel polynomials [76], various authors [74], [75], [89] introduced the divergent series of delta functions into this problem.

Let $\{\mu_n\}$ be a sequence that satisfies the condition

$$\triangle_n = \det [\mu_{i+j}]_{i,j=0}^{n} \neq 0, \ n = 0, 1, 2, \cdots. \tag{6.3.19}$$

Then a sequence of polynomials is introduced by setting $p_0 = 1$ and

$$p_n(x) = \frac{1}{\triangle_{n-1}} \begin{bmatrix} \mu_0 & \cdots & \mu_n \\ \vdots & \ddots & \vdots \\ \mu_{n-1} & \cdots & \mu_{2n-1} \\ 1 & \cdots & x^n \end{bmatrix}, \ n \geq 1. \tag{6.3.20}$$

The polynomials $p_n(x)$ are monic of degree n. The polynomials $\{p_n(x)\}$ can be made orthogonal by a weight $w(x)$ that satisfies $< w(x), p_n(x)p_m(x) >$

$= 0$, $n \neq m$, $< w(x), p_n^2(x) > $, $\neq 0$. Actually, if a dual Taylor series is allowed for the weight then the solution is immediate:

$$w(x) = \sum_{n=0}^{\infty} \frac{(-1)^n \mu_n \delta^{(n)}(x)}{n!}. \tag{6.3.21}$$

The weight $w(x)$ can also be obtained by solving an ordinary differential equation [75]. If the polynomials satisfy a differential equation of the second order, it should be of the form

$$p(x)y''(x) + q(x)y'(x) = \lambda_n y(x), \tag{6.3.22}$$

where

$$p(x) = ax^2 + bx + c, \quad q(x) = dx + e. \tag{6.3.23}$$

If $w(x)$ is a solution of the equation

$$-(p(x)w(x))' + q(x)w(x) = 0, \tag{6.3.24}$$

that vanishes rapidly at $\pm\infty$, then $w(x)$ is a weight for the polynomials $\{p_n(x)\}$.

Example 113. The Hermite polynomials satisfy the equation

$$y'' - 2xy' + 2ny = 0. \tag{6.3.25}$$

In this case the equation corresponding to (6.3.24) is

$$w' + 2xw = 0, \tag{6.3.26}$$

with solution $w(x) = ce^{-x^2}$, that give the well-known weight.

We can also solve (6.3.26) by a series $w(x) = \sum_{n=0}^{\infty} a_n \delta^{(n)}(x)$. Proceeding formally,

$$0 = w' + 2xw = \sum_{n=0}^{\infty} a_n \delta^{(n+1)}(x) - 2\sum_{n=1}^{\infty} a_n \delta^{(n-1)}(x),$$

and this yields

$$w(x) = \sum_{n=0}^{\infty} \frac{a_0 \delta^{(n)}(x)}{4^n n!}. \tag{6.3.27}$$

Example 114. For the Bessel polynomials the equation is

$$x^2 y'' + (2x + 2)y' + n(n+1)y = 0. \tag{6.3.28}$$

The equation for the weight is

$$x^2 w' - 2w = 0, \tag{6.3.29}$$

but the classical solution $e^{\frac{-2}{x}}$, $x \neq 0$, cannot be regularized at $x = 0$ to give a distribution. Furthermore, $e^{\frac{-2}{x}}$ does not vanish at infinity.

On the other hand, substitution of a dual Taylor series $\sum_{n=0}^{\infty} a_n \delta^{(n)}(x)$ gives

$$w(x) = \sum_{n=0}^{\infty} \frac{2^{n+1} \delta^{(n)}(x)}{(n+1)! n!}, \tag{6.3.30}$$

Although (6.3.30) does not define a distribution, Kim and Kwon [73] established that the series defines a hyperfunction concentrated at $\{0\}$.

The next class of problems we would like to consider is *the solution of ordinary differential equations.* The fact that dual Taylor series can be differentiated and multiplied by polynomials makes them candidates to solve differential equations with polynomial coefficients. We already met examples of this kind in the previous examples. Many classical equations, such as Bessel's or the hypergeometric, have been solved by this method [71], [81]. Wiener [114] has shown that series of Dirac delta functions also arise in the solution of functional differential equations, particularly differential equations with deviations in the argument.

Let us introduce some notation. Given an equation, we denote by S_c the set of classical solutions. Similarly, S_d are the distributional solutions, S_h the hyperfunction solutions and S_δ are the dual Taylor series solutions. Clearly $S_c \subseteq S_d \subseteq S_h$, but the relationship with the space S_δ is not so obvious.

Example 115. For a homogeneous equation with constant coefficients,

$$y^{(k)} + \alpha_{k-1} y^{(k-1)} + \cdots + \alpha_0 y = 0, \tag{6.3.31}$$

there are no solutions of the form $\sum_{n=0}^{\infty} a_n \delta^{(n)}(x)$, as follows by direct substitution. In this case the dimension of the space of classical solutions S_c is k while the dimension of S_δ is 0.

Recall that a normal equation with smooth coefficients does not have distributional solutions other than the classical ones [71]. Thus, non-classical distributional solutions of ordinary differential equations arise only for non-normal equations. A similar situation is encountered with hyperfunction solutions, since according to Komatsu's theorem [78], a differential equation with analytic coefficients

$$a_k(x) y^{(k)}(x) + \cdots + a_0(x) y(x) = 0,$$

has a space of hyperfunction solutions S_h of dimension $k + \sum_x \text{ord } a_k(x)$, where ord $a_k(x)$ is the order of the zero x at a_k, so that ord $a_k(x) = 0$ if $a_k(x) \neq 0$. Thus new hyperfunction solutions arise only for non-normal equations.

The previous example seems to indicate that this is also the case with dual Taylor series solutions. However, the situation is more complicated.

Example 116. The equation

$$y' = x^2 y, \tag{6.3.32}$$

has as classical solutions $y = ce^{\frac{x^3}{3}}$. Thus $S_c = S_d = S_h$ has dimension 1. However, if we substitute $y = \sum_{n=0}^{\infty} a_n \delta^{(n)}(x)$ into (6.3.32) we obtain

$$\sum_{n=0}^{\infty} a_n \delta^{(n+1)}(x) = \sum_{n=2}^{\infty} n(n-1) a_n \delta^{(n-2)}(x),$$

and thus $a_2 = 0$ while

$$(n+2)(n+1) a_{n+2} = a_{n-1}, \quad n \geq 1.$$

Therefore

$$y(x) = a_0 \sum_{m=0}^{\infty} \frac{4 \cdot 7 \cdots (3m-2)}{(3m)!} \delta^{(3m)}(x) + a_1 \sum_{m=0}^{\infty} \frac{2 \cdot 5 \cdots (3m-1)}{(3m+1)!} \delta^{(3m+1)}(x),$$

$$\tag{6.3.33}$$

where a_0 and a_1 are arbitrary. Thus $\dim S_\delta = 2$.

Actually, for the equation

$$y' = x^k y, \tag{6.3.34}$$

we have $\dim S_\delta = k$ while $\dim S_c = \dim S_d = \dim S_h = 1$.

Also series of Dirac delta functions arise in *singular perturbations*. Differential equations containing a small parameter $\varepsilon << 1$,

$$f(x, y, \cdots, y^{(k)}, \varepsilon) = 0 \tag{6.3.35}$$

are usually referred to as perturbation problems. In the regular perturbation problems the solution can be expressed as a power series in ε, $y = \sum_{n=0}^{\infty} y_n(x) \varepsilon^n$, where the $y_n(x)$ satisfy suitable equations and boundary conditions. The first approximation $y_0(x)$ is obtained by solving (6.3.35) with $\varepsilon = 0$.

Problems where the parameter ε multiplies the highest order derivative, such as the initial value problem

$$\varepsilon y'' + ay' + by = 0, \quad y(0) = y_0, \quad y'(0) = y_1, \tag{6.3.36}$$

do not admit such power series solutions. Here setting $\varepsilon = 0$ gives an equation of lower order and the initial or boundary conditions cannot all be satisfied. However, if the y_n are allowed to contain Dirac delta terms, Glizer and Dimitriev [49] have shown that a power series representation is still possible in many cases.

6.4 Dual Taylor Series as Asymptotics of Solutions of Equations

As we have seen, series of Dirac delta functions arise in the solution of equations while they also form the basic block in the asymptotic expansion of generalized functions. Thus, we can expect that there should be a connection between the asymptotics of the distributional solutions and the formal dual Taylor series solutions of those equations. The examples we have given show that this relationship is not so evident since, in general, there is no relationship between the dimensions of the spaces S_d and S_δ. We now give some results in this direction [62].

Theorem 58. *Let $y(x)$ be a solution of ordinary differential equation with polynomial coefficients*

$$a_k(x)y^{(k)}(x) + \cdots + a_0(x)y(x) = 0. \tag{6.4.1}$$

If y belongs to $\mathcal{A}'$, where $\mathcal{A}$ is any of the spaces $\mathcal{E}, \mathcal{P}, \mathcal{O}_M, \mathcal{O}_C$ or another where the moment asymptotic expansion holds,

$$y(\lambda x) \sim \sum_{n=0}^{\infty} \frac{(-1)^n \mu_n \delta^{(n)}(x)}{n! \lambda^{n+1}}, \quad \text{as } \lambda \to \infty, \tag{6.4.2}$$

then the dual Taylor series

$$u(x) = \sum_{n=0}^{\infty} \frac{(-1)^n \delta^{(n)}(x)}{n!} \tag{6.4.3}$$

satisfies (6.4.1).

Thus, there is a map from the space $S_d \cap \mathcal{A}'$ to S_δ. For instance, for the equation

$$y' + 2xy = 0 \tag{6.4.4}$$

we have $S_d = \{ce^{-x^2} : c \in \mathbb{R}\} \subseteq \mathcal{P}'$ and the map from S_d to S_δ is a bijection

$$ce^{-x^2} \leftrightarrow c\sqrt{\pi} \sum_{n=0}^{\infty} \frac{\delta^{(2n)}(x)}{4^n n!}.$$

For the equation

$$y'' + y = 0 \tag{6.4.5}$$

we have $S_d = \{c_1 \cos x + c_2 \sin x : c_1, c_2 \in \mathbb{R}\} \subseteq \mathcal{K}'$. But $S_\delta = \{0\}$ and the map of S_d to S_δ is trivial.

The equation

$$y' = -4x^3 y \tag{6.4.6}$$

has $S_d = \{ce^{-x^4} : c \in \mathbb{R}\}$ of dimension 1, but S_δ has dimension 3. This shows that not every series of deltas that solves an ordinary differential equation arises from the asymptotic expansion of a distributional solution: the converse of the previous theorem is not true.

However, we have the following result [62].

Theorem 59. *Let* $u(x) = \sum_{n=0}^{\infty} a_n \delta^{(n)}(x)$ *be a solution of the ordinary differential equation with polynomial coefficients*

$$a_k(x)y^{(k)}(x) + \cdots + a_0(x)y(x) = 0. \tag{6.4.7}$$

Then if $\mathcal{A}$ *is any of the spaces* $\mathcal{P}, \mathcal{O}_M, \mathcal{O}_C$ *or* $\mathcal{K}$, *there exists* $f \in \mathcal{A}'(\mathbb{R})$ *such that*

$$f(\lambda x) = o(\lambda^{-\infty}), \quad as \ \lambda \to \infty \tag{6.4.8}$$

and a solution $y \in \mathcal{A}'(\mathbb{R})$ *of the equation*

$$a_k(x)y^{(k)}(x) + \cdots + a_0(x)y(x) = f(x), \tag{6.4.9}$$

such that

$$y(\lambda x) \sim u(\lambda x) \sim \sum_{n=0}^{\infty} \frac{a_n \delta^{(n)}(x)}{\lambda^{n+1}}, \tag{6.4.10}$$

as $\lambda \to \infty$.

Summarizing, the dual Taylor series solutions of an ordinary differential equation with polynomial coefficients, $Ly = 0$, are the asymptotics of the distributional solutions of the equations $Ly = f$, where $f(\lambda x) = o(\lambda^{-\infty})$ as $\lambda \to \infty$. Observe that for these generalized functions f the associated dual Taylor series is $0 \cdot \delta(x) + 0 \cdot \delta'(x) + 0 \cdot \delta''(x) + \cdots$ and thus from the point of view of dual Taylor series it is not possible to distinguish the equations $Ly = 0$ and $Ly = f$.

In many cases, the introduction of the auxiliary function f does not cause any complications. For instance, for polynomials satisfying a second order equation

$$py'' + qy' = \lambda_n y, \tag{6.4.11}$$

if we can find a rapidly decreasing at infinity solution w of the equation

$$-(pw)' + qw = f, \tag{6.4.12}$$

where $f(\lambda x) = o(\lambda^{-\infty})$, as $\lambda \to \infty$, then w is a weight for the polynomials.

6.5 Singular Perturbations

As we mentioned in Section 6.3, series of Dirac delta functions also appear in the solution of singular perturbation problems that do not admit solutions in the form of power series with classical solutions as coefficients.

Let us start with the initial value problem

$$\varepsilon y'(x) = -y(x), \quad x > 0, \tag{6.5.1a}$$

$$y(0) = 1, \tag{6.5.1b}$$

where $\varepsilon << 1$.

Let $z(x) = H(x)y(x)$, where $H(x)$ is the Heaviside function. Then

$$z' = Hy' + \delta(x), \tag{6.5.2}$$

thus $(6.5.1a, b)$ becomes

$$\varepsilon z' = -z + \varepsilon \delta(x). \tag{6.5.3}$$

Next, we look for a solution of the form

$$z = \sum_{n=0}^{\infty} z_n(x)\varepsilon^n. \tag{6.5.4}$$

Substitution into $(6.5.2)$ and collection of like powers of ε yields

$$z_0 = 0,$$
$$z_1 = -\delta(x),$$
$$z_n + z'_{n-1} = 0, \quad n \geq 2.$$

Since it is natural to ask that $z_n(x) = 0$ for $x < 0$, we obtain

$$z_{n+1}(x) = (-1)^n \delta^{(n)}(x), \quad n \geq 0,$$

and thus

$$z(x, \varepsilon) = \sum_{n=0}^{\infty} (-1)^n \varepsilon^{n+1} \delta^{(n)}(x). \tag{6.5.5}$$

The solution of (6.5.1) is easy to find, however, so that a comparison can be made. The solution is $H(x)e^{-\varepsilon^{-1}x}$. The asymptotic relation

$$H(x)e^{\varepsilon^{-1}x} \sim \sum_{n=0}^{\infty} (-1)^n \varepsilon^{n+1} \delta^{(n)}(x), \quad \text{as } \varepsilon \to 0, \tag{6.5.6}$$

is nothing but the distributional version of Watson's lemma of Chapter 3.

A similar analysis can be applied to the initial value problem

$$\varepsilon y'(x) = a(x)y(x), \quad x > 0, \tag{6.5.7a}$$

$$y(0) = y_0, \tag{6.5.7b}$$

where $a(x)$ is smooth and negative for $x \geq 0$. Again we introduce $z(x) = H(x)y(x)$. The explicit solution is

$$z(x, \varepsilon) = H(x)y_0 e^{\varepsilon^{-1}B(x)}, \tag{6.5.8}$$

where

$$B(x) = \int_0^x a(t)dt. \tag{6.5.9}$$

Observe that if we substitute $B(x)$ for x in (6.5.6) we obtain

$$z(x, \varepsilon) \sim y_0 \sum_{n=0}^{\infty} \varepsilon^{n+1} \delta^{(n)}(B(x)). \tag{6.5.10}$$

In particular, we have Laplace's formula

$$z(x, \varepsilon) = \frac{-\varepsilon y_0}{a(0)} \delta(x) + O(\varepsilon^2). \tag{6.5.11}$$

More generally, we have the following result.

Theorem 60. *Let $y(x, \varepsilon)$ be the solution of the initial value problem*

$$\varepsilon y'(x) = A(x, \varepsilon)y(x), \quad x > 0, \tag{6.5.12a}$$

$$y(x_0) = y_0, \tag{6.5.12b}$$

where $A(x, \varepsilon)$ is smooth for $x \geq 0$ and

$$A(x, \varepsilon) \sim A_0(x) + \varepsilon A_1(x) + \varepsilon^2 A_2(x) + \cdots, \quad \text{as } \varepsilon \to 0, \tag{6.5.13}$$

where $A_0(x)$ is negative.

Let $z(x, \varepsilon) = H(x)y(x, \varepsilon)$. Then

$$z(x, \varepsilon) \sim z_1(x)\varepsilon + z_2(x)\varepsilon^2 + z_3(x)\varepsilon^3 + \cdots, \quad \text{as } \varepsilon \to 0, \tag{6.5.14}$$

in $\mathcal{D}'(\mathbb{R})$ where z_i is supported at $\{0\}$ and has order $i - 1$ at the most.

REFERENCES

[1] Andrews GE, *The Theory of Partitions*, Addison–Wesley, Reading, Massachusetts, 1976.

[2] Antonets MA, The classical limit for Weyl quantization, *Lett. Math. Phys.*, **2** (1978), 241–245.

[3] _____, Classical limit of Weyl quantization, *Teoret. Mat. Fiz.*, **38** (1979), 331–339.

[4] Apostol TM, *Introduction to Analytic Number Theory*, Springer–Verlag, New York, 1976.

[5] Baker J, Formulation of quantum mechanics based on the quasi-probability distribution induced on phase space, *Phys. Rev.*, **109** (1958), 2198–2206.

[6] Bayen F, Flato M, Fronsdal C, Lichnerowicz A, and Sternheimer D, Deformation theory and quantization, *Ann. Phys.* , **111** (1978), 61–151.

[7] Berndt BC and Evans RJ, Extensions of asymptotic expansions from chapter 15 of Ramanujan's second notebook, *J. Reine Angew Math.*, **361** (1985), 118–134.

[8] Blaive B and Metzger J, Explicit expressions for the n-th gradient of $1/r$, *J. Math. Phys.*, **25** (1984), 1721–1724.

[9] Boas RP, The Stieljes moment problem for functions of bounded variation, *Bull. Amer. Math. Soc.*, **45** (1939), 399–404.

[10] _____, *Entire Functions*, Academic Press, New York, 1954.

[11] Borwein D and Borwein JM, A note on alternating series in several dimensions, *Amer. Math. Monthly*, **93** (1986), 531–539.

[12] Bremmerman H, *Distributions, Complex Variables and Fourier Transforms*, Addison-Wesley, Reading, Massachusetts, 1965.

[13] Brüning J, On the asymptotic expansion of some integrals, *Arch. Math.*, **58** (1984), 253–259.

[14] Brüning J and Selley R, Regular singular asymptotics, *Adv. Math.*, **58** (1985), 133–148.

[15] Brychkov YA and Prudnikov AP, *Integral Transforms of Generalized Functions*, Gordon & Breach, New York, 1987.

[16] Caboz R, Codaccioni JP, and Constantinescu F, Taylor series for the Dirac function on perturbed surfaces with applications to Mechanics, *Math. Methods Appl. Sci.*, **7** (1985), 416–425.

[17] Chandrasekharan K, *Arithmetical Functions*, Springer-Verlag, New York, 1970.

[18] Cooke KL and Wiener J, Distributional and analytical solutions of functional differentials equations , *J. Math. Anal. Appl.*, **98** (1984), 111–129.

[19] Copson ET, *Asymptotic Expansions*, Cambridge University Press, London and New York, 1965.

[20] Costen RC, Jump conditions for fields that have infinite, integrable singularities at an interface, *J. Math. Phys.*, **22** (1981), 1337–1385.

[21] De Bruijn NG, *Asymptotic Methods in Analysis*, North Holland, Amsterdam, 1970.

[22] Dixon A, Una fórmula de Ramanujan, *Master Thesis*, Universidad de Costa Rica, 1989.

[23] Donoghue W, *Distributions and Fourier Transforms*, Academic Press, New York, 1969.

[24] Durán A and Estrada R, Strong moment problems for rapidly decreasing smooth functions, *Proc. Amer. Math. Soc.*, in press.

[25] Durán AJ, The Stieljes moment problem for rapidly decreasing smooth functions, *Proc. Amer. Math. Soc.*, **107** (1989), 731–741.

[26] Durbin P, Asymptotic expansions of Laplace transforms about the origin using generalized functions, *J. Inst. Math. Appl.*, **23** (1978), 182–192.

[27] Erdelyi A, *Asymptotic Expansions*, Dover, New York, 1956.

[28] _____, Asymptotic expansions of Fourier integrals involving logarithmic singularities, *SIAM J.*, **4** (1956), 38–47.

[29] Erdös P and Fuchs WHJ, On a problem of additive number theory, *J. London Math. Soc.*, **31** (1956), 67–73.

[30] Estrada R, The asymptotic expansion of certain series considered by Ramanujan, *Appl. Anal.*, **43** (1992), 191–228.

[31] _____, Series of Dirac delta functions, in: *Partial Differential Equations*, Wiener J and Hale JK, eds., Pitman, London, 1992, 44–49.

[32] Estrada R, Gracia-Bondía J, and Várilly J, On asymptotic expansions of twisted products, *J. Math. Phys.*, **30** (1989), 2789–2796.

[33] Estrada R and Kanwal RP, Applications of distributional derivatives to wave propagation, *J. Inst. Math. Appl.*, **26** (1980), 39–63.

[34] _____, Distributional boundary values of analytic and harmonic functions, *J. Math. Anal. Appl.*, **89** (1982), 262–289.

[35] _____, Distributional analysis of discontinuous fields, *J. Math. Anal. Appl.*, **105** (1985), 478–490.

[36] _____, Distributional solutions of singular integral equations, *J. Integral Equations*, **8** (1985), 41–85.

[37] Regularization and distributional derivatives of $(x_1^2 + \cdots + x_p^2)^{\frac{-n}{2}}$ in R^p, *Proc. Roy. Soc. London Ser. A*, **401** (1985), 281–297.

[38] _____, Distributional solutions of dual integral equations of Cauchy, Abel, and Titchmarsh types, *J. Integral Equations*, **9** (1985), 277–305.

[39] _____, Some results in discontinuous fields and wave propagation, *J. Math. Anal. Appl.*, **128** (1987), 389–404.

[40] _____, Higher order fundamental forms of a surface and their applications to wave propagation and distributional derivatives, *Rend.*

Circ. Mat. Palermo, **36** (1987), 27–62.

[41] _____, Moment sequences for a class of distributions, *Complex Variables Theory Appl.*, **9** (1987), 31–39.

[42] _____, Regularization, pseudofunction and Hadamard finite part, *J. Math. Anal. Appl.*, **141** (1989), 195–207.

[43] _____, A distributional theory of asymptotic expansions, *Proc. Roy. Soc. London Ser. A*, **428** (1990), 399–430.

[44] _____, The asymptotic expansion of some multidimensional generalized functions, *J. Math. Anal. Appl.*, **163** (1992), 264–283.

[45] _____, Taylor expansions for distributions, *Math. Meth. Appl. Sci.*, **16** (1993), 297–304.

[46] _____, Asymptotic separation of variables, *J. Math. Anal. Appl.*, **178** (1993), 128–140.

[47] Flajolet P and Martin GN, Probabilistic counting algorithms for data base applications, *J. Comp. Syst. Sci.*, **31** (1985), 182–209.

[48] Focke J, Asymptotiche Entwicklungen mittels der Methods der stationären Phase, *Berichte der Sächsischen Akademie der Wissenchaften zu Leipzig*, **101** (1954).

[49] Glizer VJ and Dimitriev MG, Singular perturbations and generalized functions, *Soviet Math. Dokl.*, **20** (1979), 1360–1364.

[50] Gracia-Bondía J and Várilly J, Algebras of distributions suitable for phase space quantum mechanics, *J. Math. Phys.*, **29** (1988), 869–879.

[51] Guermond JL, A general lifting-line theory for curved and swept wings, *J. Fluid Mech.*, **221** (1990), 497–513.

[52] Grothendieck A, *Produits Tensories Topologiques et Spaces Nucléaires*, American Mathematical Society, Providence, Rhode Island, 1955.

[53] Gel'fand IM and Shilov GE, *Generalized Functions*, **vol. I**, Academic Press, New York, 1964.

[54] Hafner JL, New omega theorems for two classical lattice point problems, *Invent. Math.*, **63** (1981), 181–186.

[55] Halberstam H and Roth KF, *Sequences*, Springer-Verlag, New York, 1983.

[56] Handelsman RA and Bleinstein N, *Asymptotic Expansions of Integrals*, Holt, Rinehart and Winston , New York, 1975.

[57] Handelsman H and Lew JS, Asymptotic expansion of Laplace transforms near the origin, *SIAM J. Math. Anal*, **1** (1970), 118–130.

[58] _____, Asymptotic expansion of a class of integral transforms with algebraically dominated kernel, *J. Math. Anal. Appl.*, **35** (1971), 405–433.

[59] Hardy GH, On Dirichlet divisor problem, *Proc. London Math. Soc.*, **15** (1916), 1–25.

[60] _____, *Divergent Series*, Clarendon Press, Oxford, 1949.

[61] _____, *Ramanujan: Twelve Lectures on Subjects Suggested by his Life and Work*, Chelsea, New York, 1978.

[62] Hernández LG and Estrada R, Solutions of ordinary differential equations by series of Dirac delta functions, Preprint.

[63] Hörmander L, *Linear Differential Operators*, **vol. I**, Springer-Verlag, New York, 1983.

[64] Horvath J, *Topological Vector Spaces and Distributions*, **vol. I**, Addison-Wesley, Reading, Massachusetts, 1966.

[65] Ingham AE, On the classical lattice point problems, *Proc. Cambr. Phil. Soc.*, **36** (1940), 131–138.

[66] Ivic A, *The Theory of the Riemann Zeta-function with Applications*, Wiley, New York, 1985.

[67] Jeffery H, *Asymptotic Approximations*, Oxford University Press, London and New York, 1962.

[68] Jones DS, Generalized transforms and their asymptotic behavior, *Phil. Trans. Roy. Soc. London A.*, **265** (1969), 1–43.

[69] _____, *Generalised Functions*, Cambridge University Press, London and New York, 1982.

[70] Kaneko A, *Introduction to Hyperfunctions*, Klewer Press, Boston, 1989.

[71] Kanwal RP, *Generalized Functions: Theory and Technique*, Academic Press, New York, 1983.

[72] Karamata J, Sur un mode de croissance reguliere des functions, *Mathematica (Cluj)*, **4** (1930), 38–53.

[73] Kim SS and Kwon KN, Generalized weights for orthogonal polynomials, *Differential Integral Equations*, **4** (1991), 601–608.

[74] Krall AM, Orthogonal polynomials through moment generating functions, *SIAM. J. Math. Anal.*, **9** (1978), 600–603.

[75] Krall AM, Kanwal RP, and Littlejohn LL, Distributional solutions of ordinary differential equations, *Canad. Math. Soc. Conf. Proc.*, **8** (1987), 227–246.

[76] Krall HL and Frink O, A new class of orthogonal polynomials: the Bessel polynomials, *Trans. Amer. Math. Soc.*, **65** (1949), 100–115.

[77] Kolesnik G, On the order of $\zeta\left(\frac{1}{2}+it\right)$ and $\triangle(R)$, *Pacific J. Math.*, **98** (1982), 107–122.

[78] Komatsu H, On the index of ordinary differential operators, *J. Fac. Sci. Univ. Tokyo Sect. 1A*, **18** (1971), 379–398.

[79] Landau E, *Handbuch der Lehre von der Verteilung der Primzahlen*, Chelsea, New York, 1953.

[80] Lighthill MJ, *Introduction to Fourier Analysis and Generalized Functions*, Cambridge University Press, London and New York, 1958.

[81] Littlejohn LL and Kanwal RP, Distributional solutions of the hypergeometric differential equation, *J. Math. Anal. Appl.*, **122** (1987), 325–345.

[82] Lojasiewicz S, Sur la valeur et la limite d'une distribution en un point, *Studia Math.*, **16** (1957), 1–36.

[83] Lützen J, *The Prehistory of the theory of Distributions*, Springer Verlag, New York, 1982.

[84] McClure, JP and Wong R, Explicit error terms for asymptotic expansions of Stieljes transforms, *J. Inst. Math. Appl.*, **22** (1978), 129–145.

[85] _____, Generalized Mellin convolutions and their asymptotic expansions, *Canad. J. Math.*, **36** (1984), 924–960.

[86] Meyer B, On the convergence of alternating double series, *Amer. Math. Monthly*, **60** (1953), 402–404.

[87] Milnor J, *Morse Theory*, Princeton University Press, Princeton, New Jersey, 1963.

[88] Moricz F, Some remarks on the notion of regular convergence of multiple series, *Acta Math. Hungar.*, **41** (1983), 161–168.

[89] Morton RD and Krall AM, Distributional weight functions for orthogonal polynomials, *SIAM J. Math. Anal.*, **9** (1978), 604–626.

[90] Murray JD, *Asymptotic Analysis*, Clarendon Press, Oxford, 1974.

[91] Olver FWJ, *Asymptotics and Special Functions*, Academic Press, New York, 1974.

[92] Pilipovic S, Quasiasymptotic expansion and the Laplace transform, *Appl. Anal.*, **35** (1990), 247–261.

[93] Pool JCT, Mathematical aspects of the Weyl correspondence, *J. Math. Phys.*, **7** (1966), 66–76.

[94] Rademacher H, *Topics in Analytic Number Theory*, Springer-Verlag, New York, 1973.

[95] Ramanujan S, *Notebooks*, 2 volumes, Tata Institute of Fundamental Research, Bombay, 1957.

[96] Richert HE, Zur multiplikatien Zahlentheorie, *J. Reine Angew. Math.*, **206** (1961), 31–38.

[97] Schwartz L, *Théorie des Distributions*, Hermann, Paris, 1966.

[98] Seneta E, *Regularly Varying Functions*, Springer-Verlag, Berlin, 1976.

[99] Shah SM and Wiener J, Distributional and entire solutions of ordinary differential and functional diferential equations, *Internat. J. Math. Sci.*, **6** (1983), 243–270.

[100] Shivakumar PN and Wong R, Asymptotic expansions of multiple Fourier transforms, *SIAM J. Math. Anal.*, **10** (1979), 1095–1104.

[101] Shohat, JA and Tamarkin, JD, *The Problem of Moments*, Amer. Math. Soc, Providence, 1943.

[102] Sirovich, L, *Techniques of Asymptotic Analysis*, Springer Verlag, New York, 1971.

[103] Soni, K and Soni RP, A note on asymptotic expansion of finite K_ρ and related transforms with explicit remainder, *J. Math. Anal. Appls.* **79** (1981), 163-177.

[104] _____, A note on summability and asymptotics, *SIAM J. Math. Anal.* **16** (1985), 392-404.

[105] Stankovic, B, S-asymptotics of distributions, in *Generalized Functions, Convergence Structures and Their Applications*, Plenum Press, New York, 1988, 71-78.

[106] Stein, EM, Oscillatory integrals in Fourier analysis, Bejing Lectures in Harmonic Analysis, *Annals of Math.Stud.* **112** Princeton University Press, 1986.

[107] _____, Problems in harmonic analysis related to curvature and oscillatory integrals, *Proc. Int. Cong. Math.*, Berkeley, CA, USA, 1986.

[108] Stein, EM and Weiss, G, *Introduction to Fourier Analysis on Euclidean Spaces*, Princeton University Press, Princeton, NJ, 1971.

[109] Titchmarsh, EC, *The Theory of the Riemann Zeta-function*, Clarendon Press, Oxford, 1951.

[110] Treves, F, *Topological Vector Spaces, Distributions and Kernels*, Academic Press, New York, 1967.

[111] Vladimirov, VS, *Generalized Functions in Mathematical Physics*, Mir, Moscow, 1979.

[112] Vladimirov, VS, Drozhinov, YN and Zavyalov, BI, *Multidimensional Tauberian Theorems for Generalized Functions*, Nauka, Moscow, 1986.

[113] Wasow, W, *Asymptotic Expansions for Ordinary Differential Equations*, John Wiley, New York, 1965.

[114] Wiener, J, Generalized function solutions of differential and functional differential equations, *J. Math. Anal. Appls.* **88** (1982), 170-182.

[115] Wigert, S, Sur la serie de Lambert et son application à la theorie des nombres, *Acta Math.* **41** (1916), 197-218.

[116] Whittaker, ET and Watson, GN, *A course in Modern Analysis*, AMS edition, AMS Press , New York, 1979.

[117] Wong, R, Explicit error bounds for asymptotic expansions of Mellin transforms, *J. Math. Anal. Appls.* **72** (1979), 740-756.

[118] _____, Error bounds of asymptotic expansions of integrals, *SIAM Rev.* **22** (1980), 401-435.

[119] _____, *Asymptotic Approximation of Integrals*, Academic Press, New York, 1989.

Additional Readings

Berndt, B. C., *Ramanujan's Notebooks, Part II*, Springer-Verlag, New York, 1989.

Frank, L. S., *Singular Perturbations*, North Holland, New York, 1990.

Kowalenko, V., Frankel, N.E. and Glasser, M. L., Generalized Euler-Jacobi inversion formula, preprint.

Zayed, A. I., Asymptotic expansions of some integral transforms by using generalized functions, Trans. Amer. Math. Soc., **272** (1982), 785–802.

Index